Mössbauer Spectroscopy

Mössbauer Spectroscopy

Applications in Chemistry and Materials Science

Edited by Yann Garcia, Junhu Wang, and Tao Zhang

WILEY-VCH

Editors

Prof. Yann Garcia
Université Catholique de Louvain
Institute of Condensed Matter and Nanosciences
Place Louis Pasteur 1
1348 Louvain-la-Neuve
Belgium

Prof. Junhu Wang
Mössbauer Effect Data Center (MEDC)
Dalian Institute of Chemical Physics (DICP)
Chinese Academy of Sciences
457 Zhongshan Road
116023 Dalian
China

Prof. Tao Zhang
Mössbauer Effect Data Center (MEDC)
Dalian Institute of Chemical Physics (DICP)
Chinese Academy of Sciences
457 Zhongshan Road
116023 Dalian
China

Cover Image: Courtesy of Norimichi Kojima

Library of Congress Card No.: applied for

British Library Cataloguing-in-Publication Data
A catalogue record for this book is available from the British Library.

Bibliographic information published by the Deutsche Nationalbibliothek
The Deutsche Nationalbibliothek lists this publication in the Deutsche Nationalbibliografie; detailed bibliographic data are available on the Internet at <http://dnb.d-nb.de>.

Print ISBN: 978-3-527-34691-2
ePDF ISBN: 978-3-527-82494-6
ePub ISBN: 978-3-527-82493-9
oBook ISBN: 978-3-527-82495-3

Typesetting Straive, Chennai, India
Printing and Binding CPI Group (UK) Ltd, Croydon, CR0 4YY

C9783527346912_020625

The manufacturer's authorized representative according to the EU General Product Safety Regulation is Wiley-VCH GmbH, Boschstr. 12, 69469 Weinheim, Germany, e-mail: Product_Safety@wiley.com.

Contents

Preface

After the discovery of recoil-free resonance absorption in the solid state by R. L. Mössbauer in 1958 [1], a physical method called Mössbauer spectroscopy was established. This technique finds nowadays a number of applications in various research fields including biology, physics, material science, metallurgy, mineralogy, and chemistry. It is thus not surprising that many textbooks and essays with engaging titles are available in the literature. Basically, these contain an introduction to the technique, with principal hyperfine parameters along with their theoretical description, followed by chapters about various applications, most often focused on physics, which are of little value for chemists, in particular synthetic chemists.

Thus, this book's sole focus on chemical applications constitutes a unique reference. It shows how relevant Mössbauer spectroscopy can be to giving useful information on electronic properties of matter, structural insights, and solid-state effects of chemical systems. Readers interested in learning about the theoretical and experimental grounds of the technique are invited to refer to excellent textbooks, e.g. by Gütlich et al. [2].

The book offers a wide range of chemical applications and concepts as well as numerous examples. It is written by leading specialists in Mössbauer spectroscopy who are active in various fields of chemistry at the international level. Such high-quality book is thus beneficial to readers working in different areas of chemistry and material science covered by Mössbauer spectroscopy such as energy, environment, life sciences, molecular electronics. It is intended for researchers in academia (universities, research institutes) and industry, as well as occasional readers and newcomers to the field. It could be used for educational purposes at the master's and PhD levels too. Readers at the bachelor level are invited to refer to the textbook by Yoshida and Langouche [3]. This is an aspect given that Mössbauer spectroscopy is nowadays seldomly taught in chemistry schools, with some students even ignoring what it is about or, at worst, how useful it can be.

Nowadays, the topic of energy is gaining importance due to the current increase in energy prices. The funding key chapter by Lippens introduces us to how can ^{119}Sn and ^{57}Fe operando Mössbauer spectroscopy be useful for the study of electrode materials for batteries, as well as specific catalysts. This is followed by a chapter by Long and Grandjean on ^{57}Fe Mössbauer spectral studies of iron phosphate-containing minerals, some of which are part of batteries used in the industry.

Next comes a chapter by Chen, Wang, and Liang on ^{57}Fe Mössbauer studies of iron-based silicides, which can be useful in optoelectronic and thermoelectric applications, as well as for catalytic conversion.

What would chemistry be without catalysis? Lázár gives a general overview of the use of Mössbauer spectroscopy applied to catalysts, considering both heterogeneous and homogeneous catalysts (via frozen solutions). It is followed by two chapters: (i) by Liu, Wang, and Zhang on the application of the technique in studying catalysts for preferential oxidation of CO in the presence of H_2. (ii) by Xu, Wang, Wang, and Sun on the application of ^{57}Fe Mössbauer spectroscopy in studying Fe–N–C catalysts for oxygen reduction reactions in polymer electrolyte membrane fuel cells.

But Mössbauer spectroscopy is not restricted to ^{57}Fe and ^{119}Sn isotopes. Indeed, Kojima, Kobayashi, and Seto introduce us to the field of thiolate-protected gold nanoclusters investigated by ^{197}Au Mössbauer spectroscopy, followed by the ^{197}Au Mössbauer spectroscopy study of gold mixed valence complexes, $Cs_2[Au^{I}X_2][Au^{III}Y_4]$ (X, Y = Cl, Br, I) and $[NH_3(CH_2)nNH_3]_2[(Au^{I}I_2)(Au^{III}I_4)(I_3)_2]$ ($n = 7, 8$).

Whereas chemistry readers are invited to refer to seminal reviews by Gütlich with Mössbauer applications in coordination chemistry and spin crossover phenomena [4, 5], the current book develops novel applications in the field of molecular electronics. Tokoro, Imoto, and Ohkoshi invite us to discover temperature- and photo-induced spin crossover occurring in some molecule-based magnets, studied by ^{57}Fe Mössbauer spectroscopy. Kumar and Garcia report on a new way to photo-induce the spin state in coordination compounds, e.g. by irradiating the counter-anion (AD-LISC effect), whereas Sun, Li, and Garcia comment on the possibility to use ^{57}Fe Mössbauer spectroscopy as a prime tool to explore a new family of colorimetric sensors, which could find application in food packaging freshness control.

The book project is supported by the Mössbauer Effect Data Center (MEDC), located in Dalian (China). MEDC already demonstrated its interest in the field by publishing five special issues on “chemical applications” [6–10]. The project was, of course, delayed by the COVID-19 crisis, affecting authors, reviewers, as well as the publishing house at several degree levels. Nevertheless, the editors wish to recognize the involvement and determination of each in the process of producing this book.

At the end of this preface, we wish to dedicate this book to the memories of Prof. Dr. Ph. Gütlich (1934–2022) and to Dr. Eckhard Bill (1953–2022), who did so much for chemical applications of Mössbauer spectroscopy.

We wish you all readers an enjoyable reading!

Yann Garcia, Junhu Wang, Tao Zhang

References

1 Mössbauer, R.L. (1958). *Z. Physik* 151: 124.
2 Gütlich, P., Bill, E., and Trautwein, A.X. (2011). *Mössbauer Spectroscopy and Transition Metal Chemistry: Fundamentals and Applications*. Springer.

3 Yoshida, Y. and Langouche, G. (2013). *Mössbauer Spectroscopy – Tutorial Book*. Springer.
4 Gütlich, P. and Garcia, Y. (2010). *J. Phys. Conf. Ser.* 217: 012001.
5 Gütlich, P. and Garcia, Y. (2013, Chapter 2). *Chemical Applications of Mössbauer Spectroscopy in Mössbauer Spectroscopy – Tutorial Book* (ed. Y. Yoshida and G. Langouche), 23–89. Springer.
6 Garcia, Y. (2012). Chemical applications. *Möss. Eff. Data Ref. J.* 35 (6): 111.
7 Garcia, Y. (2012). Chemical applications. *Möss. Eff. Data Ref. J.* 35 (8): 191.
8 Garcia, Y. (2013). Chemical applications. *Möss. Eff. Data Ref. J.* 36 (4): 47.
9 Garcia, Y. (2013). Chemical applications. *Möss. Eff. Data Ref. J.* 36 (6): 103.
10 Garcia, Y. (2014). Chemical applications. *Möss. Eff. Data Ref. J.* 37 (3): 49.

1

Application of Mössbauer Spectroscopy to Energy Materials

Pierre-Emmanuel Lippens, Jean-Claude Jumas, and Josette Olivier-Fourcade

Institut Charles Gerhardt de Montpellier, UMR 5253 CNRS, Université de Montpellier, ENSCM, Pôle Chimie Balard Recherche 1919 route de Mende, Montpellier Cedex 5, France

1.1 Introduction

The increasing demand for energy and the environmental problems caused by fossil fuels make necessary the fast development of renewable and clean energy sources such as solar or wind. However, their intermittent nature requires highly efficient energy storage systems. Furthermore, clean transportation, such as electric or hybrid vehicles, needs power sources of high energy density. In this regard, lithium-ion batteries can be considered as one of the most promising technologies. Although lithium-ion batteries are used in electronic devices, power tools, electric bikes, and more, there are still many challenges to overcome for new applications that require higher energy and power densities, improved safety, and lower cost. This explains the current intensive research on electrode materials [1, 2].

The performance of currently used carbon and layered lithium transition metal oxides as negative and positive electrode materials, respectively, has almost reached the upper theoretical limits. New negative electrode materials such as silicon and tin have high theoretical capacities, but they suffer from strong volume variations during cycling, which leads to capacity fading and strongly reduces the cycle life [3, 4]. Among the different approaches proposed to improve the performance of such materials, downsizing the particle size or dispersing particles within an electrochemically inactive matrix have been proposed [5, 6]. For the positive electrodes, new Fe-containing materials such as $LiFePO_4$, $LiFe_{1-x}Mn_xPO_4$, or $LiFePO_4F$ appear to be very promising for environmental and economic aspects [7, 8]. Finally, the potential problems of cobalt and lithium supply in a near future have recently led to the development of alternative technologies such as Na-ion batteries [9, 10].

Improving the performance of electrode materials requires a good knowledge of the electrochemical reaction mechanisms in batteries. Mössbauer spectroscopy is often used for the characterization of materials, but is also a unique tool to follow such reactions at the atomic scale from the analysis of the Mössbauer parameters: isomer shift (given in this chapter relative to α-Fe and $BaSnO_3$ for

Mössbauer Spectroscopy: Applications in Chemistry and Materials Science,
First Edition. Edited by Yann Garcia, Junhu Wang, and Tao Zhang.

^{57}Fe and ^{119}Sn isotopes, respectively), quadrupole splitting and hyperfine magnetic field [11, 12]. Different examples are considered to illustrate the application of Mössbauer spectroscopy to electrode materials for Li-ion and Na-ion batteries. This includes tin-based negative electrode materials: β-Sn, tin oxides, tin borophosphates, tin intermetallics, and tin-silicon composites, as well as iron-based positive electrode materials: $LiFePO_4$, $LiFe_{1-x}Mn_xPO_4$, $Fe_{1.19}PO_4(OH)_{0.57}(H_2O)_{0.43}$, and $Na_{1.5}Fe_{0.5}Ti_{1.5}(PO_4)_3$.

Reducing emission and pollution of existing transportation based on fossil fuels is also a great challenge. Reforming catalysis is a major petroleum refining process for the production of hydrogen or high-octane gasoline. In particular, catalytic reforming is a chemical process used to convert naphtha, produced during petroleum refining, into high-octane number gasoline. If Pt/Al_2O_3 was the first naphtha-reforming catalyst, a great progress has been achieved with supported bimetallic-reforming catalysts, in which Pt is promoted by another metal such as Sn, that offer high selectivity at low pressure [13]. Improved selectivity can be obtained by the addition of a second promoter as In [14]. Mössbauer spectroscopy has been widely used in catalysis [15, 16]. Some examples are presented here to show how complex redox processes in Sn-Pt based catalysts involving many different tin species can be elucidated.

1.2 Mössbauer Spectroscopy for Li-ion and Na-ion Batteries

1.2.1 Characterization of Electrode Materials and Electrochemical Reactions

The main specifications of electrode materials for Li-ion or Na-ion batteries are specific and volumetric capacities, nominal potential, cycle life, calendar life, rate capability, safety, environmental impact, and recycling. It is of course impossible to optimize all these features at the same time. For example, layered lithium transition metal oxides, as commonly used positive electrode materials for Li-ion batteries, have high capacity and high potential vs. Li^+/Li. Lithium iron phosphate ($LiFePO_4$) has lower capacity but provides better safety and higher electric power. Thus, new electrode materials are needed to improve the performance of Li-ion and Na-ion batteries. This requires a better knowledge of the electrochemical reaction mechanisms that take place during the charge–discharge cycles.

Different techniques have been used for the characterization of electrode materials and to follow the reactions within the batteries. This includes ex situ experiments, i.e. the electrode material is extracted from the battery, in situ experiments, i.e. the electrode material is in the battery, and operando experiments as electrochemical reactions proceed [17]. X-ray diffraction (XRD) and Mössbauer spectroscopy have often been combined to obtain complementary information about long range and local properties, respectively [18]. It is thus possible to determine the formed species even if they are amorphous, to follow changes in local structure or oxidation state in order to elucidate the reaction mechanisms.

In situ and operando Mössbauer and XRD measurements require specific electrochemical cells that allow transmission of γ-rays and reflection of X-rays, respectively, while voltage or electric current is imposed. Such a cell is based on the usual positive electrode/separator/electrolyte/negative electrode configuration [19, 20]. To investigate the electrochemical reactions for a given electrode material, a half-cell is commonly used, where the negative electrode is metallic lithium (or sodium) and the positive electrode is the formulated electrode material under investigation. The formulation consists of mixing the electrochemically active material (nanoparticles or microparticles) with an electronic conductive additive, like carbon black, and a binder to form a slurry that is casted onto the metal current collector [21]. In the case of an electrochemical half-cell, the discharge corresponds to the lithiation (or sodiation) of the electrode material under investigation and the charge to delithiation (or desodiation).

1.2.2 Tin-Based Negative Electrode Materials for Li-ion Batteries

1.2.2.1 Electrochemical Reactions of Lithium with Tin

Metallic and semi-metallic elements can be used as negative electrode materials for high energy Li-ion batteries. For example, each Sn atom in β-Sn can react with a maximum of 4.4 Li, which corresponds to specific and volumetric capacities of 992 $mAh{\cdot}g^{-1}$ and 2111 $mAh{\cdot}cm^{-3}$, respectively [3]. This is more than 2.6 times the capacity of currently used graphite (372 $mAh{\cdot}g^{-1}$ and 719 $mAh{\cdot}cm^{-3}$). In addition, the average potential of β-Sn in a Li half-cell is higher than that of carbon, which reduces lithium plating and improves safety. Unfortunately, the formation of $Li_{x<4.4}Sn$ phases during lithiation induces a strong increase of the particle volume (>300%) that causes cracks, while delithiation leads to the pulverization of the particles. These two effects are responsible for mechanical and electrical instabilities of the negative electrode film, which drastically reduces the cycle life of the battery. XRD and ^{119}Sn Mössbauer spectroscopy have been combined to obtain deeper insights into Li–Sn alloying reactions.

The experimental voltage curves of β-Sn in a Li half-cell, obtained during the first cycle in galvanostatic regime at low current density, show different well-defined plateaus that can be attributed to two-phase reactions (Figure 1.1) [22].

The voltage plateaus are observed for both lithiation and delithiation processes, showing the reversibility of the mechanism. The following alloying reactions have been proposed by considering the different crystalline phases of the commonly accepted Li–Sn phase diagram [24]:

$$5\ \beta\text{-}Sn + 2\ Li \rightarrow Li_2Sn_5 \tag{1.1}$$

$$Li_2Sn_5 + 3\ Li \rightarrow 5\ LiSn \tag{1.2}$$

$$3\ LiSn + 4\ Li \rightarrow Li_7Sn_3 \tag{1.3}$$

$$2\ Li_7Sn_3 + Li \rightarrow 3\ Li_5Sn_2 \tag{1.4}$$

$$5\ Li_5Sn_2 + Li \rightarrow 2\ Li_{13}Sn_5 \tag{1.5}$$

$$2\ Li_{13}Sn_5 + 9\ Li \rightarrow 5\ Li_7Sn_2 \tag{1.6}$$

$$5\ Li_7Sn_2 + 9\ Li \rightarrow 2\ Li_{22}Sn_5 \tag{1.7}$$

First-principles calculations of the cell voltage were performed with different methods based on density functional theory (DFT) for reactions (1.1)–(1.6) [22, 23].

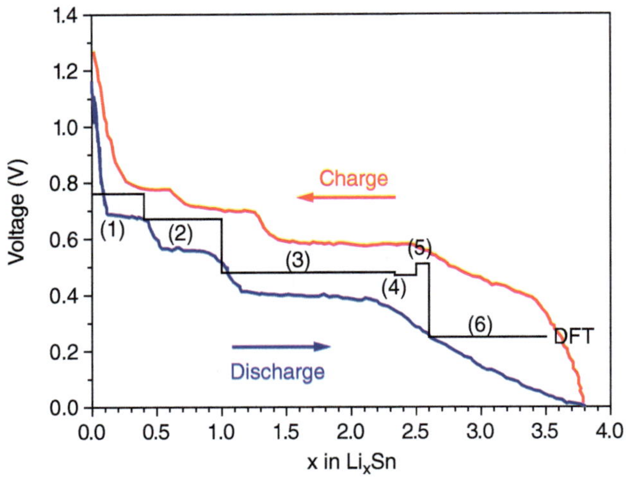

Figure 1.1 Experimental voltage profile for the first cycle of β-Sn in Li half-cell [22] and voltage plateaus evaluated with density functional theory (DFT) for the reactions (1.1)–(1.6) described in the text [23]. Source: Adapted from Refs. [22, 23].

The comparison between the experimental and theoretical voltage profiles suggests that the two first plateaus at average experimental voltages of 0.75 V and 0.65 V reflect the formation of Li_2Sn_5 (reaction 1.1) and LiSn (reaction 1.2) crystalline phases, respectively. The following plateau, at about 0.5 V, could be attributed to the formation of Li_7Sn_3 (reaction 1.3), Li_5Sn_2 (reaction 1.4), and/or $Li_{13}Sn_5$ (reaction 1.5) that have closed chemical compositions and formation energies. The expected plateau for the formation of Li_7Sn_2 (reaction 1.6) is not observed experimentally since the voltage curve shows a continuous decrease in the range 2.2–3.8 Li per Sn. This was interpreted by the formation of a metastable phase with a body-centered cubic disordered structure showing the same short-range order as $Li_{22}Sn_5$ [22].

Operando XRD was used to follow the structural changes of β-Sn small particles in Li half-cell during the first cycle in galvanostatic regime (Figure 1.2).

During the discharge, the successive XRD patterns show the main peaks of β-Sn, Li_2Sn_5, and LiSn, confirming that the two first voltage plateaus correspond to the reactions (1.1) and (1.2). Then, there is one broad band around $2\theta = 25°$ and a peak at $2\theta = 45°$ that can be attributed to one or more Li-rich Li_xSn phases, but it is not possible to determine the chemical composition from the XRD patterns. The large linewidth and the small intensity of the peaks indicate that the electrochemically formed Li_xSn phases are poorly crystallized and/or of small size. The nanostructuration of the pristine material is typical of alloying reactions and explains the inability of XRD to give accurate information about such electrochemical mechanisms, especially for the highly lithiated electrodes. The mechanism is clearly reversible during the charge, but it is still difficult to distinguish all the intermediate Li_xSn phases. In such a case, ^{119}Sn Mössbauer spectroscopy is of particular interest since it is sensitive to the local environment of the Sn atoms and not to long-range order.

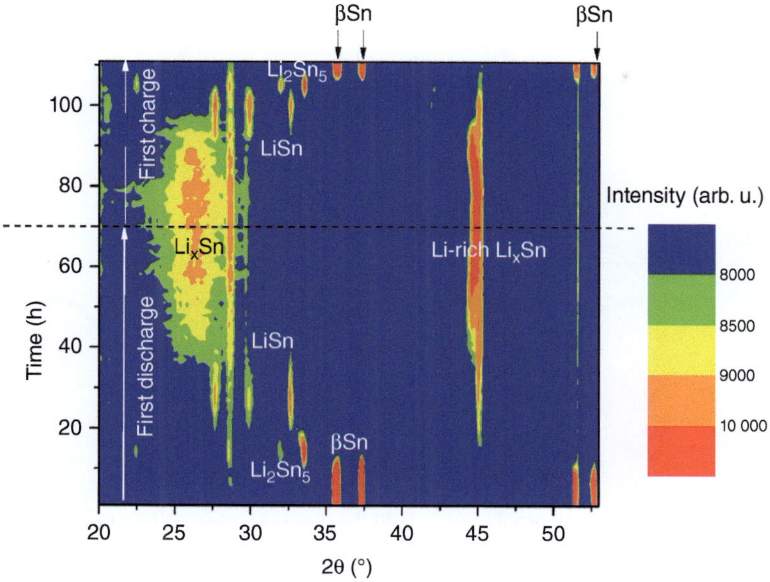

Figure 1.2 Operando XRD patterns obtained during the first discharge and the first charge of β-Sn in Li half-cell (galvanostatic regime) where colors reflect X-ray intensity.

To have some reference compounds that can help in the interpretation of the ^{119}Sn Mössbauer experiments during the charge–discharge reactions, the spectra of the seven Li_xSn crystalline phases, Li_2Sn_5, LiSn, Li_7Sn_3, Li_5Sn_2, $Li_{13}Sn_5$, Li_7Sn_2 and $Li_{22}Sn_5$, were measured at room temperature [24]. These phases were obtained in two different ways, namely high temperature solid-state reactions [25] and mechanosynthesis followed by annealing [26]. The crystal structures determined by XRD show the existence of one (Li_5Sn_2), two (Li_2Sn_5, LiSn, Li_7Sn_2), three (Li_7Sn_3, $Li_{13}Sn_5$), and four ($Li_{22}Sn_5$) crystallographic sites for Sn. The Mössbauer spectra obtained for the two series of synthesized materials give similar Mössbauer parameters. The observed small differences are due to the existence of impurities and the difficulties to fit unresolved spectra. One set of spectra and the corresponding parameters are reported in Figure 1.3 and Table 1.1, respectively.

The values of the quadrupole splitting, Δ, can be related to the different local environments of Sn resulting from the existence of different types of nearest neighbors (Li, Sn), polyhedral geometries, and bond lengths [26]. The average value of the isomer shift, δ_{av}, is of about 2.4 mm·s^{-1} for the two Sn-rich phases and decreases from 2.1 to 1.8 mm·s^{-1} with increasing number of Li per Sn for the Li-rich phases. These values are typical of the Sn(0) oxidation state. They reflect the existence of a Sn based sublattice for the Sn-rich phases and the decrease in the number of Sn—Sn bonds per Sn for the Li-rich phases.

The values of δ_{av} are plotted for the Li_xSn references as a function of x (Figure 1.4). The linear correlation shows that $\delta_{av}(x)$ mainly depends on the average chemical composition of Li_xSn and can be used to evaluate x for amorphous phases or small particles as often encountered in electrochemical reactions. The linear function

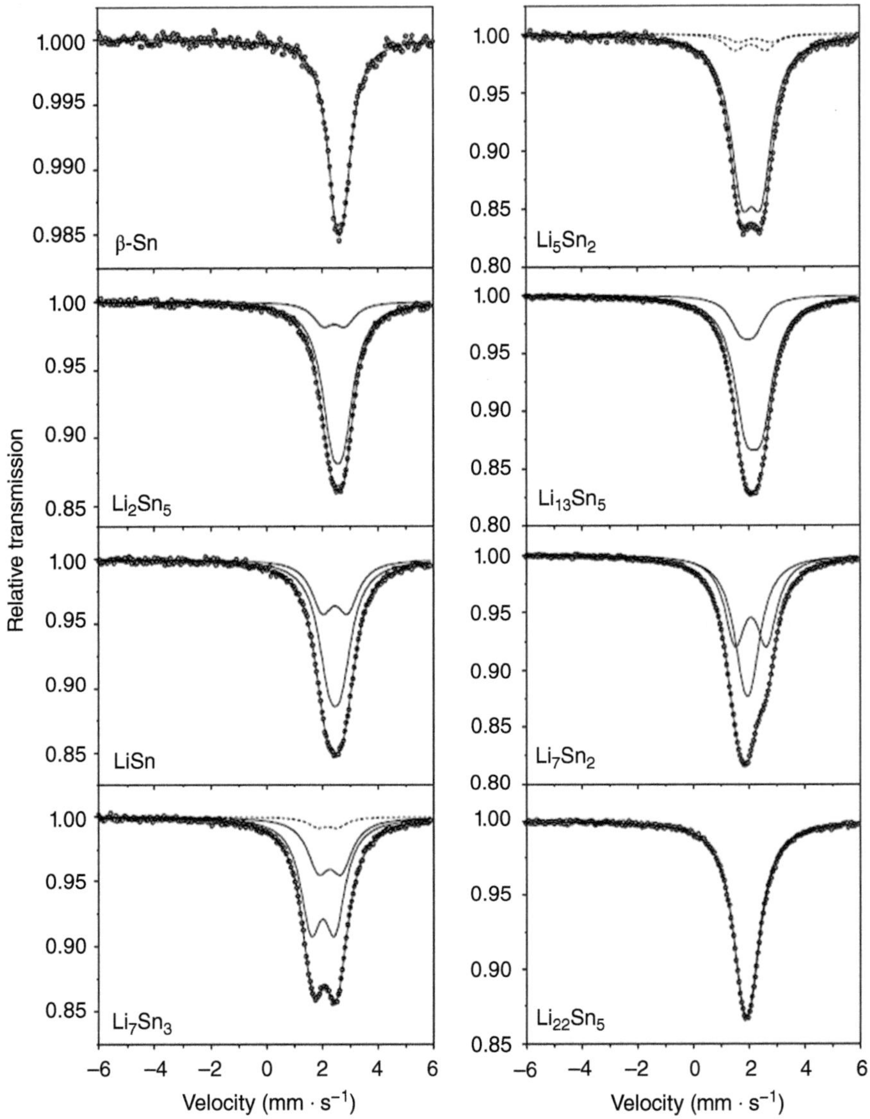

Figure 1.3 ^{119}Sn Mössbauer spectra of Li_xSn crystalline phases measured at room temperature. Source: Reproduced from Ref. [26]/with permission from Elsevier.

obtained for the regression line shown in Figure 1.4 can be written as:

$$\delta_{av}(\text{mm} \cdot \text{s}^{-1}) = 2.55 - 0.20x \tag{1.8}$$

Other empirical correlation rules have been derived from the Li_xSn Mössbauer references, such as δ–Δ correlation diagrams used to predict the electrochemical activity of Sn based negative electrode materials for Li-ion batteries (Figure 1.5) [27].

The ^{119}Sn Mössbauer spectroscopy was used for ex situ measurements at different stages of the first discharge of β-Sn in Li half-cells [28]. The shape of the voltage curve

Table 1.1 Values of the ^{119}Sn Mössbauer parameters of Li_xSn crystalline phases obtained by Robert et al. [26]. The values of the isomer shift, δ, relative to $BaSnO_3$ and quadrupole splitting, Δ, are reported for Sn in different crystallographic sites.

	β-Sn	Li_2Sn_5	LiSn	Li_7Sn_3	Li_5Sn_2	$Li_{13}Sn_5$	Li_7Sn_2	$Li_{22}Sn_5$
δ/Δ (mm·s^{-1}) (crystallographic sites)	2.56/0.29 (4a)	2.49/042 (8i) 2.36/0.78 (2d)	2.38/0.43 (2m) 2.38/0.91 (1a)	2.19/0.82 (2e) 1.94/0.86 (2e, 2e)	2.01/0.69 (6c)	1.86/0.48 (1a) 2.07/0.58 (2d, 2d)	1.84/0.28 (4i) 1.96/1.13 (4h)	1.83/0.31 (16e, 16e, 24f, 24g)

Source: Adapted from Ref. [26].

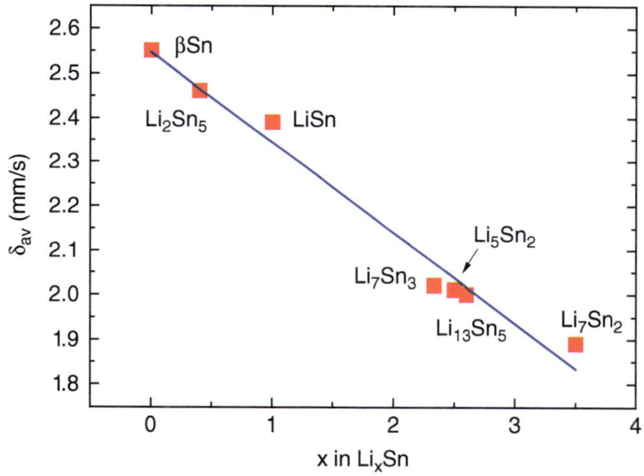

Figure 1.4 Average experimental values of the isomer shift of the Li_xSn crystalline references vs. number of Li per Sn and linear regression line (blue).

is similar to that of Figure 1.1, except at the very beginning of lithiation due to the existence of tin oxides as impurities. The spectra recorded for the insertion of 0.5, 1, and 3.4 Li per Sn were fitted to two doublets and the values of the Mössbauer parameters are similar to those of Li_2Sn_5, LiSn, and Li_7Sn_2 references, respectively, although there are some small differences that can be attributed to the poor crystallinity and to variations in the chemical composition of the electrochemically lithiated species.

Although β-Sn has a higher capacity than graphite, it suffers from large volume variations during cycling that affect the mechanical and electrical integrity of the electrode film. Tin oxide composites were proposed to reduce such effects [29], and the electrochemical mechanisms are described in the following two paragraphs by considering the application of Mössbauer spectroscopy.

1.2.2.2 Tin Oxides

Stannic oxide (SnO_2) and stannous oxide (SnO) have tetragonal $P4_2/mnm$ and $P4/nmm$ structures, respectively. The Mössbauer spectrum of SnO_2 is formed by

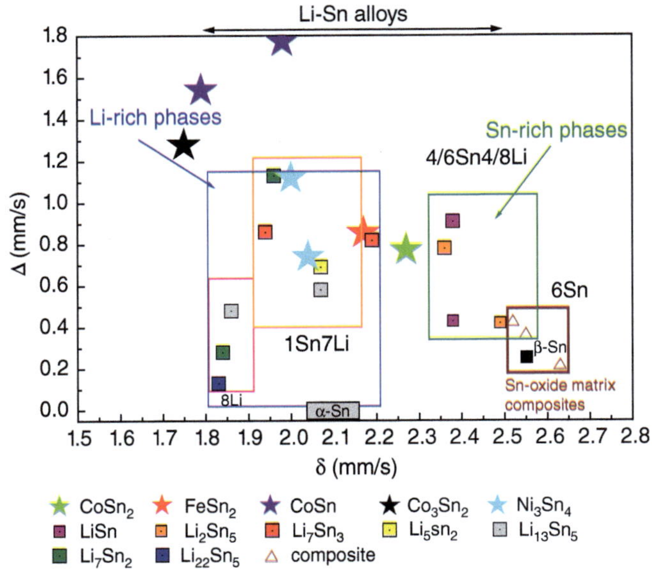

Figure 1.5 Correlation diagram for Sn-based materials based on experimental values of isomer shift, δ, and quadrupole splitting, Δ.

a single peak reflecting the Sn(IV) oxidation state ($\delta = 0\,\mathrm{mm \cdot s^{-1}}$) and a slightly distorted SnO_6 octahedral environment ($\Delta = 0.5\,\mathrm{mm \cdot s^{-1}}$). SnO can be described as parallel Sn–O–Sn layers, and each Sn atom is bonded to four O atoms to form a SnO_4 square based pyramid. The Mössbauer spectrum consists of an asymmetric doublet characteristic of the Sn(II) oxidation state ($\delta = 2.63\,\mathrm{mm \cdot s^{-1}}$), while the quadrupole splitting ($\Delta = 1.33\,\mathrm{mm \cdot s^{-1}}$) reflects the anisotropy of the Sn p-type electron density arising from the Sn 5p lone pair along the fourfold pyramid axis.

The voltage profiles of SnO_2 and SnO based electrodes obtained in galvanostatic regime show similar trends [30–33]. The first discharge of SnO is formed by a plateau at ~1 V for 2 Li per Sn (Region R1), followed by a continuous voltage decrease showing different pseudo-plateaus (Region R2) (Figure 1.6). Then, the voltage profiles of the two tin oxides are similar to that of β-Sn for the subsequent charge–discharge cycles. The electrochemical mechanism was studied by ex situ [31] and operando [32] ^{119}Sn Mössbauer spectroscopy. As lithiation proceeds in R1, there is a progressive broadening of the spectra at the low velocity side, ending with a large and unresolved peak at the end of the first plateau (A–D in Figure 1.7). The experimental spectra were successfully fitted to three components corresponding to SnO, β-Sn, and a Sn(IV) oxide, respectively, and an additional component centered at about $1.2\,\mathrm{mm \cdot s^{-1}}$. The contribution of SnO decreases during the lithiation while that of β-Sn increases, which corresponds to the conversion reaction:

$$\mathrm{SnO + 2\,Li \rightarrow \beta\text{-}Sn + Li_2O} \tag{1.9}$$

The subspectrum at $1.2\,\mathrm{mm \cdot s^{-1}}$ can been attributed to Sn bonded to both Sn and O atoms resulting from interactions between β-Sn and Li_2O small particles or to the existence of Li–Sn–O amorphous phases. In region R2, the ex situ spectra E and F

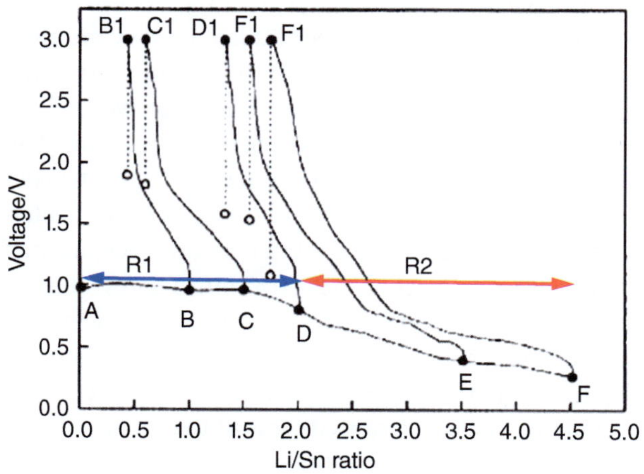

Figure 1.6 Voltage curves of SnO in Li half-cell with the points of measurements for the first discharge (A–F) and full charges at 3 V (B1–F1). Source: Reproduced from Ref. [31]/with permission from Elsevier.

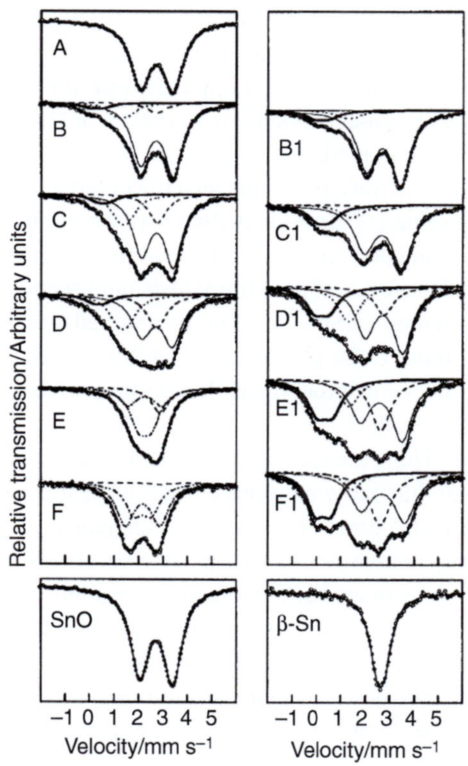

Figure 1.7 Ex situ ^{119}Sn Mössbauer spectra at different stages of the first discharge and of the end of charges. Source: Reproduced from Ref. [31]/with permission from Elsevier.

(Figure 1.7) are similar to published in situ Mössbauer spectra [32]. However, more spectra were recorded in the latter case due to a deeper lithiation. They are close to the spectra of the reference materials, confirming the formation of Sn-rich and Li-rich Li_xSn equilibrium phases, except around Li_5Sn_2 as discussed above for β-Sn. The Li_xSn dealloying reactions operate during the charge, but a back reaction of Sn with O, arising from the delithiation of Li_2O, is observed at high voltage. The later reaction is out of the usual voltage range for negative electrodes in Li-ion batteries and should not be considered. In that case, the electrochemical mechanism of SnO consists of the irreversible conversion reaction of Sn(II)O into β-Sn(0) followed by reversible Li_xSn alloying reactions. The Li_2O particles form an electrochemically inactive matrix that maintains the dispersion of the Li_xSn particles and buffers the volume variations, improving the cycling behavior compared to β-Sn. However, the in situ formation of the Li_2O matrix leads to a capacity loss at the first cycle that requires a compensation by the lithiated positive electrode in a Li-ion full-cell. The situation is even more critical for SnO_2 since four Li per Sn are required for its transformation into β-Sn/Li_2O nanocomposite.

1.2.2.3 Tin Borophosphates

Tin composite oxides (TCO) were proposed by Fuji in the late 1990s as negative electrode materials for Li-ion batteries [29]. Among the different TCOs, the tin borophosphate glass $SnO(B_2O_3)_{0.25}(P_2O_5)_{0.25}$ (Sn_2BPO_6 in the text) was studied by in situ ^{119}Sn Mössbauer spectroscopy [32]. The voltage profile of the first cycle of Sn_2BPO_6 in a Li half-cell is similar to that of SnO, except that the voltage of the first plateau (1.75 V) is higher (Figure 1.8). The Mössbauer spectrum of $Sn(II)_2BPO_6$, similar to that of SnO, is transformed into a broad band of weak intensity at the end of the first plateau. This reflects the transformation of the Sn(II) based pristine material into β-Sn small clusters embedded in a mixed borophosphate and lithium oxide matrix (Figure 1.8). Then, the Mössbauer spectra are similar to the Li_xSn references except around Li_5Sn_2. The spectra obtained along the first charge until 1 V were attributed to the Li_xSn phases resulting from dealloying reactions, which is consistent with a reversible process. As the charge proceeds until 2.5 V, there is a partial reformation of Sn_2BPO_6.

Thus, the electrochemical mechanism of Sn_2BPO_6 glass consists in the formation of Sn(0) tiny particles embedded in a more complex matrix than that obtained for SnO, followed by alloying–dealloying Li_xSn reactions. The value of the voltage cutoff for charge (≈0.8 V) is crucial to avoid back reaction of Sn with O. The electrochemical performance of TCO, such as cyclability, shows that a borophosphate matrix is more efficient than a Li_2O matrix, as obtained for tin oxide compounds, to maintain the dispersion of the Li_xSn particles and buffer the volume variations during cycling.

Tin borophosphate composites were also considered as possible negative electrode materials for Li-ion batteries, such as $Sn/(BPO_4)_{0.2}$ that shows a reversible capacity of about 500 mAh·g^{-1} [33, 34]. The operando Mössbauer measurements were performed in galvanostatic regime at C/10 in the voltage range 0.1–1.2 V [35]. The Mössbauer spectrum of the pristine material shows the existence of β-Sn but also of an amorphous Sn(II)-borophosphate phase not observed by XRD

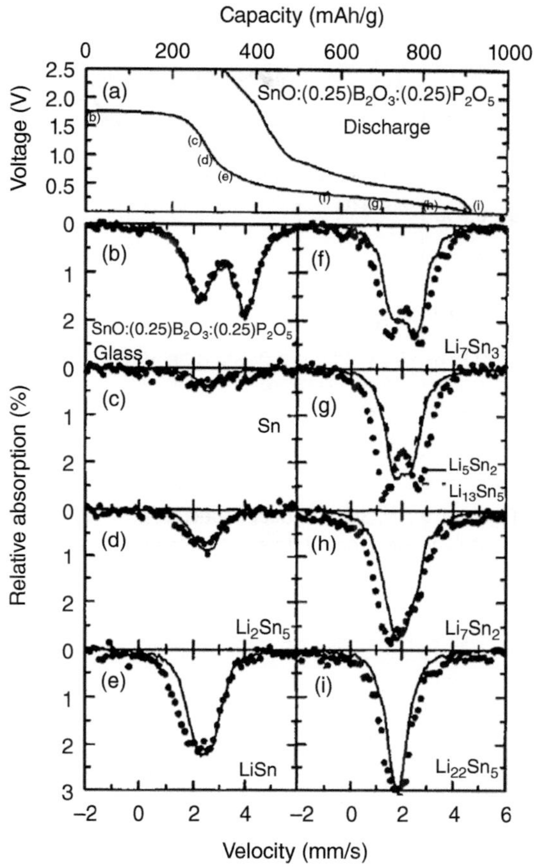

Figure 1.8 In situ ^{119}Sn Mössbauer spectra at different stages of the first discharge of Sn_2BPO_6. Source: Reproduced from Ref. [32]/with permission from Elsevier.

(Figure 1.9). At the beginning of discharge, the voltage curve shows a plateau at 1.5 V, significantly shorter than the plateaus of tin oxides or TCO discussed above, followed by a voltage decrease until 0.4 V for about 0.6 Li per Sn (Figure 1.10a). The intensity of the Sn(II) Mössbauer absorption strongly decreases in the range 0–0.2 Li per Sn, which indicates that the plateau corresponds to the reduction of the Sn(II) amorphous phase into Sn(0) (Figure 1.10b). Then, there is no more change in the spectra until 0.9 V, in line with the irreversible reaction of lithium with the electrolyte, leading to the formation of a solid electrolyte interphase (SEI) at the particle surface. As discharge proceeds further, the voltage curve shows a plateau at 0.4 V and then decreases until 0.1 V in the range from 0.6 to 3.2 Li per Sn. The average Mössbauer isomer shift decreases from 2.45 to 2.0 mm·s^{-1}. By considering the $\delta_{av}(x)$ linear function for Li_xSn (see Eq. 1.8), the latter value suggests that $x \approx 2.5$ (Figure 1.10c and 1.10d), in agreement with the 2.6 Li reacting with β-Sn in this region. This low value of x, compared to the expected $Li_{22}Sn_5$ phase, is due to the voltage cutoff of 0.1 V used in this experiment, which prevents deep lithiation. During the first charge, the average isomer shift linearly increases, as expected from the delithiation of Li_xSn, ending with $\delta_{av} \approx 2.35$ mm·s^{-1}, which corresponds to $x \approx 1$. This confirms the reversibility of the alloying process although

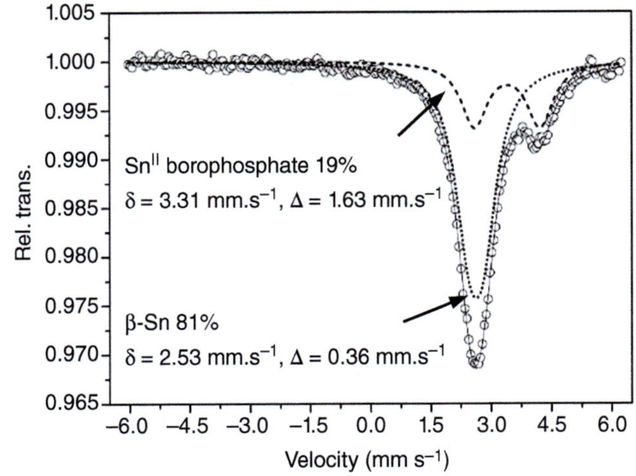

Figure 1.9 ^{119}Sn Mössbauer spectrum of Sn/$(BPO_4)_{0.2}$ composite (relative transmission: rel. trans.) showing the relative amounts of Sn^{II} and Sn^0 species. Source: Adapted from Ref. [35].

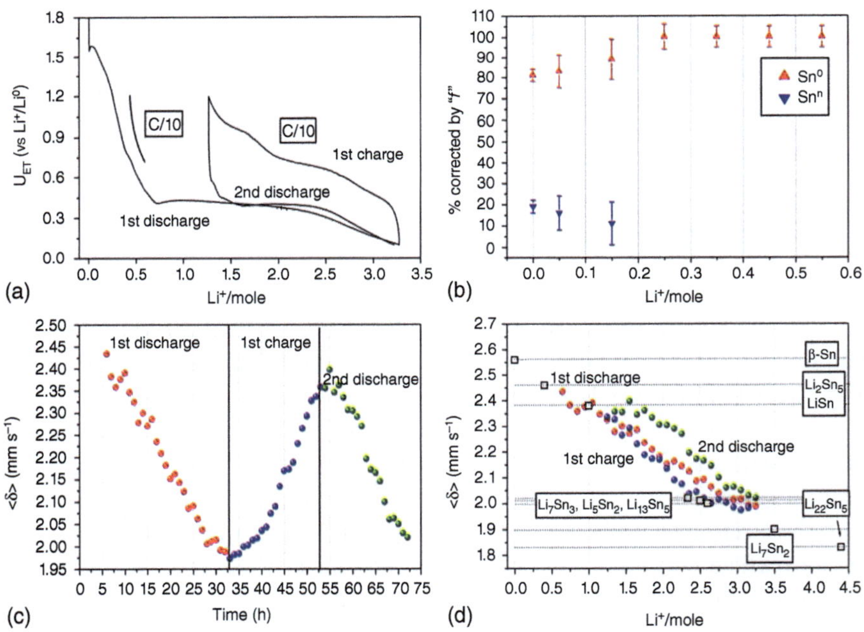

Figure 1.10 Voltage profile of Sn/$(BPO_4)_{0.2}$ in Li half-cell for the galvanostatic regime of C/10 (a), amounts of Sn(II) and Sn(0) at the beginning of first discharge (b), evolution of the Sn(0) average isomer shift, δ, as a function of time reaction (c), and as a function of the number of Li per Sn (d): first discharge in red, first charge in blue, and second discharge in green.

there is no full delithiation. The voltage profile observed for the second discharge is similar to the first one but is in the range 1.3–3.2 Li per Sn. The average isomer shift follows the same trend as during the first discharge, corresponding to the progressive change in the average composition of Li_xSn from x = 1 to about 2.5.

The results reported here for tin oxides as negative electrode materials for Li-ion batteries show that the first discharge consists in the restructuration of the electrode material, ending with Li-rich Li_xSn nanoparticles dispersed in an electrochemically inactive matrix composed of oxide, borate, or phosphate based particles. This matrix helps to buffer the strong volume variations of Li_xSn during alloying–dealloying reactions, reducing the capacity fading and improving the cycle life. However, there are still some issues regarding the capacity loss at first cycle due to the irreversible consumption of Li and the large voltage polarization due to the poor electronic conductivity of the oxide based matrix. Significant improvements have been obtained with tin intermetallics as described in the following section.

1.2.2.4 Tin-Based Intermetallics

Advantages over Tin Oxides The use of tin intermetallics as negative electrode materials for Li-ion batteries is based on the same concept as tin oxides, which is to form a nanocomposite during the first discharge, except that the matrix is composed of metallic particles in order to improve the electronic conductivity. In addition, the microstructure of metal based matrices is different from that of oxide matrices, which could be advantageous for buffering the volume changes resulting from Li_xSn alloying reactions. But the most interesting aspect of tin intermetallics is the existence of Sn(0) oxidation state. There is no reduction of tin prior to alloying lithiation during the first discharge, as Sn(IV) or Sn(II) in tin oxides, which reduces the capacity loss of the first cycle. Most of the tin-based intermetallics, MSn_x, combine Sn with a transition metal, M, that does not react with Li. The first discharge is expected to extrude the M atoms from MSn_x to form metallic nanoparticles (the matrix) that maintain the dispersion of Li_xSn nanoparticles. Although carbon additives and a binder are added to MSn_x particles to form the conductive film casted on the metal current collector, carbon can also be introduced during the synthesis process to form a MSn_x/C composite. Such composites have been widely studied in relation with the commercialization by Sony of the Nexelion Li-ion batteries [36–40]. The specific capacity was about twice as high than that of graphite. A detailed study of $FeSn_2$ is presented in the following part. This compound is of particular interest to follow lithiation–delithiation reactions since it contains both ^{57}Fe and ^{119}Sn Mössbauer isotopes.

Electrochemical Reactions of Lithium with $FeSn_2$ $FeSn_2$ has a tetragonal structure (*I4/mcm*), and the unit cell contains one crystallographic site for Fe at the center of a Sn based square antiprism. There is also one crystallographic site for Sn bonded to four Fe atoms to form a $SnFe_4$ square based pyramid. $FeSn_2$ microparticles were synthesized by solid-state reaction as electrode material for Li-ion batteries and nanostructured by an additional ball-milling step [41]. The $FeSn_2$ crystal is antiferromagnetic below 378 K [42], and the ^{57}Fe Mössbauer spectrum is formed

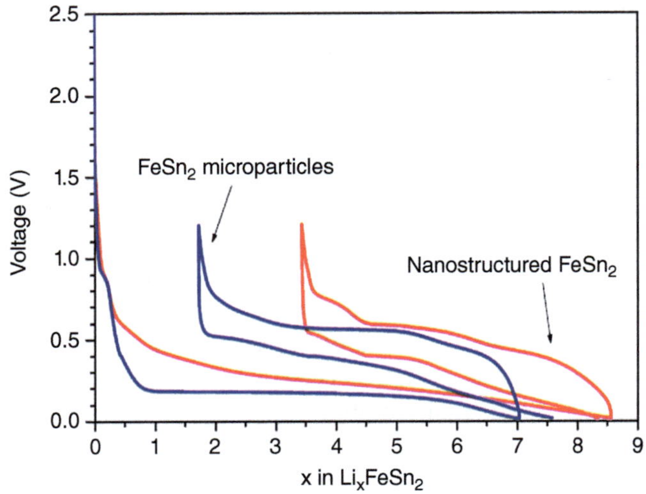

Figure 1.11 Voltage profiles of $FeSn_2$ based electrodes at C/50. Source: Reproduced from Ref. [41]/with permission from Elsevier.

by a sextet at room temperature. The ^{119}Sn Mössbauer spectrum shows broad structures in the range 0–5 $mm{\cdot}s^{-1}$, reflecting a transferred hyperfine magnetic field [43]. The Mössbauer parameters: $\delta = 0.5\,mm{\cdot}s^{-1}$, $\Delta = 0\,mm{\cdot}s^{-1}$, $B = 11\,T$ for Fe and $\delta = 2.18\,mm{\cdot}s^{-1}$, $\Delta = 0.83\,mm{\cdot}s^{-1}$, $B = 2.4\,T$ for Sn are typical of Fe(0) and Sn(0) oxidation states with a Sn asymmetrical environment. Similar values of the Mössbauer parameters were obtained for nanostructured $FeSn_2$ except there is no hyperfine magnetic field because of the small size and the poor crystallinity of the $FeSn_2$ ground particles [41].

The voltage profile of the electrode containing $FeSn_2$ microparticles shows a low voltage plateau for the first discharge and a voltage hysteresis at the average value of 0.5 V for the following cycles (Figure 1.11). The voltage curve of the first discharge differs from that of tin oxides and reflects a two-phase reaction. This is confirmed by operando XRD that shows the direct transformation of $FeSn_2$ into a Li-rich Li_xSn phase as lithiation proceeds. However, the chemical composition and the structure of the formed Li_xSn phase cannot be determined due to unresolved Bragg peaks. The operando ^{119}Sn Mössbauer spectra show strong changes from the magnetic spectrum of $FeSn_2$ to an asymmetrical peak at the end of discharge (Figure 1.12a). The Mössbauer parameters of the fully lithiated electrode are close to those of the Li_7Sn_2 crystalline reference except for one value of the quadrupole splitting ($\Delta = 0.72\,mm{\cdot}s^{-1}$) smaller than the reference ($\Delta = 1.13\,mm{\cdot}s^{-1}$). This was attributed to the small size and/or poor crystallinity of the Li_7Sn_2 particles [44]. The ^{57}Fe Mössbauer spectra change from a sextet for $FeSn_2$ to a doublet with Mössbauer parameters ($\delta = 0.2\,mm{\cdot}s^{-1}$, $\Delta = 0.5\,mm{\cdot}s^{-1}$) that can be attributed to α-Fe nanoparticles in paramagnetic state (Figure 1.12b).

All the intermediate ^{57}Fe and ^{119}Sn Mössbauer spectra were successfully fitted to the spectra of $FeSn_2$/α-Fe(nano) and $FeSn_2/Li_7Sn_2$(nano) phases, respectively, showing that the first discharge can be described by the two-phase reaction:

$$FeSn_2 + 7\,Li \rightarrow Li_7Sn_2 + Fe \tag{1.10}$$

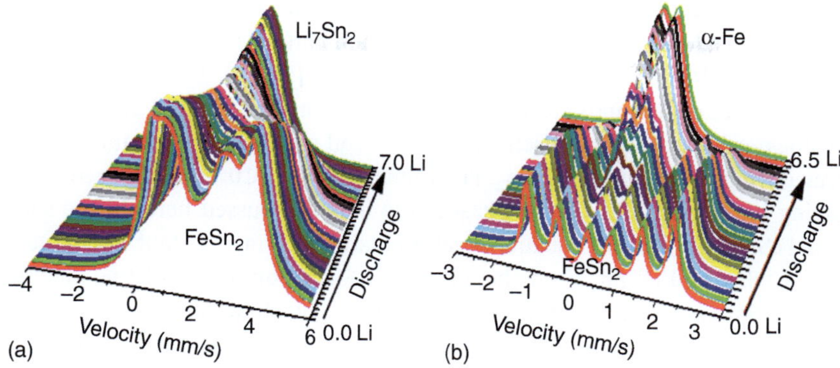

Figure 1.12 Operando ^{119}Sn (a) and ^{57}Fe (b) Mössbauer spectra obtained during the first galvanostatic discharge of $FeSn_2$ (microparticles) electrode in Li half-cell.

The relative amounts of the different species were evaluated from the variations of the subspectrum areas and the recoil-free fractions. They are found to vary linearly with the number of inserted lithium ions in line with the reaction given by Eq. (1.10). The value of the saturation magnetization of the fully lithiated electrode material at 4 K is close to that of bulk α-Fe. The zero-field cooling and field cooling curves obtained at different stages of discharge show the progressive formation of superparamagnetic α-Fe nanoparticles with a constant average diameter of about 3 nm. This result is consistent with the extrusion of Fe atoms from $FeSn_2$ during the conversion reaction and the formation of α-Fe nanoparticles.

For nanostructured $FeSn_2$, the voltage profile is a smooth curve decreasing as lithiation proceeds, instead of the plateau observed for the $FeSn_2$ microparticles (Figure 1.11). However, the observed changes in ^{57}Fe and ^{119}Sn Mössbauer spectra are also consistent with the conversion reaction given by Eq. (1.10) [44]. This indicates that nanostructured and crystalline $FeSn_2$ are both transformed into a Li_7Sn_2/α-Fe nanocomposite during the first discharge. Thus, the first discharge should be considered as a restructuring step for the electrode material, while α-Fe/Li_7Sn_2 nanocomposite is the real starting material for the reversible cycles.

During the first charge of $FeSn_2$ microparticles, the operando ^{119}Sn Mössbauer spectra change from the asymmetrical peak of Li_7Sn_2 into a broad doublet and the average isomer shift increases linearly from 1.8 to 2.3 $mm\cdot s^{-1}$. According to Eq. (1.8), this corresponds to changes in chemical composition from Li_7Sn_2 to LiSn. The observed reverse trend for the second discharge is consistent with the reversibility of the mechanism but strongly differs from the first discharge. The ^{57}Fe Mössbauer spectra do not change significantly during this first charge and the second discharge, showing that the α-Fe nanoparticles do not react with Sn during the reversible cycles. Thus, the reversible mechanism of $FeSn_2$ based electrodes consists of Li_xSn alloying–dealloying reactions. The α-Fe nanoparticles and carbon additives form a composite matrix that buffers the volume variations of Li_xSn during cycling and improves the electronic conductivity of the electrode.

MSn_x Intermetallic Compounds (M = Mn, Co, Ni, Cu) Other tin-based intermetallics have been considered as anode materials for Li-ion batteries. This includes $MnSn_2$ [45, 46], $CoSn_2$ [47, 48], Ni_3Sn_4 [49–51], and Cu_6Sn_5 [52–54]. The voltage profiles of the first three compounds are similar to that of $FeSn_2$ (see Figure 1.11) except for the voltage value of the first discharge plateau and the existence of an additional plateau during the charge for $MnSn_2$. The voltage profile of Cu_6Sn_5 is more complex, involving intermediate Li–Cu–Sn phases, and is not discussed here. In the same way as $FeSn_2$, the ^{119}Sn Mössbauer spectra obtained during the first discharge of $MnSn_2$, $CoSn_2$, and Ni_3Sn_4 show the direct transformation of the pristine material into M/Li_7Sn_2 (M = Mn, Co, Ni) nanocomposites. The situation is more complex for the first charge. The average isomer shift increases as delithiation proceeds for $CoSn_2$ [47] and nanostructured $MnSn_2$ [46] but the spectra are broadened, reflecting the formation of Li–Co–Sn and Li–Mn–Sn ternary phases, respectively, or strong chemical bonds between Li_xSn and metallic nanoparticles. In addition, the voltage curve of $MnSn_2$ shows an additional plateau at the end of charge. The ^{119}Sn Mössbauer spectra obtained at this voltage reflect magnetic relaxation. This was explained by the existence of magnetic $MnSn_2$ particles within the electrode, resulting from partial back reaction of Sn-rich Li_xSn with Mn. For Ni_3Sn_4, the Mössbauer spectrum obtained at the end of the first charge is close to that of the pristine material and can be attributed to the reaction of Sn atoms with Ni nanoparticles [49].

Thus, depending on the transition metal M (Mn, Fe, Co, Ni), back reactions of Sn with M atoms take place at different levels during the delithiation process. Such back reactions that prevent Sn coalescence could improve the cycle life. Although the first and second discharges differ, they both end with the formation of a M/Li_7Sn_2 nanocomposite. The nanostructuration of the first discharge is irreversible, while the cycle formed by the first charge and the second discharge is typical of the reversible cycles. They reflect Li–Sn alloying reactions with possible back reactions between M and Sn atoms. The M/C matrix formed during the first discharge improves the cycle life compared to β-Sn. However, the capacity remains too low for the next generation of high energy Li-ion batteries, and addition of Si was proposed to improve the electrochemical performance.

MSn_x/Si Composites Si can electrochemically react with up to 3.75 Li at room temperature, leading to a theoretical capacity of 3580 $mAh \cdot g^{-1}$, which is about four times higher than that of β-Sn. The electrochemical mechanism of crystalline Si with Li differs from that of β-Sn and is based on Li_xSi alloying–dealloying reactions with the formation of amorphous phases, but still at a low voltage [55]. Thus, the addition of silicon to tin intermetallics is expected to significantly increase the capacity of the negative electrode materials for Li-ion batteries. Better performance was indeed obtained for the Ni_3Sn_4/Si/C composite that shows higher specific capacity than Ni_3Sn_4 [56]. Both Ni_3Sn_4 and Si are electrochemically active in the composite and the operando ^{119}Sn Mössbauer spectra obtained during the charge/discharge cycles are similar to those found for Ni_3Sn_4 negative electrode material. They show the formation of a Ni/Li_7Sn_2 nanocomposite during the first discharge, followed by reversible cycles of Li_xSn alloying–dealloying reactions including the back reaction of Ni with Sn atoms during the charge processes. Both Li_xSi and Li_xSn alloying reactions go along with strong volume variations, but the Ni/C matrix maintains the electronic conductivity and the dispersion of the electrochemically active particles, while the gradual formation of porosity in the electrode film upon cycling helps to buffer the volume variations and the electrolyte impregnation. As a result, a capacity of 800 $mA.g^{-1}$ over 200 cycles was obtained with a capacity loss at first cycle lower than 20%. Other Si-enriched Sn-intermetallic based composites have been proposed, showing improved performance [57–59].

1.2.3 Iron-Based Electrode Materials

1.2.3.1 $LiFePO_4$ as Positive Electrode Material for Li-ion Batteries

Lithium metal oxides with layered structure are commonly used as positive electrode materials for Li-ion batteries. The lithium ions are intercalated between the layers, leading to rather high specific capacity and operating voltage. However, there are some issues regarding the cycle life, rate capability, and safety due to the layered structure. $LiFePO_4$ with olivine structure is an interesting positive electrode material that overcomes these limitations, but has a lower energy density [60]. The voltage curves obtained for $LiFePO_4$ in a Li half-cell and galvanostatic regime are formed by similar plateaus for charge and discharge at 3.4 V that are typical of reversible two-phase reactions. At room temperature, the ^{57}Fe Mössbauer spectrum of $LiFePO_4$ is a doublet ($\delta = 1.22\,mm{\cdot}s^{-1}$, $\Delta = 2.96\,mm{\cdot}s^{-1}$), reflecting high spin Fe^{2+}. The intensity of this doublet decreases during delithiation (charge) while the intensity of another doublet ($\delta = 0.45\,mm{\cdot}s^{-1}$, $\Delta = 1.52\,mm{\cdot}s^{-1}$) increases. The latter doublet is typical of high spin Fe^{3+} in $FePO_4$ [61]. The $LiFePO_4$–$FePO_4$ two-phase reaction is found to be reversible for the lithiation (discharge) and the relative amounts of $LiFePO_4$ and $FePO_4$, evaluated from the relative contributions of the Fe^{2+} and Fe^{3+} subspectra, vary linearly with the number of inserted lithium ions [62, 63].

The substitution of Mn (or Co) for Fe in $LiFePO_4$ has been proposed to increase the electrode potential and, as a result, the energy density of the battery. The partial substitution of Mn for Fe leads to a voltage profile with two plateaus at 3.4 and 4 V instead of only one plateau at 3.4 V for $LiFePO_4$ [64]. The operando ^{57}Fe Mössbauer spectra (Figure 1.13) and the XRD patterns were measured for the first cycle of $Li_xMn_{0.25}Fe_{0.75}PO_4$ in Li half-cell.

For $0.55 < x < 0.95$, the Mössbauer spectra show the progressive oxidation of Fe(II) ($\delta = 1.3\,mm{\cdot}s^{-1}$, $\Delta = 3.0\,mm{\cdot}s^{-1}$) into Fe(III) ($\delta = 0.42\,mm{\cdot}s^{-1}$, $\Delta = 1.10\,mm{\cdot}s^{-1}$) with a quadrupole splitting lower than that of $FePO_4$, and another doublet attributed to Fe(II) ($\delta = 1.23\,mm{\cdot}s^{-1}$, $\Delta = 2.75\,mm{\cdot}s^{-1}$), suggesting the existence of a new iron-based phase. The intensity of the Bragg peaks of $LiMn(II)_{0.25}Fe(II)_{0.75}PO_4$ obtained by operando XRD [64] decreases as charge proceeds, while the intensity increases for the new phase, attributed to $Li_{0.55}Fe(II)_{0.3}Fe(III)_{0.45}Mn(II)_{0.25}PO_4$. For $0.25 < x < 0.55$, there is a sloppy voltage increase up to 4.0 V and a systematic shift of the (020) diffraction peak of $Li_{0.55}Fe(II)_{0.3}Fe(III)_{0.45}Mn(II)_{0.25}PO_4$. This indicates that the reaction with lithium is monophasic, ending with $Li_{0.25}Fe(III)_{0.75}Mn(II)_{0.25}PO_4$. The Fe(II) isomer shift and quadrupole splitting

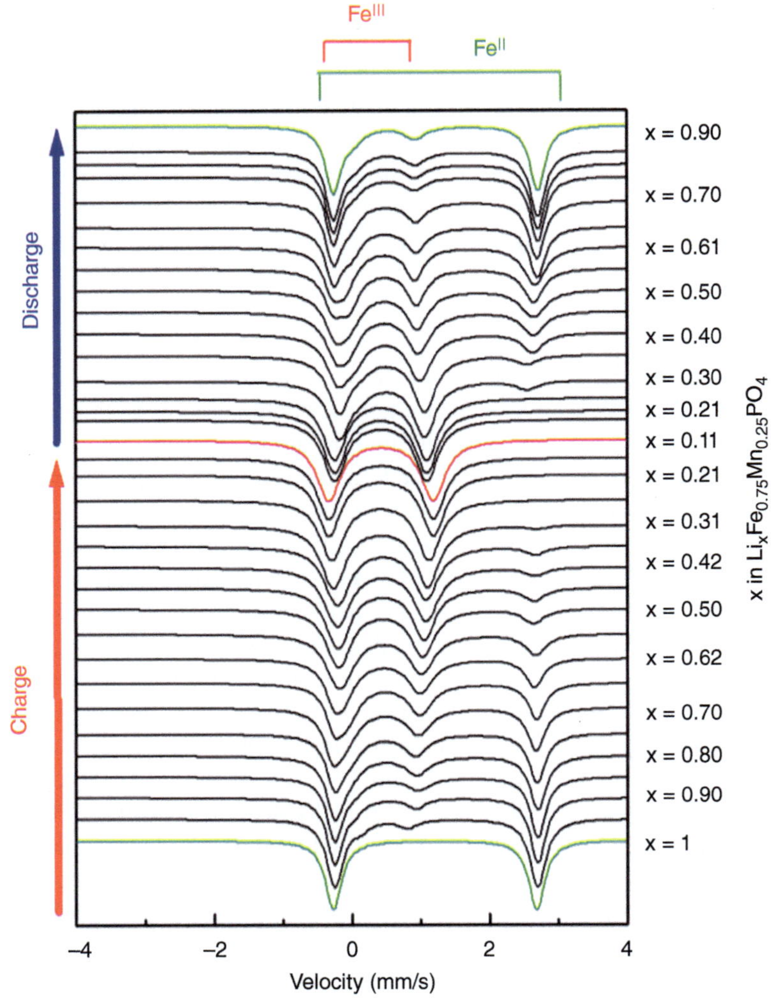

Figure 1.13 ^{57}Fe Mössbauer spectra collected during the first galvanostatic charge–discharge cycle of $LiMn_{0.25}Fe_{0.75}PO_4$ in Li half-cell. Source: Reproduced from Ref. [64]/with permission from Springer Nature.

and the Fe(III) isomer shift remain unchanged while the quadrupole splitting of Fe(III) increases gradually. This was attributed to the increasing disorder around Fe(III). For $0.1 < x < 0.25$, corresponding to the voltage plateau at 4 V, Mn(II) is oxidized into Mn(III). The Mössbauer spectra were fitted to the doublet of $Li_{0.25}Fe(III)_{0.75}Mn(II)_{0.25}PO_4$ and an additional Fe(III) doublet ($\delta = 0.42\,mm{\cdot}s^{-1}$, $\Delta = 1.50\,mm{\cdot}s^{-1}$) attributed to $Fe(III)_{0.75}Mn(III)_{0.25}PO_4$.

1.2.3.2 $Fe_{1.19}PO_4(OH)_{0.57}(H_2O)_{0.43}$/C as Positive Electrode Material for Li-ion Batteries

$Fe_{1.19}PO_4(OH)_{0.57}(H_2O)_{0.43}$/Cnt (FPHH/Cnt, Cnt: carbon nanotubes) has been proposed as a positive electrode material for Li-ion batteries [65]. This composite, obtained by hydrothermal synthesis, has a specific capacity of 120 mAh·g^{-1} at 1 C for

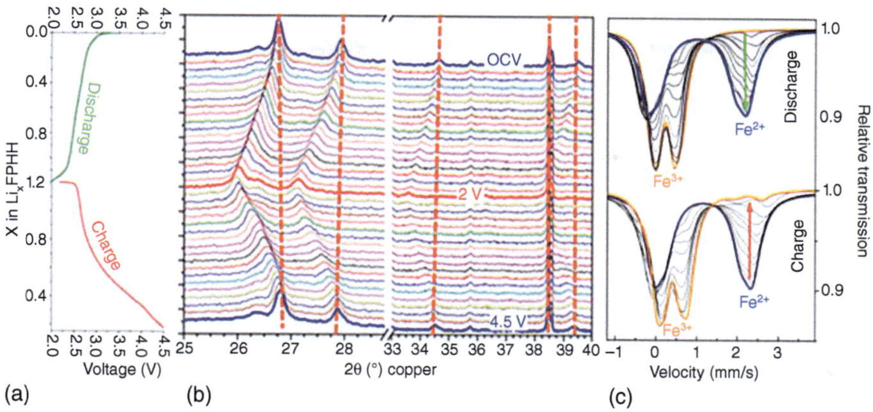

Figure 1.14 First cycle of FPHH based electrode material in Li half-cell: voltage curves (a), operando XRD patterns in the range $2\theta = 25-28.5°$ and 33–40° (b), operando ^{57}Fe Mössbauer spectra (c). Source: Reproduced from Ref. [65]/with permission from American Chemical Society.

500 cycles and shows unusual lithium insertion mechanism. The structure of FPHH ($I4_1/amd$) can be described by perpendicular chains of face sharing iron octahedra along 100 and 010 directions and connected by PO_4 tetrahedra. There is one crystallographic site for Fe with 59% occupation, but the existence of Fe vacancies and O—H bonds are responsible for Fe position disorder. The ^{57}Fe Mössbauer spectrum of FPHH was fitted to two doublets with the same isomer shift: $\delta = 0.4\,mm{\cdot}s^{-1}$ (high spin Fe^{3+}) but different quadrupole splittings: $\Delta = 0.33\,mm{\cdot}s^{-1}$ and $0.67\,mm{\cdot}s^{-1}$, reflecting the distribution of Fe—O bonds and Fe vacancies. The Li^+ diffusion paths are along 100 and 010 channels interconnected by Fe vacancies.

The voltage profile of the first cycle is typical of a one-phase reaction (Figure 1.14). This is confirmed by operando XRD that shows continuous variations in the position of the Bragg peaks of FPHH, in line with the increase and the decrease of the cell volume (~10%) during the discharge and charge, respectively. The Mössbauer spectra change as lithiation proceeds from one doublet for FPHH with an average isomer shift, $\delta_{av} = 0.4\,mm{\cdot}s^{-1}$, typical of Fe^{3+}, to a spectrum formed by two asymmetrical peaks with the average Mössbauer parameters: $\delta_{av} = 1.04\,mm{\cdot}s^{-1}$, $\Delta_{av} = 2.2–2.6\,mm{\cdot}s^{-1}$, typical of high spin Fe^{2+}. The shape and the linewidth of the Mössbauer peaks reflect quadrupole splitting distributions. The evolution of these distributions during the charge–discharge cycle is due to changes in the local environment of Fe atoms. This evolution and the variations of the lattice parameters with Li content reflect a more complex mechanism than a random occupation of the vacant sites by Li. A mechanism based on the selective occupation of iron and channel vacancies that affects both Fe^{2+} and Fe^{3+} environments has been proposed [65].

1.2.3.3 $Na_{1.5}Fe_{0.5}Ti_{1.5}(PO_4)_3/C$ as Electrode Material for Na-ion Batteries

Phosphate compounds with Nasicon structure (sodium super ionic conductor) form an attractive family of electrode materials for Na-ion batteries due to the flexibility of the structure and the existence of a lot of vacant sites for Na^+ insertion. Carbon coated $Na_{1.5}Fe_{0.5}Ti_{1.5}(PO_4)_3/C$ can be used as negative or positive electrode material

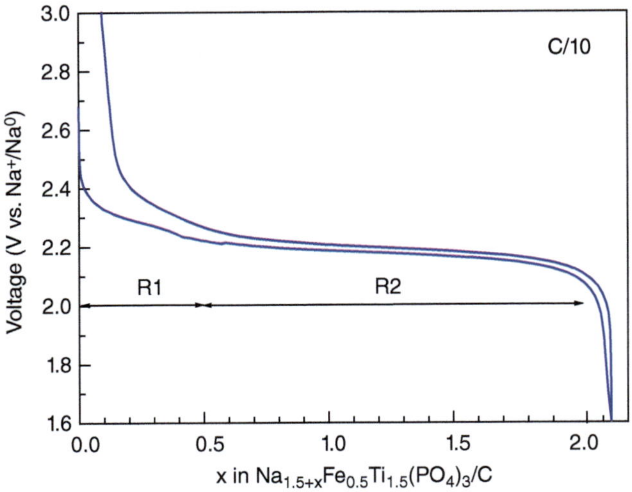

Figure 1.15 Voltage curve of the first cycle of $Na_{1.5}Fe_{0.5}Ti_{1.5}(PO_4)_3$/C in Na half-cell and galvanostatic regime (0.1 Na per hour). Two different mechanisms are observed in regions R1 and R2. Source: Reproduced from Ref. [66]/with permission from American Chemical Society.

for Na-ion batteries due to the existence of two voltage plateaus. $Na_{1.5}Fe_{0.5}Ti_{1.5}(PO_4)_3$ has a rhombohedral structure (*R*-3*c*) with a random occupation of the metal sites by Fe and Ti. The ^{57}Fe Mössbauer spectrum is formed by a doublet due to high spin Fe^{3+} while that of $Na_{1.5}Fe_{0.5}Ti_{1.5}(PO_4)_3$/C shows an additional doublet due to high spin Fe^{2+} in impurities arising from pyrolysis (9 at%) [66, 67].

The voltage curve of the first cycle of $Na_{1.5}Fe_{0.5}Ti_{1.5}(PO_4)_3$/C in a Na half-cell is shown in Figure 1.15. During the first discharge, the voltage decreases from 2.4 to 2.2 V in the range 0–0.5 Na per f.u. (formula unit), suggesting a one-phase reaction (Region R1), and then forms a plateau until the end of sodiation at 2.1 Na per f.u. (Region R2), reflecting a two-phase reaction. This two-step mechanism is reversible for the first charge. The plateau at 2.2 V is also observed for $NaTi_2(PO_4)_3$ and can be attribute to the redox couple Ti^{4+}/Ti^{3+} [68].

The operando ^{57}Fe Mössbauer spectra obtained during the first discharge show the reduction of high spin Fe^{3+} into high spin Fe^{2+} ($\delta = 1.2\ \mathrm{mm{\cdot}s^{-1}}$, $\Delta = 2.7\ \mathrm{mm{\cdot}s^{-1}}$) in region R1. Then, the spectra do not change anymore in region R2 (Figure 1.16).

Deeper insight in the mechanism was obtained by operando XRD [66], confirming that the first discharge is a solid solution reaction coming with the reduction of Fe^{3+} into Fe^{2+}, followed by a two-phase reaction, leading to the reduction of Ti^{4+} into Ti^{3+}:

$$Na_{1.5}Fe^{3+}_{0.5}Ti^{4+}_{1.5}(PO_4)_3\ [R\bar{3}c] + 0.5\ Na = Na_2Fe^{2+}_{0.5}Ti^{4+}_{1.5}(PO_4)_3\ [R\bar{3}c] \tag{1.11}$$

$$Na_2Fe^{2+}_{0.5}Ti^{4+}_{1.5}(PO_4)_3\ [R\bar{3}c] + 1.5\ Na = Na_{3.5}Fe^{2+}_{0.5}Ti^{3+}_{1.5}(PO_4)_3\ [P\bar{1}c] \tag{1.12}$$

These two reactions are reversible for the first charge and form the basic mechanism for the charge–discharge cycles. The observed better cyclability and

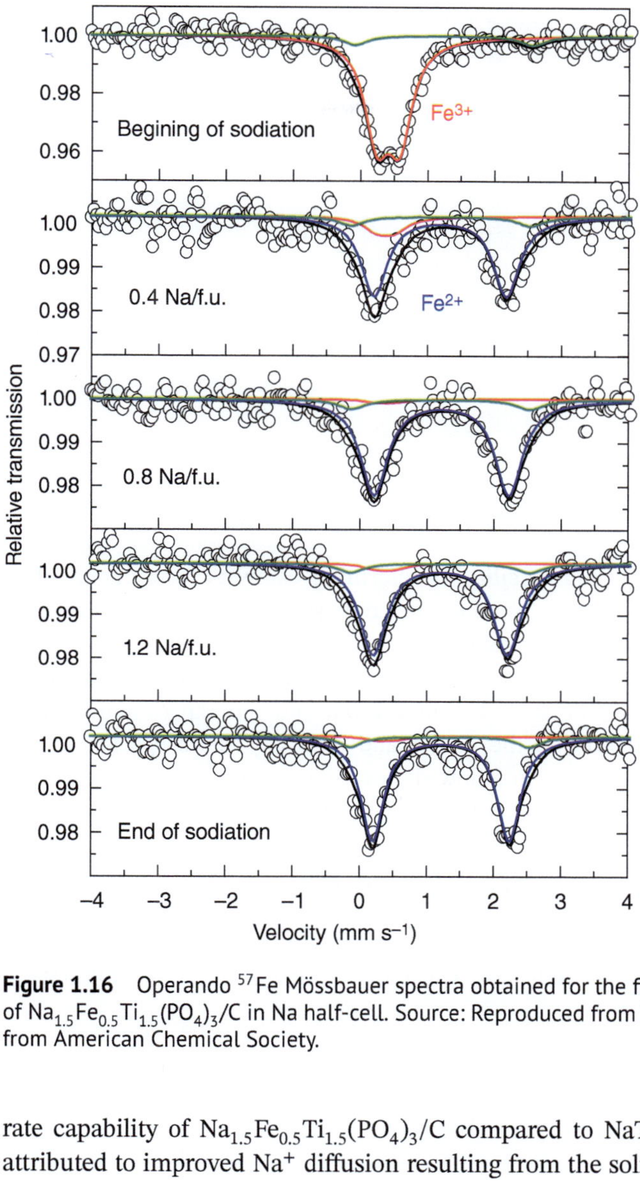

Figure 1.16 Operando ^{57}Fe Mössbauer spectra obtained for the first discharge (sodiation) of $Na_{1.5}Fe_{0.5}Ti_{1.5}(PO_4)_3$/C in Na half-cell. Source: Reproduced from Ref. [66]/with permission from American Chemical Society.

rate capability of $Na_{1.5}Fe_{0.5}Ti_{1.5}(PO_4)_3$/C compared to $NaTi_2(PO_4)_3$/C have been attributed to improved Na^+ diffusion resulting from the solid solution mechanism in R1.

1.3 Mössbauer Spectroscopy of Tin-Based Catalysts

1.3.1 Reforming Catalysis

Alumina-supported bimetallic Pt-Sn catalysts have been proposed to accelerate dehydrogenation, isomerization, and dehydrocyclization reactions in order to increase the gasoline octane number. Addition of tin to monometallic Pt/Al_2O_3

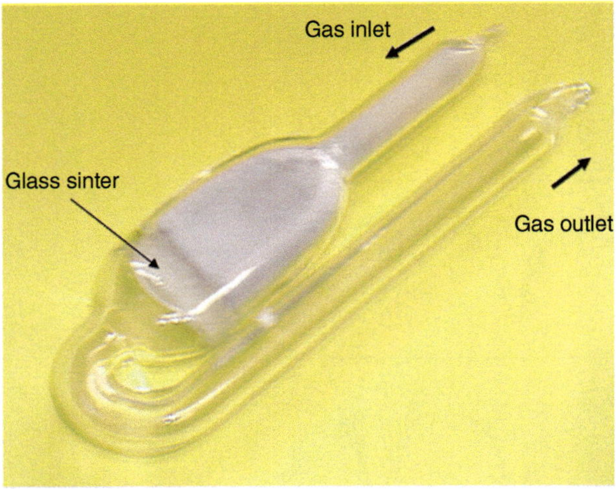

Figure 1.17 Specific cell for in situ ^{119}Sn Mössbauer measurements of Sn-based catalysts.

improves the selectivity and stability of reforming catalysts by reducing their deactivation caused by coke formation (reaction step) and the coalescence of metallic particles (regeneration step). The redox reactions with Pt-Sn/Al_2O_3 were studied by ^{119}Sn Mössbauer spectroscopy in order to evaluate the effects of the synthesis method, the sample composition, and the experimental conditions of reduction/oxidation on the nature and reactivity of tin species. In situ measurements were performed to follow the evolution of the catalyst during the oxidation/reduction cycles with a lab-made cell (Figure 1.17).

1.3.2 Redox Properties of Pt-Sn Based Catalysts

A Mössbauer δ–Δ correlation diagram has been established from the characterization of Sn/Al_2O_3 and Pt-Sn/Al_2O_3 samples (Figure 1.18) [69]. The tin oxidation states

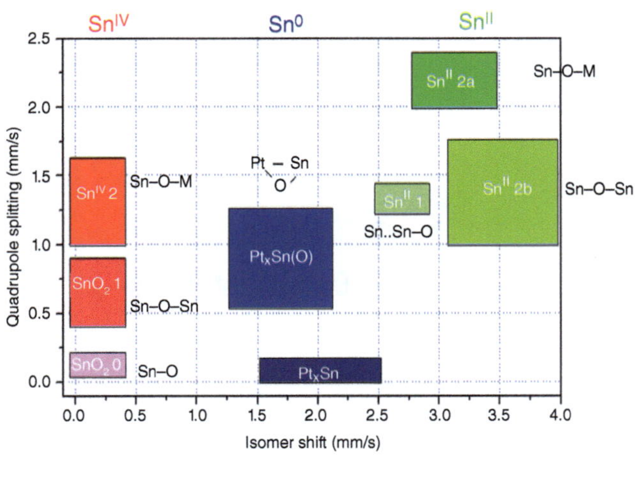

Figure 1.18 Correlation diagram for Sn species in Pt_xSn phases and Sn-based catalysts. Source: Reproduced from Ref. [69]/with permission from Springer Nature.

Sn(IV), Sn(II), and Sn(0) were identified from the values of the isomer shift, while the values of the quadrupole splitting have allowed to distinguish different Sn local environments. This includes three types of Sn(IV): "SnO_2 0" in SnO_6 free units, "SnO_2 1" in SnO_2-like lattices, and "Sn(IV) 2" in Sn(IV)–O–M bridges with M = Al, Pt; two types of Sn(0): "Pt_xSn" and "$Pt_xSn(O)$" close to oxygen; and three types of Sn(II): "Sn(II) 1" in SnO-like structure, "S(II) 2a" in Sn(II)–O–M bridges, and "Sn(II) 2b" in Sn(II)–O–Sn(II) bridges.

The correlation diagram was used to explain the phase transformations occurring in the Pt-Sn particles of Pt-Sn/Al_2O_3 samples prepared by sol-gel or impregnation methods, after oxidation and reduction [70]. The reaction mechanisms between Sn–(n–C_4H_9)$_4$ and alumina surface sites were also identified in order to optimize the controlled preparation of Pt-Sn/Al_2O_3 catalysts [71]. A structural model was given that describes the morphology of the bimetallic particles and their evolution under

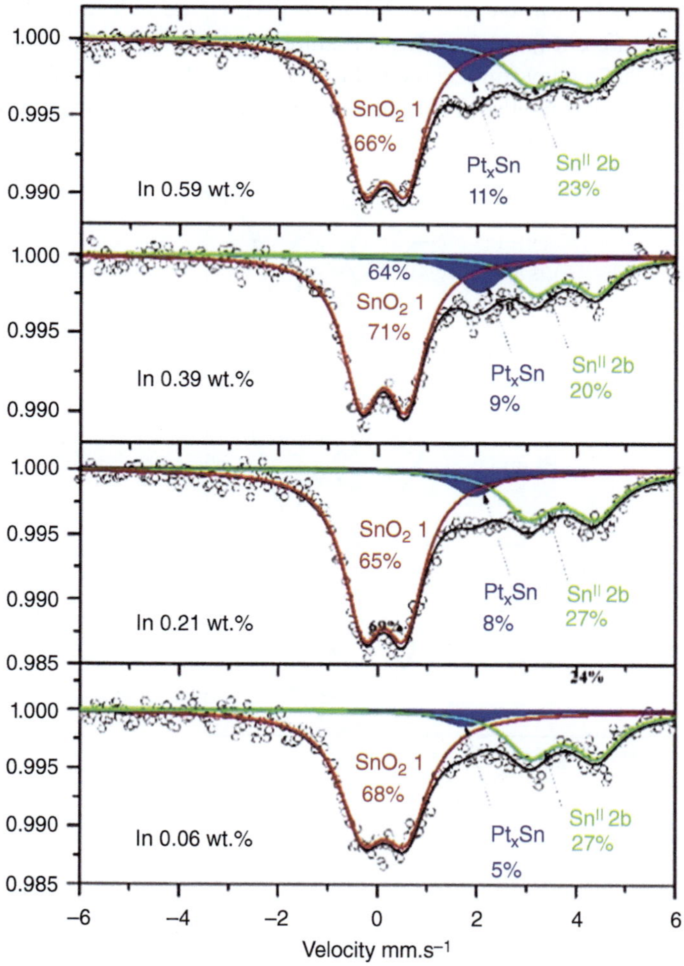

Figure 1.19 ^{119}Sn Mössbauer spectra of reduced catalysts Pt-Sn-In/Al_2O_3–Cl (CP) with different indium loadings. Source: Reproduced from Ref. [69]/with permission from Springer Nature.

various heat treatments at the atomic scale. This model was used to optimize the preparation method of Pt-Sn/Al_2O_3 catalysts in order to preferentially anchor Pt next to Sn particles and improve the catalytic activity. The results are very promising for applications [72, 73].

1.3.3 Trimetallic Pt-Sn-In Based Catalysts

Trimetallic naphtha-reforming catalysts comprising Pt-Sn modified by the addition of indium (Pt/Al_2O_3SnIn–Cl) show high selectivity due to the addition of In as second promoter [14]. Different samples synthesized by coprecipitation (CP), successive impregnations (SI), and surface organostannic chemistry of metals (SOMC) were investigated by ^{119}Sn Mössbauer spectroscopy. The correlation diagram of Figure 1.18 was used to identify various Sn-based species. The presence of In in Pt/Al_2O_3SnIn–Cl obtained by CP of Sn and In with alumina precursor, followed by sol-gel method, favors the formation of Pt_xSn alloys [74]. Figure 1.19 shows the Mössbauer spectra of different Pt/Al_2O_3SnIn$_y$–Cl samples reduced at 500 °C under H_2. The formation of Pt_xSn alloys during the reduction step is observed for all the samples and the increase of isomer shift with increasing indium content y indicates

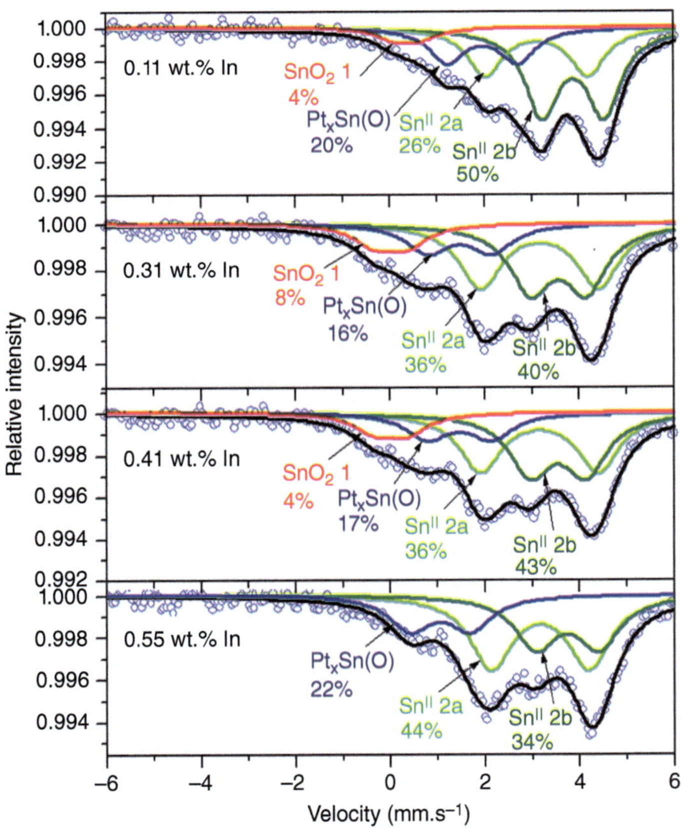

Figure 1.20 ^{119}Sn Mössbauer spectra of reduced catalysts Pt-Sn-In/Al_2O_3–Cl (SI) with different indium loadings.

changes in the chemical composition of Pt_xSn from Pt_3Sn to PtSn. As y increases, more Sn oxides are in a weak interaction with alumina, and Pt-Sn alloying becomes possible with higher tin atomic concentration in the formed alloys.

The presence of indium in Pt-Sn-In/Al_2O_3–Cl catalysts, obtained by SI of metals, leads to the formation of $Pt_xSn(O)$ oxo-metallic phases [75]. The Mössbauer spectrum obtained for 0.11 wt% of In can be fitted to four subspectra that can be assigned to unreduced "Sn(IV) 1" oxide, "Sn(II) 2a" and "Sn(II) 2b" oxides, and oxo-metallic $Pt_xSn(O)$ phase (Figure 1.20). The other spectra obtained for 0.31, 0.41, and 0.55 wt% of In also show oxo-metallic phases that represent 16, 17, and 22% of the Sn species, respectively. The isomer shift of $Pt_xSn(O)$ decreases with increasing In content, which indicates that x increases. Such a decrease in the Sn atomic concentration agrees with the substitution of Sn by In and proves the close Pt–In proximity in these catalysts.

Improving the formation of Pt_xSn alloyed clusters on γ-alumina with a high Sn(0)/Pt ratio represents a crucial step toward more selective heterogeneous

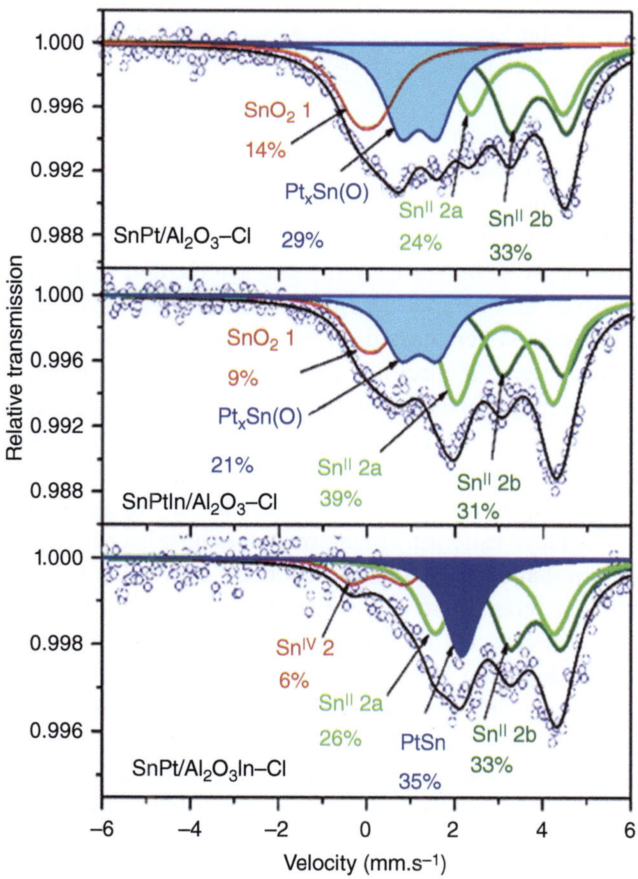

Figure 1.21 ^{119}Sn Mössbauer spectra of reduced catalysts obtained by SOMC: Sn-Pt/Al_2O_3–Cl, Sn-Pt-In/Al_2O_3–Cl (In by SI), and Sn-Pt/Al_2O_3In–Cl (In by CP). Source: Reproduced from Ref. [76]/with permission from American Chemical Society.

catalysts [76]. For that purpose, bimetallic Sn-Pt and trimetallic Sn-Pt-In-based catalysts were prepared by SOMC on monometallic Pt and bimetallic Pt-In-based catalysts, respectively. The ^{119}Sn Mössbauer spectra of reduced Sn-Pt/Al_2O_3–Cl, Sn-Pt-In/Al_2O_3–Cl (In added by impregnation), and Sn-Pt/Al_2O_3In–Cl (In added by CP with alumina source) are shown in Figure 1.21. The spectra of Sn-Pt/Al_2O_3–Cl and Sn-Pt-In/Al_2O_3–Cl show the contribution of Pt_xSn(O) species but not of Pt_xSn alloys, in contrast to Sn-Pt/Al_2O_3In–Cl. Pt-Sn alloying is favored in the latter case, when In is introduced in the support via CP. This can be explained by the existence of In(III) species to stabilize Pt_xSn at the interface with the alumina support.

The application of ^{119}Sn Mössbauer spectroscopy to ternary Pt-Sn-In based systems provides some explanations about the effect of indium as a function of its introduction method, its loading, and the introduction method of the other elements (Pt, Sn). The CP catalysts have higher Sn(0)/Pt ratios than SI catalysts where the elements were initially thought to be closer to Pt due to the preparation method. This is attributed to the presence of Pt_xSn alloys in CP catalysts only and to the substitution of Sn in Pt_xSn(O) by surface indium in SI catalysts. The observed differences in the activity and selectivity of the two differently prepared Pt-Sn-In systems can be related to the Sn(0)/Pt ratio [77].

1.4 Conclusion

In this chapter, we have presented different applications of ^{57}Fe and ^{119}Sn Mössbauer spectroscopies to energy materials, including electrode materials for Li-ion and Na-ion batteries and tin-based catalysts.

The Mössbauer spectroscopy can be used for the characterization of pristine electrode materials but also as an operando technique to follow electrochemical reactions. The examples considered here have been selected to illustrate the three main types of reactions encountered in electrode materials for batteries. The first one is referred to as an alloying reaction where x Li atoms react with a p-block atom M to reversibly form Li_xM. About 4 Li can react with Sn ($x \approx 4$), which explains the high specific capacity of Sn-based electrode materials although they contain a heavy element. However, alloying reactions are associated with strong volume variations that limit the cycle life of batteries. Tin intermetallics have been proposed to overcome this problem. In that case, the first lithiation is a conversion reaction that transforms the pristine material into a nanocomposite for cycling. This is the second type of reaction that also occurs for tin oxides and composites. For both alloying and conversion reactions, the Mössbauer spectroscopy has been mainly used to identify the electrochemically formed species that are often nanosized and poorly crystallized. Finally, the insertion reactions involve Li^+ diffusion on the vacant sites of a host material as often encountered in positive electrode materials containing transition metals. Such reactions go along with the reduction and oxidation of metal ions due to lithiation and delithiation, respectively. In the case of iron-based electrode materials, the ^{57}Fe Mössbauer spectroscopy has been mainly used to detect changes in the oxidation state of Fe or of other metal elements in its environment.

Some examples of supported alumina bimetallic Pt-Sn and trimetallic Pt-Sn-In catalysts have been discussed. In that case, in situ Mössbauer spectroscopy has been used for the identification of tin species formed during the oxidation and reduction steps. The proposed approach, based on the use of an isomer shift – quadrupole splitting correlation diagram, has allowed to identify the most interesting tin species with improved performance and to optimize the synthesis methods.

Mössbauer spectroscopy provides reliable information on complex reaction mechanisms observed in electrochemistry and catalysis. However, an accurate analysis of such mechanisms often requires to combine different in situ measurements, as shown in this chapter for XRD and Mössbauer spectroscopy. Undoubtedly, the use of synchrotron radiation could also be of great interest to expand the field of applications to other elements apart from Fe and Sn and to reduce the measurement time.

Acknowledgments

The authors are grateful to CNES (Toulouse, France), SAFT (Bordeaux, France), UMICORE (Olen, Belgium), and IFPEN (Vernaison, France) for their supports.

References

1 Goodenough, J.B., Whittingham, M.S., and Yoshino, A. (2019). The Nobel Prize in Chemistry 2019. NobelPrize.org. Nobel Prize Outreach AB 2023. www.nobelprize.org/prizes/chemistry/2019/summary. (accessed 4 February 2023).
2 Larcher, D. and Tarascon, J.M. (2015). Towards greener and more sustainable batteries for electrical energy storage. *Nat. Chem.* 7 (1): 19–29.
3 Obrovac, M.N. and Chevrier, V.L. (2014). Alloy negative electrodes for Li-ion batteries. *Chem. Rev.* 114 (23): 11444–11502.
4 Besenhard, J.O., Yang, J., and Winter, M. (1997). Will advanced lithium-alloy anodes have a chance in lithium-ion batteries? *J. Power Sources* 68 (1): 87–90.
5 Larcher, D., Beattie, S., Morcrette, M. et al. (2007). Recent findings and prospects in the field of pure metal as negative electrodes for Li-ion batteries. *J. Mater. Chem.* 17 (36): 3759–3772.
6 Winter, M. and Besenhard, J.O. (1999). Electrochemical lithiation of tin and tin based intermetallics and composites. *Electrochim. Acta* 45 (1–2): 31–50.
7 Zhang, W.J. (2011). Structure and performance of $LiFePO_4$ cathode materials: a review. *J. Power Sources* 196 (6): 2962–2970.
8 Barpanda, P., Ati, M., Melot, B.C. et al. (2011). A 3.90 V iron-based fluorosulphate material for lithium-ion batteries crystallizing in the triplite structure. *Nat. Mater.* 10 (10): 772–779.
9 Palomares, V., Serra, P., Villaluenga, I. et al. (2012). Na-ion batteries, recent advances and present challenges to become low cost energy storage systems. *Energy Environ. Sci.* 5 (3): 5884–5901.

10 Hu, Y.S., Komaba, S., Forsyth, M. et al. (2019). A new emerging technology: Na-ion batteries. *Small Methods* 3 (4): 1900184.

11 Tarascon, J.M. (2010). The greatness of Mössbauer spectroscopy to the design of better Li-ion batteries. *Möss. Eff. Ref. Data J.* 33 (2): 53–54.

12 Lippens, P.E., El Khalifi, M., Chamas, M. et al. (2012). How Mössbauer spectroscopy can improve Li-ion batteries. *Hyperfine Interact.* 206 (1–3): 35–46.

13 Dautzenberg, F.M., Helle, J.N., Biloen, P., and Sachtler, H. (1980). Conversion of n-hexane over mono-functional supported and unsupported PtSn catalysts. *J. Catal.* 63 (1): 119–128.

14 Le Goff, P.Y., Le Peltier, F., Giraud, J. et al. (2010). Bimetallic or multi-metallic catalyst that has an optimized bimetallicity index and an optimized hydrogen adsorption capacity. Fr Patent 2910347, filed 7 December 2007 and issues 25 February 2010.

15 Berry, F. (1984). *Mössbauer Spectroscopy Applied to Inorganic Chemistry* (ed. G.J. Long), 391–442. New York and London: Plenum Press.

16 Li, X., Zhu, K., Pang, J. et al. (2018). Unique role of Mössbauer spectroscopy in assessing structural features of heterogeneous catalysts. *Appl. Catal. B Environ.* 224: 518–532.

17 Harks, P.P.R.M.L., Mulder, F.M., and Notten, P.H.L. (2015). In situ methods for Li-ion battery research: a review of recent developments. *J. Power Sources* 288 (1): 92–105.

18 Jumas, J.C., Fullenwarth, J., Sougrati, M.T. et al. (2014). Analysis of the features of an electrode material of an electrochemical cell. WO Patent 2014033402-A1, filed 28 August 2013 and issued 6 March 2014.

19 Lippens, P.E. and Jumas, J.C. (2010). It's all about the battery. *Möss. Eff. Ref. Data J.* 33 (2): 31–54.

20 Leriche, J.B., Hamelet, S., Shu, J. et al. (2010). An electrochemical cell for operando study of lithium batteries using synchrotron radiation. *J. Electrochem. Soc.* 157 (5): A606–A610.

21 Sivasankaran, V., Marino, C., Chamas, M. et al. (2011). Improvement of intermetallics electrochemical behavior by playing with the composite electrode formulation. *J. Mater. Chem.* 21 (13): 5076–5082.

22 Courtney, I.A., Tse, J.S., Mao, O. et al. (1998). *Ab initio* calculation of the lithium-tin voltage profile. *Phys. Rev. B* 58 (23): 15583–15588.

23 Lippens, P.E., Aldon, L., Ionica, C.M. et al. (2005). Characterization of Li insertion mechanisms in negative electrode materials for Li-ion batteries by Mössbauer spectroscopy and first-principles calculations. In: *Solid State Ionics-2004*, vol. 842 (ed. P. Knauth, C. Masquelier, E. Traversa, et al.), 249–260. Warrendale.

24 Sangster, J. and Bale, C.W. (1998). The Li-Sn (lithium-tin) system. *J. Phase Equilib.* 19 (1): 70–75.

25 Dunlap, R.A., Small, D.A., MacNeil, D.D. et al. (1999). A Mössbauer effect investigation of the Li–Sn system. *J. Alloys Compd.* 289 (1–2): 135–142.

26 Robert, F., Lippens, P.E., Olivier-Fourcade, J. et al. (2007). Mössbauer spectra as a fingerprint in tin-lithium compounds: applications to Li-ion batteries. *J. Solid State Chem.* 180 (1): 339–348.

27 Naille, S., Jumas, J.C., Lippens, P.E. et al. (2009). ^{119}Sn Mössbauer parameters as predictive tool for future Sn-based negative electrode materials. *J. Power Sources* 189 (1): 814–817.

28 Chouvin, J., Olivier-Fourcade, J., Jumas, J.C. et al. (1999). ^{119}Sn Mössbauer study of Li_xSn alloys prepared electrochemically. *Chem. Phys. Lett.* 308 (5–6): 413–420.

29 Idota, Y., Kubota, T., Matsufuji, A. et al. (1997). Tin-based amorphous oxide: a high-capacity lithium-ion-storage material. *Science* 276 (5317): 1395–1397.

30 Sandu, I., Brousse, T., Schleich, D.M. et al. (2004). SnO_2 negative electrode for lithium ion cell: *in situ* Mössbauer investigation of chemical changes upon discharge. *J. Solid State Chem.* 177 (11): 4332–4340.

31 Chouvin, J., Olivier-Fourcade, J., Jumas, J.C. et al. (2000). SnO reduction in lithium cells: study by X-ray absorption, ^{119}Sn Mössbauer spectroscopy and X-ray diffraction. *J. Electroanal. Chem.* 494 (2): 136–146.

32 Courtney, I.A., Dunlap, R.A., and Dahn, J.R. (1999). *In-situ* 119Sn Mössbauer effect studies of the reaction of lithium with SnO and SnO: 0.25 B_2O_3: 0.25 P_2O_5 glass. *Electrochim. Acta* 45 (1–2): 51–58.

33 Jumas, J.C., Lippens, P.E., Olivier-Fourcade, J. et al. (2006). Negative electrode composite material, production method, negative electrode and lithium-ion battery. Fr Patent 2873855, Eu *794993*, filed 28 July 2005 and issued 3 January 2008.

34 Aboulaich, A., Womes, M., Olivier-Fourcade, J. et al. (2010). Sn-0.4BPO_4 composite as a promising negative electrode for rechargeable lithium batteries. *Solid State Sci.* 12 (1): 65–72.

35 Conte, D.E., Mouyane, M., Jumas, J.C. et al. (2012). A combined Mössbauer spectroscopy and X-ray diffraction operando study of Sn-based composite anode materials for Li-ion accumulators. *J. Solid State Electrochem.* 16 (12): 3837–3848.

36 SONY (2005). Sony's new Nexelion hybrid Li-ion batteries to have thirty percent more capacity than conventional offering. Sony Group, News Releases, 15 February 2005. www.sony.com/en/SonyInfo/News/Press/200502/05-006E. (accessed 4 February 2023).

37 Hassoun, J., Panero, S., Mulas, G., and Scrosati, B. (2007). An electrochemical investigation of a Sn–Co–C ternary alloy as a negative electrode in Li-ion batteries. *J. Power Sources* 171 (2): 928–931.

38 Mao, O., Dunlap, R.A., and Dahn, J.R. (1999). Mechanically alloyed Sn–Fe(–C) powders as anode materials for Li-ion batteries: I. The Sn_2Fe–C system. *J. Electrochem. Soc.* 146 (2): 405–413.

39 Ferguson, P.P., Martine, M.L., George, A.E., and Dahn, J.R. (2009). Studies of tin–transition metal–carbon and tin–cobalt–transition metal–carbon negative electrode materials prepared by mechanical attrition. *J. Power Sources* 194 (2): 794–800.

40 Ferguson, P.P., Dunlap, R.A., and Dahn, J.R. (2010). An *in situ* study of the electrochemical reaction of Li with nanostructured $Sn_{30}Co_{30}C_{40}$. *J. Electrochem. Soc.* 157 (3): A326–A332.

41 Chamas, M., Lippens, P.E., Jumas, J.C. et al. (2011). Comparison between microparticles and nanostructured particles of $FeSn_2$ as anode materials for Li-ion batteries. *J. Power Sources* 196 (16): 7011–7015.

42 Le Caer, G., Malaman, B., Venturini, G., and Kim, I.B. (1982). ^{119}Sn hyperfine fields in $MnSn_2$. *Phys. Rev. B* 26 (9): 5085–5096.

43 Le Caer, G., Malaman, B., Venturini, G. et al. (1985). A Mossbauer study of $FeSn_2$. *J. Phys. F* 15 (8): 1813–1827.

44 Chamas, M., Sougrati, M.T., Reibel, C., and Lippens, P.E. (2013). Quantitative analysis of the initial restructuring step of nanostructured $FeSn_2$-based anodes for Li-ion batteries. *Chem. Mater.* 25 (12): 2410–2420.

45 Mahmoud, A., Chamas, M., Jumas, J.-C. et al. (2013). Electrochemical performances and mechanisms of $MnSn_2$ as anode material for Li-ion batteries. *J. Power Sources* 244: 246–251.

46 Philippe, B., Mahmoud, A., Ledeuil, J.B. et al. (2014). $MnSn_2$ electrodes for Li-ion batteries: mechanisms at the nano scale and electrode/electrolyte interface. *Electrochem. Acta* 123 (20): 72–83.

47 Ionica-Bousquet, C.M., Lippens, P.E., Aldon, L. et al. (2006). *In situ* ^{119}Sn Mössbauer effect study of Li–$CoSn_2$ electrochemical system. *Chem. Mater.* 18 (26): 6442–6447.

48 Wang, S., He, M., Walter, M. et al. (2018). Monodisperse $CoSn_2$ and $FeSn_2$ nanocrystals as high-performance anode materials for lithium-ion batteries. *Nanoscale* 10 (15): 6827–6831.

49 Ehinon, K.K.D., Naille, S., Dedryvère, R. et al. (2008). Ni_3Sn_4 electrodes for Li-ion batteries: Li–Sn alloying process and electrode/electrolyte interface phenomena. *Chem. Mater.* 20 (16): 5388–5398.

50 Naille, S., Dedryvère, R., Zitoun, D., and Lippens, P.E. (2009). Atomic-scale characterization of tin-based intermetallic anodes. *J. Power Sources* 189 (1): 806–808.

51 Naille, S., Lippens, P.E., Morato, F. et al. (2006). ^{119}Sn Mössbauer study of nickel–tin anodes for rechargeable lithium-ion batteries. *Hyperfine Interact.* 167 (1–3): 785–790.

52 Fransson, L., Nordström, E., Edström, K. et al. (2002). Structural transformations in lithiated η'-Cu_6Sn_5 electrodes probed by *in situ* Mössbauer spectroscopy and X-ray diffraction. *J. Electrochem. Soc.* 149 (6): A736–A742.

53 Naille, S., Dedryvère, R., Martinez, H. et al. (2007). XPS study of electrode/electrolyte interfaces of η-Cu_6Sn_5 electrodes in Li-ion batteries. *J. Power Sources* 174 (2): 1086–1090.

54 Tan, X.F., McDonald, S.D., Gu, Q. et al. (2019). Characterisation of lithium-ion battery anodes fabricated via *in-situ* Cu_6Sn_5 growth on a copper current collector. *J. Power Sources* 415 (1): 50–61.

55 Li, J. and Dahn, J.R. (2009). An *in situ* X-ray diffraction study of the reaction of Li with crystalline Si. *J. Electrochem. Soc.* 154 (3): A156–A161.

56 Edfouf, Z., Sougrati, M., Fariaut-Georges, C. et al. (2013). Reactivity assessment of lithium with the different components of novel $Si/Ni_{3.4}Sn_4/Al/C$ composite anode for Li-ion batteries. *J. Power Sources* 238 (15): 210–217.

57 Jumas, J.C., Lippens, P.E., Olivier-Fourcade, J. et al. (2017). Nouveau matériau composite enrichi en silicium, son procédé de fabrication et son utilisation à titre d'électrode. Fr Patent 3051483, filed 19 May 2017 and issued 24 November 2017.

58 Azib, T., Bibent, N., Latroche, M. et al. (2020). Ni–Sn intermetallics as efficient buffering matrix of Si anodes in Li-ion batteries. *J. Mater. Chem. A* 8 (35): 18132–18142.

59 He, Y., Ye, Z., Chamas, M. et al. (2022). Porous $Si/Cu_6Sn_5/C$ composite containing native oxides as anode material for Li-ion batteries. *J. Mater. Sci. Mater. Electron.* 33: 235–243.

60 Pahdi, A.K., Nanjundaswamy, K.S., and Goodenough, J.B. (1997). Phospho-olivines as positive-electrode materials for rechargeable lithium batteries. *J. Electrochem. Soc.* 144 (4): 1188–1194.

61 El Khalifi, M. and Lippens, P.E. (2016). First-principles investigation of the ^{57}Fe Mössbauer parameters of $LiFePO_4$ and $FePO_4$. *J. Phys. Chem. C* 120 (45): 28375–28389.

62 Andersson, A.S., Kalska, B., Häggström, L. et al. (2000). Lithium extraction/insertion in $LiFePO_4$: an X-ray diffraction and Mössbauer spectroscopy study. *Solid State Ionics* 130 (1–2): 41–52.

63 Aldon, L., Perea, A., Womes, M. et al. (2010). Determination of the Lamb-Mössbauer factors of $LiFePO_4$ and $FePO_4$ for electrochemical *in situ* and operando measurements in Li-ion batteries. *J. Solid State Chem.* 183 (1): 218–222.

64 Jumas, J.C., Sougrati, M.T., Perea, A. et al. (2013). Combined operando studies of new electrode materials for Li-ion batteries. *Hyperfine Interact.* 217 (1): 107–115.

65 Mahmoud, A., Karegeya, C., Sougrati, M.T. et al. (2018). Electrochemical mechanism and effect of carbon nanotubes on the electrochemical performance of $Fe_{1.19}(PO_4)(OH)_{0.57}(H_2O)_{0.43}$ cathode material for Li-ion batteries. *ACS Appl. Mater. Interfaces* 10 (40): 34202–34211.

66 Difi, S., Saadoune, I., Sougrati, M.T. et al. (2015). Mechanisms and performances of $Na_{1.5}Fe_{0.5}Ti_{1.5}(PO_4)_3/C$ composite as electrode material for Na-ion batteries. *J. Phys. Chem. C* 119 (45): 25220–25234.

67 Difi, S., Saadoune, I., Sougrati, M.T. et al. (2016). Role of iron in $Na_{1.5}Fe_{0.5}Ti_{1.5}(PO_4)_3/C$ as electrode material for Na-ion batteries studied by operando Mössbauer spectroscopy. *Hyperfine Interact.* 237 (1): 61–68.

68 Delmas, C., Cherkaoui, F., Nadiri, A., and Hagenmuller, P. (1987). A nasicon-type phase as intercalation electrode: $NaTi_2(PO_4)_3$. *Mater. Res. Bull.* 22 (5): 631–639.

69 Jumas, J.C., Sougrati, M.T., Olivier-Fourcade, J. et al. (2013). Identification and quantification of Sn-based species in trimetallic Pt-Sn-In/Al_2O_3-Cl naphtha-reforming catalysts. *Hyperfine Interact.* 217: 137–144.

70 Olivier-Fourcade, J., Womes, M., Jumas, J.C. et al. (2004). Investigation of redox properties of different PtSn/Al_2O_3 catalysts. *ChemPhysChem* 5 (11): 1734–1744.

71 Womes, M., Le Peltier, F., Morin, S. et al. (2007). Study of the reaction mechanisms between Sn-(n-$C_4H_9)_4$ and alumina surface sites. Application to the controlled preparation of PtSn/Al_2O_3 catalysts. *J. Mol. Catal. A Chem.* 266 (1–2): 55–64.

72 Le Peltier, F., Didillon, B., Jumas, J.C. et al. (2003). Supported bimetallic catalyst with a strong interaction between a group VIII metal and tin and its use in a catalytic reforming process. Patent FR2813209, US2002045544, EP1181978, 23 August 2001 and issues 12 August 2003.

73 Le Peltier, F., Didillon, F., Jumas, J.C. et al. (2003). Process for dehydrogenating organic compounds in the presence of a supported bimetallic catalyst with a strong interaction between a group VIII metal and tin. Patent FR2813304, US20020445787, EP1182180, filed 23 August 2001 and issues 29 July 2003.

74 Jahel, A., Avenier, P., Lacombe, S. et al. (2010). Effect of indium in trimetallic Pt/Al_2O_3SnIn–Cl naphtha-reforming catalysts. *J. Catal.* 272 (2): 275–286.

75 Jahel, A., Lacombe, S., Avenier, P. et al. (2012). Mössbauer spectroscopy: an elegant tool for the active sites identification and quantification in Pt-Sn-In based naphtha reforming catalysts. *Hyperfine Interact.* 207 (1–3): 37–40.

76 Jahel, A., Moizan-Baslé, V., Chizallet, C. et al. (2012). Effect of indium doping of γ-alumina on the stabilization of PtSn alloyed clusters prepared by surface organostannic chemistry. *J. Phys. Chem. C* 116 (18): 10073–10083.

77 Avenier, P., Lacombe, S., Jumas, J.C., Olivier-Fourcade, J. (2012). Multi-metallic catalyst with a strong metallic interaction. FR Patent 2947464, filed 15 June 2010 and issues 17 May 2012.

2

Mössbauer Spectral Studies of Iron Phosphate Containing Minerals and Compounds

Gary J. Long and Fernande Grandjean

University of Missouri, Department of Chemistry, Missouri University of Science and Technology, 400 West 11th Street, Rolla, MO 65409-0010, USA

2.1 Introduction

The possibility of forming compounds with one or more phosphate anions, PO_4^{3-}, or the related HPO_4^{2-}, hydrogen phosphate, and $H_2PO_4^{-}$, dihydrogen phosphate, and the various alkali and alkaline-earth phosphate compounds with iron(II) and/or iron(III) cations, yields a rich structural variety of minerals and compounds. In addition, the possible structural range is extended by the possible presence of mixed valence iron(II) and iron(III) in minerals and compounds with various Robin and Day [1] class electron transfer between the iron(II) and iron(III) cations. Further, as a result of the range of structural possibilities, many iron phosphate compounds, including the simplest compound, α-$FePO_4$ [2, 3], exhibit both structural phase transitions or unusual magnetic exchange interactions or sometimes both.

Further, many of the alkali and alkaline-earth iron phosphate containing minerals and compounds also exhibit valuable practical applications, such as crop fertilization, heterogeneous catalysis, ion exchange media, and, more recently, cathodes in lithium-ion type batteries. These applications have led to extensive Mössbauer spectral work of these various minerals and compounds. In addition, the possibility of finding phosphate minerals on the surface of Mars, in part through Mössbauer spectroscopy, has opened another field of research.

This chapter will survey a good portion of the recent iron-57 Mössbauer spectral work on the above mentioned iron phosphate containing natural minerals and compounds. The chapter will also briefly discuss a few of the problems we have encountered in the published Mössbauer-effect literature on iron phosphate compounds. Finally, during the course of preparing this chapter, the authors have come upon two papers with intriguing titles for anyone interested in the properties of iron phosphate compounds and minerals. These two papers have led to the next two sections of this chapter.

Mössbauer Spectroscopy: Applications in Chemistry and Materials Science, First Edition. Edited by Yann Garcia, Junhu Wang, and Tao Zhang.

2.2 Thermodynamic Properties of Iron Phosphate Containing Compounds

The first paper [4] mentioned above and titled "Estimating the Thermodynamic Properties of Phosphate Minerals at High and Low Temperature from the Sum of Constituent Units" is especially interesting because the standard thermodynamic properties, i.e. the standard enthalpy of formation, ΔH_f°, and the standard Gibbs free energy of formation, ΔG_f°, of only a few phosphate minerals and compounds have been reported. Hence, the polyhedral units model approach presented in this paper may prove very useful in the field of iron phosphate research. Because relatively few iron phosphate containing minerals or compounds are included in the results given in reference [4], Table 2.1 [5–9] yields the calculated ΔH_f° and ΔG_f° values that can be obtained with the polyhedral units model along with a comparison with the limited number of experimental values available.

Unfortunately, the polyhedral units model approach is limited to compounds that can be formulated in terms of a polyhedral units basis set, see table 2 in reference [4], a set that is limited to P_2O_5, Li_2O, Na_2O, K_2O, $(NH_4)_2O$, $H_2O(H)$, MgO, CaO, FeO, CoO, NiO, ZnO, CuO, PbO, UO_3, Al_2O_3, crystalline H_2O, $H_2O(OH^-)$, F—O, and Cl—O. Further, Li_2O cannot be used in a calculation of ΔG_f° and CoO, NiO, and UO_3 cannot be used in a calculation of ΔH_f° because their basis unit factors are not included in table 2 in reference [4]. Further, we have come upon a small unexpected but acceptable difference [10] in the example calculation included on page 110 of reference [4]. Also acceptable are the results reported [4] for berlinite, α-AlPO4, i.e. calculated $\Delta H_f^\circ = -1752(5)$ kJ/mol and $\Delta G_f^\circ = -1625(6)$ kJ/mol from reference [4] table 4, which agree well with the experimental values of $\Delta H_f^\circ = -1733.80$ kJ/mol and $\Delta G_f^\circ = -1601.20$ kJ/mol reported by Wagman et al. [11].

Thus, at this point, the computational method presented in reference [4] seems to be viable and so the authors tried to apply the method to calculate the properties of the mineral strengite, $FePO_4 \cdot 2H_2O$, a calculation that is not reported in reference [4] and involves the following equations:

$$\begin{aligned}\Delta H_f^\circ(FePO_4 \cdot 2H_2O) &= h(FeO) + ½\,h(P_2O_5) + ½\,h(H_2O(OH^-)) + 2h(H_2O(cryst.))\\ &= -319.16 - 1726.84/2 - 267.20/2 - 2 \times 299.22\\ &= -1915(3)\ kJ/mol\end{aligned}$$

$$\begin{aligned}\Delta G_f^\circ(FePO_4 \cdot 2H_2O) &= g(FeO) + ½g(P_2O_5) + ½g(H_2O(OH^-)) + 2g(H_2O(cryst.))\\ &= -269.53 - 1636.94/2 - 255.04/2 - 2 \times 239.10\\ &= -1694(3)\ kJ/mol,\end{aligned}$$

where h_i and g_i are the basic unit values given in table 2 of reference [4]. These calculated values agree rather well with the experimental values of $\Delta H_f^\circ = -1888.24$ kJ/mol and $\Delta G_f^\circ = -1657.70$ kJ/mol reported in reference [12]. The user of this method

Table 2.1 Standard 298.15 K thermodynamic parameters for several iron-bearing phosphate minerals and compounds.[a)]

Natural mineral or compound	Ideal composition	Calc./obs.	ΔH_f°, kJ/mol	% Uncertainty	% Difference	ΔG_f°, kJ/mol	% Uncertainty	% Difference	Dana number	References
Berlinite	α-$AlPO_4$	Calc.	−1754(5)	0.285	−1.164	−1625(6)	0.369	−0.613	38.4.2.1	[5]
		Obs.	−1733.82(3)[b)]	—	—	−1615(11)[b)]	—	—		
Arrojadite	$K_2Na_5Fe^{2+}{}_{14}Fe^{3+}(PO_4)_{12}(OH)_2$	Calc.	−18 277(52)	0.28	—	−16 775(71)	0.42	—	41.7.2.1	[5]
Arsenocrandallite	$CaAl_3(OH)_6[PO_3(O_{1/2}(OH)_{1/2})]_3$	Calc.	−8161(34)	0.42	—	−7463(45)	0.60	—	42.7.4.1	[5]
Barbosalite	$Fe^{2+}Fe^{3+}{}_2(PO_4)_2(OH)_2$	Calc.	−3219(15)	0.47	—	−2956(21)	0.71	—	41.10.1.4	[5]
Bobierrite	$Mg_3(PO_4)_2{\cdot}8H_2O$	Calc.	−6199(16)	0.26	—	−5435(15)	0.28	0.22	40.3.7.1	[5]
		Obs.	—	—	—	−5447(4)[b)]	0.07	—		
Ferristrunzite	$[Fe^{3+}Fe^{3+}{}_2(PO_4)_2(OH)_3{\cdot}5(H_2O)]$	Calc.	−4982(20)	0.40	—	−4406(27)	0.61	—	42.11.9.3	[6]
Ferrolaueite	$[Fe^{2+}Fe^{3+}{}_2(PO_4)_2(OH)_2{\cdot}8(H_2O)]$	Calc.	−5612(16)	0.29	—	−4868(21)	0.43	—	42.11.10.7	[6]
Graftonite	$[Fe_3{}^{2+}(PO_4)_2]$	Calc.	−2684(9)	0.38	—	−2445(13)	0.53	—	38.3.3.1	[5]
Heterosite	η-$FePO_4$	Calc.	−1316(5)	0.38	−0.01	−1216(7)	0.58	−0.06	38.4.1.1	[7]
		Obs.	−1297.5			−1148.4				[7]
Ludlamite	$Fe_3{}^{2+}(PO_4)_2(H_2O)_4$	Calc.	−3881(9)	0.23	—	−3402(13)	0.38	—	40.3.5.1	[5]
Maricite	$NaFe^{2+}PO_4$	Calc.	−1542(4)	0.26	—	−1421(6)	0.42	—	38.1.2.1	[5]
Merrilite	$Ca_9NaMg(PO_4)_7$	Calc.	−14 232(56)	0.39	—	−13 373(44)	0.33	—	38.3.4.4	[6]
Phosphosiderite	$Fe^{3+}PO_4(H_2O)_2$	Calc.	−1915(5)	0.26	—	−1694(7)	0.41	—	40.4.3.2	[5]
Strengite	$Fe^{3+}PO_4{\cdot}2H_2O$	Calc.	−1915(5)	0.26	−1.40	−1694(7)	0.41	−2.14	40.4.1.2	[5]
		Obs.	−1888.20	—		−1657.70	—			[7]

(Continued)

Table 2.1 (Continued)

Natural mineral or compound	Ideal composition	Calc./obs.	ΔH_f°, kJ/mol	% Uncertainty	% Difference	ΔG_f°, kJ/mol	% Uncertainty	% Difference	Dana number	References
Tarbuttite	$Zn_2(PO_4)(OH)$	Calc.	−1936(9)	0.46		−1736(8)	0.46	—	41.6.7.1	[5]
Triphylite	$LiFePO_4$	Calc.	−1591(12)	0.75	1.57	—	—	—	38.1.1.1	[5]
		Obs.	−1616(2)	0.12	—	—	—	—		[8]
Vivianite	$Fe_3(PO_4)_2{\cdot}8H_2O$	Calc.	−5079(20)	0.39	—	−4358(26)	0.60	0.44	40.3.6.1	[5]
						−4377.2				[7]
Whitlockite	$Ca_9Mg(PO_4)_6(HPO_4)$	Calc.	−14 006(56)	0.40	—	−13 167(44)	0.33	—	38.3.4.1	[5]
α-$FePO_4$	α-$FePO_4$	Calc.	−1316(5)	0.38	0.08	−1216(7)	0.58		—	
		Obs.	−1317(1)[c]						—	
$Fe_2(PO_4)O$	$Fe_2(PO_4)O$	Calc.	−1635(6)	0.37	—	−1485(9)	0.61	—	—	
$Fe_2P_2O_7$	$Fe_2P_2O_7$	Calc.	−2046(7)	0.34	—	−1906(9)	0.47	—	—	
$Fe_3(PO_4)O_3$	$Fe_3(PO_4)O_3$	Calc.	−1968(11)	0.56	—	−2283(7)	0.31	—	—	
$Fe_7(PO_4)_6$	$Fe_7(PO_4)_6$	Calc.	−7682(26)	0.33	—	−7053(35)	0.50	—	—	
$Fe_9(PO_4)O_8$	$Fe_9(PO_4)O_8$	Calc.	−3869(16)	0.41	—	−3372(29)	0.86	—	—	
			<-3474			<-3284				[9]
$Pb_3Fe_2(PO_4)_4$	$Pb_3Fe_2(PO_4)_4$	Calc.	−5228(17)	0.33	—	−4795(22)	0.46	—	—	
$Na_4CaFe_4(PO_4)_6$	$Na_4CaFe_4(PO_4)_6$	Calc.	−9223(27)	0.29	—	−8572(34)	0.40	—	—	

a) Calculated parameters are obtained with the polyhedral units model approach given in reference [4].
b) The average of the literature values given in reference [4].
c) The experimental value obtained from strengite, $FePO_4{\cdot}2H_2O$, after the removal of the two waters of crystallization. An observed value of −1267(1) kJ/mol has been reported in reference [8].

should note that, to reach this satisfactory agreement, one *must* be careful in expressing ΔH_f° and ΔG_f° in terms of the "basic units" given in table 2 of reference [4]. Specifically in separating the strengite, $FePO_4 \cdot 2H_2O$, into its "basic unit" components, one must account for all the oxygens and include the "basic unit" $H_2O(OH^-)$ as indicated in reference [4]. The uncertainties in the above calculated values were obtained as given in equation (6) in reference [4]; in table 2 of reference [4] no uncertainty for *h*(FeO), i.e. the error ± in *h*(FeO), is missing – the correct value is 1.75 kJ/mol.

2.3 Room Temperature Mössbauer Spectra of Iron Phosphate Containing Minerals

The second paper [6] mentioned above is titled "Mössbauer Parameters of Iron in Phosphate Minerals: Implications for Interpretation of Martian Data." Further recent papers [13] titled "Synthesis and Characterization of the Mars-relevant Phosphate Minerals Fe- and Mg-whitlockite and Merrillite and a Possible Mechanism that Maintains Charge Balance During Whitlockite to Merrillite Transformation," "Mössbauer Spectroscopy: Theory and Laboratory Spectra of Geologic Materials," and "Mössbauer Spectroscopy at Gusev Crater and Meridiani Planum: Iron Mineralogy, Oxidation State, and Alteration on Mars," may also be of interest. Some of these papers, especially the first, will be discussed in detail below.

In their new, highly useful, and extensive room temperature iron-57 Mössbauer spectral study of 65 different iron-containing phosphate minerals, Dyar et al. [6] have published an extensive "fingerprint" collection of 64 spectra; the paper also includes 62 additional related mineral studies found in the literature. This study provides an extensive catalogue of Mössbauer spectra along with analyses of their paramagnetic spectral hyperfine parameters; the goal of the study is identifying which, if any, iron phosphate minerals may be found in the Mössbauer spectra measured on the surface of Mars. To achieve this goal, the authors have presented extensive plots comparing all the iron-57 isomer shifts as a function of the associated quadrupole splitting. This plot, figure 9A in reference [6], is then combined for comparison with the analogous plots for the iron sulfate and iron silicate minerals that may be present on Mars.

Unfortunately in studying the work of Dyar et al. [6], the authors have come upon several small problems. The first problem is that their figure 9A plot reveals a number of data points that seem quite unexpected, data points that, if reliable, would be quite interesting, but that could also be rather misleading. The second problem was the difficulty in identifying which mineral corresponds to the hyperfine parameters associated with a given plotted point, including the unexpected data points. Fortunately, the hyperfine parameters are given in table 2 of reference [6], an extensive table that is not included in the published paper but is available as supplementary information. Although in a few cases the naming on the mineral may be different, the original numerical data corresponding to the spectral results found in table 2 of reference [6] and in Table 2.2 of this chapter may be found at the website: nemo.cs.umass.edu:51515/group/Phosphate.

Table 2.2 Mössbauer spectral parameters for iron-bearing phosphate minerals.[a)]

		Fe^{2+}					Fe^{3+}							
Natural mineral	**Ideal composition**	δ, mm/s[b)]	ΔE_Q, mm/s	Γ, mm/s	% area	Site assign.	δ, mm/s[b)]	ΔE_Q, mm/s	Γ, mm/s	% area	Site assign.	Dana number	Code in reference [6]	References
Alluaudite	$NaCaFe^{2+}(Mn,Fe^{2+},Fe^{3+},Mg)_2(PO_4)_3$						0.43	0.44	0.26	100	M2	38.2.3.6	ML-P40(2)	[5, 14]
Ferroalluaudite	$NaCaFe^{2+}(Fe^{2+},Mn,Fe^{3+},Mg)_2(PO_4)_3$	1.27	2.75	0.3	10	M2	0.44	0.45	0.26	90	M2	38.2.3.5	ML-P14	[15]
Arrojadite	$K_2Na_5Fe^{2+}{}_{14}Fe^{3+}(PO_4)_{12}(OH)_2$	1.19	1.13	0.3	6	M4,7	0.33	0.79	0.3	4	Imp.	41.7.2.1	ML-P25	[5, 16]
		1.22	1.73	0.3	16	M4,7								
		1.2	2.15	0.3	31	M3,5								
		1.18	2.65	0.3	28	M3,5								
		1.21	3.08	0.3	16	M2								
Arsenocrandallite	$CaAl_3(OH)_6[PO_3(O_{1/2}(OH)_{1/2})]_3$						0.39	1.17	0.55	51	Al	42.7.4.1	DD150	[5, 17]
							0.4	0.46	0.32	49	Al			
Barbosalite	$Fe^{2+}Fe^{3+}{}_2(PO_4)_2(OH)_2$	1.1	3.52	0.23	25	M^{2+}	0.29	0.56	0.3	28	M^{3+}	41.10.1.4	DD248-A	[5, 18]
							0.44	0.41	0.26	47				
Barbosalite	$Fe^{2+}Fe^{3+}{}_2(PO_4)_2(OH)_2$	1.1	3.56	0.27	27	M^{2+}	0.43	0.38	0.29	56	M^{3+}	41.10.1.4	DD134	[5, 18]
							0.39	1.53	0.38	17				
Bobierrite	$Mg_3(PO_4)_2\cdot 8H_2O$	1.21	3.24	0.31	49	M2	0.26	0.85	0.61	11	Imp.	40.3.7.1	MDD11	[5, 6]
		1.18	2.41	0.28	40	M1								
Chalcosiderite	$CuFe_6(PO_4)_4(OH)_8\cdot 4H_2O$						0.38	1.09	0.25	62	M1,2	42.9.3.4	ML-P39	[5, 6]
							0.41	0.43	0.26	38	M3			
Childrenite	$(FeMn)Al(PO_4)(OH)_2\cdot H_2O$	1.24	1.66	0.25	100	M						42.7.1.1	ML-P29	[5, 19]

Fairfieldite	$Ca_2(Mn,Fe^{2+})(PO_4)_2·2H_2O$	1.23	2.94	0.3	71	M	0.41	0.94	0.54	25	M?	40.2.2.1	DD171	[6]
		1.06	2.45	0.4	4	M								
Ferrisicklerite	$Li(Fe,Mn)PO_4$	1.22	2.75	0.28	3	Imp.	0.42	0.47	0.28	51	Mn2	39.1.4.11	DD162-A	[6]
							0.44	1.44	0.32	46	Mn2			
Ferristrunzite	$Fe^{3+}Fe^{3+}{}_2(PO_4)_2(OH)_3·5H_2O$						0.39	0.76	0.48	72		42.11.9.3	ML-P15	[20]
							0.41	0.38	0.39	28				
Ferrolaueite	$Fe^{2+}Fe^{3+}{}_2(PO_4)_2(OH)_2·8H_2O$	1.28	3.1	0.3	19		0.39	1.31	0.3	30		42.11.10.7	MDD20	[21]
							0.37	0.32	0.3	10				
							0.25	0.66	0.3	41				
Frondelite	$MnFe_4(PO_4)_3(OH)_5$						0.4	0.76	0.46	61	M1	41.9.2.2	DD130	[5, 22]
							0.41	0.26	0.26	39	M1			
Gormanite	$Fe_3{}^{2+}Al_4(PO_4)_4(OH)_6·2H_2O$	1.19	2.02	0.27	77	Fe	0.31	1.22	0.3	6	Al	42.9.2.2	MDD13	[5, 23]
		1.22	2.54	0.3	12	Fe								
		1.04	2.98	0.3	5	Fe								
Graftonite	$Fe_3{}^{2+}(PO_4)_2$	1.11	1.84	0.3	44	M2						38.3.3.1	ML-P24	[5, 24]
		1.26	1.95	0.3	9	M1								
		1.18	2.34	0.3	26	M3								
		1.23	2.96	0.3	20	Imp.								
Hagendorfite	$(Na,Ca)MnFe_2(PO_4)_3$	1.36	2.66	0.24	21	M2	0.23	1.00	0.28	28	Mn	38.2.3.2	DD237	[5, 6]
		1.3	2.23	0.25	25	M2								
		1.2	1.68	0.41	26	M2								
Hagendorfite	$(Na,Ca)MnFe_2(PO_4)_3$	1.24	2.95	0.3	25	M2	0.35	0.69	0.3	17	Mn	38.2.3.2	DD184	[5, 25]
		1.25	2.09	0.3	29	M2								
		1.2	1.86	0.33	29	M2								

(Continued)

Table 2.2 (Continued)

Natural mineral	Ideal composition	Fe^{2+} δ, mm/s[b)]	Fe^{2+} ΔE_Q, mm/s	Fe^{2+} Γ, mm/s	Fe^{2+} % area	Fe^{2+} Site assign.	Fe^{3+} δ, mm/s[b)]	Fe^{3+} ΔE_Q, mm/s	Fe^{3+} Γ, mm/s	Fe^{3+} % area	Fe^{3+} Site assign.	Dana number	Code in reference [6]	References
Hureaulite	$Mn_5^{2+}(H_2O)_4(PO_4H)_2(PO_4)_2$	1.23	2.39	0.31	33	M3						39.2.1.1	ML-P31-A	[5, 26]
		1.22	1.23	0.24	67	M2								
Kulanite	$BaFe^{2+}{}_{1.3}Mn^{2+}{}_{0.6}Mg_{0.2}Al_2(PO_4)_3(OH)_3$	1.2	1.78	0.3	43	M1	0.36	0.56	0.33	11	Al	41.9.1.1	MDD16	[5, 6]
		1.26	1.71	0.3	24	M1								
		1.23	1.35	0.3	22	M2								
Kulanite	$BaFe^{2+}{}_{1.3}Mn^{2+}{}_{0.6}Mg_{0.2}Al_2(PO_4)_3(OH)_3$	1.2	1.96	0.3	5	M1	0.41	0.47	0.3	14	Al	41.9.1.1	ML-P32	[5, 6]
		1.22	1.72	0.3	35	M1								
		1.22	1.43	0.3	47	M2								
Lazulite	$Fe^{2+}Fe_2^{3+}PO_4)_2(OH)_2$	1.13	3.29	0.29	66	M^{2+}	0.27	0.66	0.59	21	M^{3+}	41.10.1.1	DD125	[5, 27]
		1.22	1.87	0.26	13	M^{2+}								
Lipscombite	$(Fe^{2+},Mn)Fe_2^{3+}(PO_4)_2(OH)_2$						0.4	0.85	0.51	69	M2	41.10.2.1	MDD9	[5, 28]
							0.41	0.44	0.32	31	M1			
Lithiophilite	$LiMnPO_4$-$LiFePO_4$	1.24	2.97	0.29	94	M2	0.36	0.62	0.43	6	M2	38.1.1.2	DD135	[5, 6]
Ludlamite	$Fe_3^{2+}(PO_4)_2(H_2O)_4$	1.23	2.42	0.23	41	M2	0.29	0.65	0.26	34	M2	40.3.5.1	DD219	[5, 29]
		1.27	1.87	0.41	25	Imp.								
Lulzacite	$Sr_2Fe^{2+}(Fe^{2+},Mg)_2Al_4(PO_4)_4(OH)_{10}$	1.21	2.58	0.38	18	FeM1						41.7.4.1	DD197	[30]
		1.12	2.03	0.29	15	FeM4								
		1.24	1.78	0.28	68	Imp.								
Maricite	$NaFe^{2+}PO_4$	1.17	3.01	0.4	7	Imp.	0.32	0.9	0.5	10	Imp.	38.1.2.1	MDD14	[5, 31]
		1.21	2.21	0.23	83	Fe								

Messelite	$Ca_2(Fe^{2+},Mn)(PO_4)_2 \cdot 2H_2O$	1.2	3.17	0.29	100	Fe						40.2.2.2	DD228	[5, 32]
Metavivianite	$Fe^{2+}_{3-x}Fe_x^{3+}(PO_4)_2(OH)_x \cdot (8-x)H_2O$	1.21	3.03	0.25	30	Fe1	0.43	0.77	0.41	56	Fe1,2	40.3.9.1	DD319	[5, 33]
		1.23	2.77	0.39	14	Fe2								
Phosphosiderite	$Fe^{3+}PO_4(H_2O)_2$						0.45	0.23	0.3	59	Fe	40.4.3.2	ML-P22	[5, 6]
							0.43	0.47	0.46	41	Fe			
Purpurite	$Mn^{3+}PO_4$-$Fe^{3+}PO_4$						0.43	1.71	0.29	56	Mn1	38.4.1.2	DD270	[5, 6]
							0.4	0.74	0.45	44	Mn1			
Purpurite	$Mn^{3+}PO_4$-$Fe^{3+}PO_4$						0.43	1.48	0.5	60	Mn1	38.4.1.2	DD227	[5, 6]
							0.41	0.58	0.43	40	Mn1			
Roscherite	$Ca(Mn,Fe)_5Be_4(PO_4)_6(OH)_4 \cdot 6H_2O$	1.14	1.77	0.31	28	M1	0.35	0.84	0.59	15	M1	42.7.7.1	DD274	[5, 6]
		1.3	1.69	0.27	58	M2								
Strengite	$Fe^{3+}PO_4 \cdot 2H_2O$						0.42	0.35	0.29	68	Fe	40.4.1.2	DD293-A	[5, 6]
							0.45	0.33	0.4	32	Fe			
Strengite	$Fe^{3+}PO_4 \cdot 2H_2O$						0.39	0.55	0.3	35	Fe	40.4.1.2	ML-P16	[5, 6]
							0.42	0.3	0.3	65	Fe			
Strunzite	$MnFe_2(PO_4)_2[(OH)_2] \cdot 6H_2O$	1.21	3.19	0.24	8	Mn	0.4	0.65	0.34	80	M2	42.11.9.1	HMM108175	[5, 6]
		1.12	2.75	0.3	3	M2								
		1.16	2.36	0.33	10	M1								
Tarbuttite	$Zn_2(PO_4)(OH)$						0.37	0.71	0.58	81	Zn1	41.6.7.1	DD306	[5, 6]
							0.37	0.48	0.24	19	Zn2			
Triphylite	$LiFePO_4$	1.21	2.98	0.23	18	Li	0.47	1.23	0.47	23		38.1.1.1	DD193-B	[5, 34]
		1.26	1.35	0.34	3	Fe	0.42	0.43	0.32	55				

(Continued)

Table 2.2 (Continued)

Natural mineral	Ideal composition	Fe^{2+}					Fe^{3+}					Dana number	Code in reference [6]	References
		δ, mm/s[b)]	ΔE_Q, mm/s	Γ, mm/s	% area	Site assign.	δ, mm/s[b)]	ΔE_Q, mm/s	Γ, mm/s	% area	Site assign.			
Triphylite	$LiFePO_4$	1.09	3.49	0.3	5							38.1.1.1	DD301B	[5, 34]
		1.23	2.97	0.27	53									
Triphylite	$LiFePO_4$	1.24	2.98	0.26	92	Fe	0.41	0.48	0.43	8	Imp.	38.1.1.1	DD305-A	[5, 34]
Triphylite	$LiFePO_4$	1.24	2.95	0.3	47	Fe						38.1.1.1	DD308-A	[5, 34]
		1.2	2.31	0.3	12	M2								
		1.12	1.73	0.3	41									
Triphylite	$LiFePO_4$	1.29	3.48	0.33	12	Fe	0.31	0.72	0.36	12		38.1.1.1	DD308-T	[5, 34]
		1.23	2.93	0.27	30	Fe								
		1.24	2.25	0.43	18	M2								
		1.11	1.66	0.29	28	M2								
Triphylite	$LiFePO_4$	1.22	2.95	0.26	93	Fe	0.42	0.41	0.32	7	Imp.	38.1.1.1	ML-P38	[5, 34]
Triplite	$(Mn,Fe)_2PO_4(F,OH)$	1.25	2.89	0.37	24	Mn1	0.31	0.94	0.55	7	Imp.	41.6.1.2	DD129	[5, 6]
		1.24	1.74	0.31	69	Mn2								
Triplite	$(Mn,Fe)_2PO_4(F,OH)$	1.23	2.76	0.26	56	Mn1						41.6.1.2	DD302	[5, 6]
		1.25	1.93	0.42	44	Mn2								
Vivianite	$Fe_3(PO_4)_2 \cdot 8H_2O$	1.22	3.15	0.28	40	M2	0.35	0.86	0.54	27	M2	40.3.6.1	DD171	[5, 6]
		1.24	2.33	0.45	34	M1								

Vivianite	$Fe_3(PO_4)_2 \cdot 8H_2O$	1.23	2.96	0.28	55	M2	0.3	0.64	0.66	13	M2	40.3.6.1	DD311	[5, 6]
		1.18	2.43	0.31	32	M1								
Vivianite	$Fe_3(PO_4)_2 \cdot 8H_2O$	1.23	2.94	0.3	53	M2	0.29	1.03	0.3	10	M2	40.3.6.1	DD312A	[5, 6]
		1.19	2.47	0.3	21	M1	0.2	0.36	0.3	9	M2			
		1.18	2.02	0.3	7	Imp.								
Vivianite	$Fe_3(PO_4)_2 \cdot 8H_2O$	1.23	2.93	0.3	29	M2	0.37	0.65	0.61	27	M2	40.3.6.1	DD315A	[5, 6]
		1.17	2.44	0.35	26	M1								
		1.23	1.83	0.29	18	M1,2								
Wagnerite	$(Mg,Fe^{2+})_2(PO_4)F$	1.23	2.65	0.43	77	M1	0.2	1.09	0.67	16	Imp.	41.6.2.1	DD330	[5, 6]
		1.24	1.8	0.3	7	M2								
Whiteite	$(Mn^{2+},Ca)(Fe^{2+},Mn^{2+})Mg_2Al_2(PO_4)_4(OH)_2 \cdot 8(H_2O)$	1.21	2.12	0.31	89	M2	0.24	0.74	0.61	11	Al	42.11.3.1	DD329	[5, 36]

a) Selected data obtained from Dyar et al. [6].
b) The isomer shifts are referred to room temperature α-iron.

The third difficulty was the apparently rather large spectral line widths associated with several of the 62 spectra shown in figures 1–8 of reference [6]; some of the apparent line widths were seemingly quite large even for natural minerals. Finally, table 1 in reference [6], the listing of the mineral samples studied, indicated that some of the samples were not single phase but sometimes contained nontrivial amounts of secondary phase iron-containing minerals. Hence, before any comparison is made of the results in Table 2.2 [5, 6, 14–36] with a new spectrum, the user must be sure to check the nominal purity of the sample as given in table 1 of reference [6].

As a consequence of some of the above noted problems, we have plotted all the 186 numerical sets of hyperfine parameters provided in table 2 of reference [6] and have generated a plot of the isomer shift versus the quadrupole splitting that is essentially similar to figure 9A in reference [6] and shows the same unexpected data points found in this figure [37]. The resulting analogous plot, prepared after the elimination of some of the unexpected data points, see discussion below, is shown in Figure 2.1. A comparison of this figure when compared with figure 9A in reference [6] makes obvious the eliminated data points. The excluded data points and the basis for their exclusion are given in detail in a footnote [38] found in the References and Notes portion of this chapter.

In order to make the iron phosphate hyperfine parameters more accessible to the reader, the iron(II) and iron(III) Mössbauer spectral isomer shifts and quadrupole splittings included in Figure 2.1 as well as the corresponding line widths and relative spectral areas are given in Table 2.2. This extensive table also gives the name of the natural mineral, its nominal composition, site assignments, its Dana classification number, and its code number used in reference [6]. As a result, this table provides a more convenient access to the Mössbauer spectral results than is offered in reference [6].

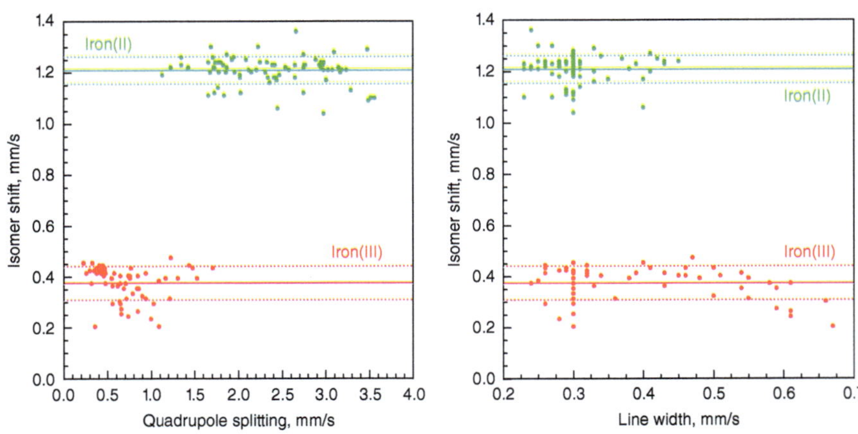

Figure 2.1 The room temperature Mössbauer spectral isomer shift vs. quadrupole splitting, left, and line width, right, for the iron(II) ions in green and the iron(III) ions in red in the phosphate minerals listed in Table 2.2. The solid lines correspond to the mean isomer shifts of 1.21(5) and 0.37(7) mm/s, and the dotted lines correspond to $\pm\sigma$, the standard deviation from the mean.

As expected Figure 2.1 shows a bimodal isomer shift distribution of its 145 data points as the result of the expected difference in the iron(II) and iron(III) Mössbauer spectral isomer shifts. The green points and the solid green line in the left portion of Figure 2.1 correspond to the average of 84 isomer shifts with a mean and (standard deviation) of 1.21(5) mm/s, a value that is consistent with that expected for high-spin, $S = 2$, iron(II) ions with a $3d^6$ electronic configuration. The red points and the solid red line in Figure 2.1 correspond to the average of 61 isomer shifts with a mean (and standard deviation) of 0.37(7) mm/s, a value that is consistent with that expected for high-spin, $S = 5/2$, iron(III) ions with a $3d^5$ electronic configuration. The analogous plot of the Mössbauer spectral isomer shifts versus their full line widths is shown in the right portion of Figure 2.1; this figure will be discussed below.

The Mössbauer spectral isomer shifts and quadrupole splittings indicate that the majority of the iron(II) mineral parameters correspond to iron(II) in a pseudooctahedral six-coordinate oxygen environment, although some of the lower isomer shifts may, as expected, be associated with a five-coordinate oxygen environment. This seems to be the case for both the $M2$ and $M3$ five-coordinate sites in graftonite, which have isomer shifts of $\delta = 1.11$ and 1.18 mm/s, respectively, whereas the six-coordinate $M1$ site has $\delta = 1.26$ mm/s, which is just above the mean $\delta = 1.21(5)$ mm/s iron(II) value. Another low iron(II) isomer shift of $\delta = 1.04$ mm/s reported for gormanite may correspond to a four- or five-coordinate site, but more likely this low value is the consequence of the high uncertainty in fitting the parameters for a weak component that is only 5% of the total spectral absorption area.

The highest iron(II) isomer shift shown in the left portion of Figure 2.1 has an isomer shift of $\delta = 1.36$ mm/s, a value that is unexpectedly high for an iron(II) ion in a pseudooctahedral oxygen coordination environment; indeed if correct this isomer shift would be one of the highest ever observed iron(II) isomer shifts in an oxidic environment; see the upper entry for hagendorfite in Tables 2.2 and 2.3 and Figure 2.1. Further, the iron(III) shift of $\delta = 0.23$ mm/s for this mineral is rather low for its oxidic coordination environment. Indeed, the experimental Mössbauer spectra of the two samples of hagendorfite, shown in Figure 2.2 herein and in figure 1 of reference [6], are rather similar, as would be expected, and their as reported spectral hyperfine parameters in Table 2.2 and in the uppermost and lowermost portions of Table 2.3 are quite different. The reason for these unexpected isomer shifts indicates that the fit shown [6] for the analysis of the Mössbauer spectrum obtained for the DD237 sample of hagendorfite may be problematic. This problem occurs because of the association of the wrong low-energy quadrupole splitting components of two doublets with their high-energy components.

Herein, the experimental Mössbauer spectra obtained [6] for the hagendorfite DD184 and DD237 samples have been refitted in an attempt to overcome the component assignment problem for DD237. The revised fits are shown in Figure 2.2, and the revised spectral parameters are both given in Table 2.3 and shown in Figure 2.3. As may be noted in this table the spectral hyperfine parameters for the two as revised sets of hagendorfite spectral isomer shifts and quadrupole splittings

Table 2.3 Reported and revised hagendorfite, $(Na,Ca)MnFe_2(PO_4)_3$, Mössbauer spectral parameters.[a)]

Mineral code	δ,[b)] mm/s	ΔE_Q, mm/s	Γ, mm/s	% area	Site Assignment	Comment
DD237 as reported[c)]	**1.36**	2.66	0.29	21	Fe(II) *M*2	**Very high** δ-value
	1.3	2.23	0.25	25	Fe(II) *M*2	High δ-value
	1.2	1.68	0.41	26	Fe(II) *M*2	Acceptable δ-value
	0.23	1.00	0.28	28	Fe(III)	**Low** δ-value
DD237 revised fit[d)]	1.235(1)	2.936(3)	0.323(2)	24.3(3)	Fe(II) *M*2	Acceptable δ-value
	1.276(1)	2.179(3)	0.323(2)	29.6(3)	Fe(II) *M*2	Acceptable δ-value
	1.207(2)	1.719(5)	0.355(2)	17.9(3)	Fe(II) *M*2	Acceptable δ-value
	0.340(1)	0.684(2)	0.323(2)	28.2(1)	Fe(III)	Acceptable δ-value
DD184 revised fit[d)]	1.236(1)	2.938(4)	0.322(3)	24.9(4)	Fe(II) *M*2	Acceptable δ-value
	1.247(1)	2.195(4)	0.322(3)	28.1(4)	Fe(II) *M*2	Acceptable δ-value
	1.204(1)	1.774(5)	0.354(2)	29.7(4)	Fe(II) *M*2	Acceptable δ-value
	0.345(3)	0.702(5)	0.322(3)	17.2(4)	Fe(III)	Acceptable δ-value
DD184 as reported[c)]	1.24	2.95	0.3	25	Fe(II) *M*2	Acceptable δ-value
	1.25	2.09	0.3	29	Fe(II) *M*2	Acceptable δ-value
	1.2	1.86	0.33	29	Fe(II) *M*2	Acceptable δ-value
	0.35	0.69	0.3	17	Fe(III)	Acceptable δ-value

a) The statistical uncertainties of the revised parameters are given in parentheses, and the actual uncertainties may be two to three times larger.
b) The isomer shifts are referred to α-iron.
c) The as-reported numerical parameters are taken from Table 2.2 and were originally obtained from table 2 in reference [6].
d) The revised parameters, obtained herein by using an alternate coupling of the low-velocity doublet absorption line with the high-velocity doublet absorption line for the DD237 spectrum, are discussed in the text.

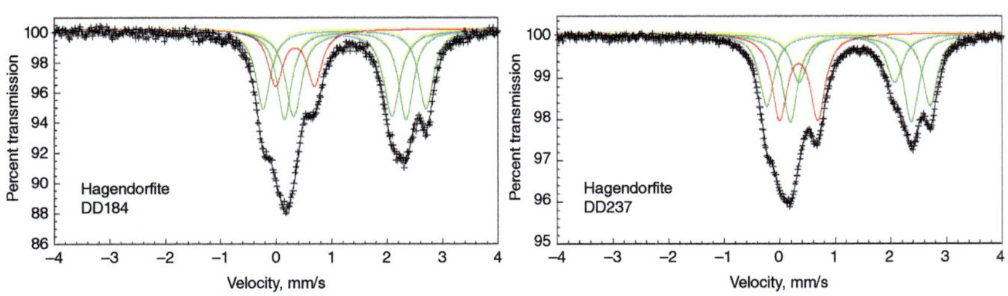

Figure 2.2 The room temperature Mössbauer spectra of the two samples of the hagendorfite mineral and fit with the spectral parameters given in Table 2.3. The black, red, and green solid lines represent the total fit, the iron(III) and iron(II) components, respectively.

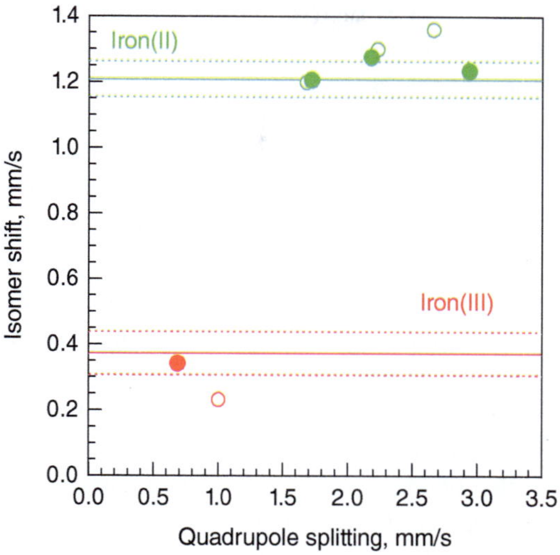

Figure 2.3 The room temperature Mössbauer spectral isomer shift vs. quadrupole splitting, for the iron(II) ions in green and the iron(III) ions in red reported for hagendorfite DD237 given in table 2 in reference [6], open symbols, and revised herein and given in Table 2.3, solid symbols. The solid lines correspond to the mean isomer shifts of 1.21(5) and 0.37(7) mm/s and the dotted lines correspond to $\pm\sigma$, the standard deviation from the mean for all the data shown in Figure 2.1. Note that all revised solid symbols are within $\pm\sigma$ of the mean values, whereas the open symbols are invalid.

are now, as expected, very similar and typical of high-spin iron(II) and iron(III) in an oxidic environment. The validity of the revised fit of the DD237 spectrum given in Table 2.3 is further confirmed by the values of the hyperfine parameter correlation coefficients given in Table 2.4 in which the largest correlation coefficient is 0.413, an acceptable value for the correlation between two adjacent independently fit quadrupole doublets. Indeed, except for the correlation coefficients between adjacent doublets, the correlation coefficients are quite small and often have a magnitude of less than 0.1. This determination of the correlation coefficients is an example in which strong support for the corresponding spectral fit is provided. Another example of such a use of the correlation coefficients is presented below

Table 2.4 Correlation coefficients for fits of the revised hagendorfite DD237 hyperfine parameters.

	δ_1	δ_2	δ_3	δ_4	ΔE_{Q1}	ΔE_{Q2}	ΔE_{Q3}	ΔE_{Q4}
δ_1	1	0.164	0.171	0.197	−0.097	−0.022	−0.087	−0.086
δ_2		1	0.086	0.007	0.097	0.095	0.251	0.084
δ_3			1	0.246	0.018	0.197	−0.020	0.019
δ_4				1	−0.378	−0.308	−0.224	−0.313
⋮								
ΔE_{Q1}					1	0.413	0.088	0.400
ΔE_{Q2}						1	0.344	0.307
ΔE_{Q3}							1	−0.012
ΔE_{Q4}								1
δ or ΔE_Q [a]	1.235(1)	1.276(1)	1.207(1)	0.340(1)	2.936(3)	2.179(3)	1.719(5)	0.684(2)

a) The corresponding observed parameter uncertainties are given in parentheses; the isomer shifts, δ, and the quadrupole splittings, ΔE_Q, are given in mm/s.

and illustrates their importance in demonstrating the use of a very poor fitting model that leads to totally unacceptable correlation coefficients and thus highly questionable spectral fits.

The reader should note that the authors could obtain a statistically satisfactory fit of the hagendorfite DD237 spectrum with the coupling of the lines used in reference [6] but with slightly different parameters from those published in this reference. It appears that depending on the initial guesses introduced in the fitting code two statistically equivalent fits may be obtained but one fit is far more physically reasonable than the other based on the values of the hyperfine parameters.

It is interesting to note that the assignment of the outermost iron(II) spectral doublets in both hagendorfite DD184 and DD237 is not associated with the hagendorfite mineral but rather is a consequence of the presence of c. 25 % of triphylite, a secondary phase whose presence is also revealed by powder x-ray diffraction results, see table 1 of reference [6]. The high-spin iron(II) Mössbauer spectral hyperfine parameters reported [6] for triphylite are $\delta = 1.218(1)$ and 1.236(1) mm/s and $\Delta E_Q = 2.949(1)$ and 2.957(1) mm/s for the essentially pure triphylite samples NL-P38 and DD305, respectively, see Table 2.2 – values that are essentially the same as those reported for the outermost iron(II) quadrupole doublet in both hagendorfite DD184 and DD237 samples, see Table 2.3.

The above classical doublet component miss-assignment problem is often found in Mössbauer spectral papers dealing with two or more quadrupole doublets. This is also the case for the lüneburgite, DD218; trolleite or dufrenite, DD157; dufrenite, ML-P34D; and perhaps for the barbosalite, DD134 and DD248-A, spectral assignments found in reference [6]; the revised hyperfine parameters given in Table 2.5 avoid this problem.

In order to investigate the apparently large full-widths at half-maximum noted above, a plot of all the spectral isomer shifts as a function of their line widths, Γ, is shown in the right portion of Figure 2.1. Fortunately, except for the $\Gamma = 0.13$ mm/s, which probably should be 0.23 mm/s, reported for a 2% area component in satterlyite, MDD18, none of the reported line widths are below 0.195 mm/s, which is the minimum possible line width for an iron-57 Mössbauer spectral absorption, the minimum value determined by the Heisenberg uncertainty principle; papers reporting smaller values either are reporting the half-width at half-maximum or are in error. The typical Mössbauer spectrometer with a rather new cobalt-57 source will typically yield for a single line absorption a line width of $\Gamma = 0.23$–0.25 mm/s and at best 0.22 mm/s.

Indeed most of the line widths reported in reference [6] are between $\Gamma = 0.24$ and 0.34 mm/s, values that are quite acceptable for minerals; many minerals exhibit larger line widths of up to $\Gamma = 0.45$ mm/s for various reasons. But spectral components with line widths above about $\Gamma = 0.45$ mm/s may possibly correspond to a distribution of local environments on a local site, differing crystallographic sites in the mineral with similar hyperfine parameters, or the presence of a secondary mineral with similar hyperfine parameters. In the case of the ca. 12 line widths observed between $\Gamma = 0.50$ and 0.67 mm/s in the minerals studied in reference [6], and shown in Figure 2.1 and Table 2.2, one or more of these possibilities may be

Table 2.5 Some mineral revised Mössbauer spectral parameters.[a)]

Mineral, code	δ,[b)] mm/s	ΔE_Q, mm/s	Γ, mm/s	*% Area*	Probable site assignment
Lüneburgite, DD218	1.224(2)	1.828(3)	0.268(4)	47.5(7)	Fe(II) *M*2
	1.140(4)	2.694(9)	0.268(4)	23(3)	Fe(II) *M*1
	0.32(2)	1.09(5)	0.46(3)	11(1)	Fe(III) *M*1
	0.22(2)	0.48(2)	0.46(3)	19(1)	Fe(III) *M*2
Trolleite or dufrenite, DD157	0.396(1)	0.725(3)	0.412(2)	42.2(5)	Fe(III) Al2
	0.396(1)	0.256(3)	0.428(5)	33.8(7)	Fe(III) Al2
	0.384(1)	1.220(3)	0.412(2)	24.0(2)	Fe(III) Al1
Dufrenite, ML-P34D[c)]	0.380(1)	0.610(5)	0.385(1)	44(2)	Fe(III) Al2
	0.390(1)	0.813(6)	0.385(1)	27(2)	Fe(III) Al2
	0.392(1)	1.510(3)	0.385(1)	14(2)	Fe(III) Al1
Lithiophilite or lithiophyllite, DD135	1.239(1)	2.964(1)	0.288(2)	94.8(4)	Fe(II) in triphylite
	0.43(2)	0.49(3)	0.38(4)	5.2(4)	Unknown Fe(III)

a) The revised parameters have been obtained by using the alternate nested coupling of the Fe(II) and Fe(III) doublets; see text. The statistical uncertainties of the revised parameters are given in parentheses. The actual uncertainties may be two to three times larger.
b) The isomer shifts are referred to α-iron.
c) The spectrum of ML-P34D also exhibits c. 4% of an Fe(II) doublet and 10% of a magnetic sextet, both of unknown origin.

operative. The reader should consult table 1 in reference [6] for the possible presence of secondary phases for a specific mineral of interest.

So now the question arises: "What is the form of the minerals that contain phosphorus known to be present on the surface of Mars?" Dyar and colleagues have initially compared the plots of the isomer shift versus the quadrupole splitting values of the iron phosphate minerals reported in their table 2 of reference [6] and in the literature with those actually observed on the surface of Mars, see figures 12 and 13 in reference [6]. The possible iron-bearing minerals known [39–41] to be on the surface of Mars are jarosite, $KFe_3(SO_4)_2)(OH)_6$; olivine group minerals, such as $(Mg,Fe)_4SiO_4$; and pyroxene group minerals, such as $(Mg,Fe)_4SiO_3$, as well as nanoparticle oxides. The goal is to see if any of the plausible iron-bearing phosphate minerals might have "complete overlap" of their isomer shifts and quadrupole splittings with these minerals. It seems that few if any of the iron phosphate minerals overlap with jarosite, but some overlap may occur, within the 2σ standard deviations reported by Morris et al. [39, 40], between the iron phosphates and the parameter regions surrounding the observed Mars spectral doublets arising from pyroxene, olivine, and nanophase oxide, see figure 13 in reference [6]. Unfortunately, as noted by Dyar et al. [6], it would be difficult to unequivocally assign any of the observed Mars's parameters to specific iron phosphate phases strictly on the basis of iron phosphate Mössbauer spectral parameters alone; a "conclusion [that] illustrates why

Mössbauer spectroscopy is rarely used in terrestrial applications for the purpose of mineral identification in paramagnetic minerals" [6].

In an effort to try to overcome the above limitation, Lane et al. [42] have reported Mössbauer spectral studies of the Paso Robles Martian light and dark soils and have combined their results with Martian visible-near-infrared reflectance spectral and thermal infrared spectral studies, termed VNIR and TIR studies, respectively, in their paper.

The observed Mössbauer spectra of the Paso Robles light and dark soils and their mixture is shown [42] in figure 7 [43]. These spectra indicate that the two major spectral components in the dark soils are high-spin iron(II) doublets, whereas the two major spectral components in the light soils are high-spin iron(III) doublets. Out of a group of 17 iron phosphate minerals, see table 2 of reference [42], only the iron(III) phosphate ferristrunzite mineral, $Fe_3(PO_4)_2(OH)_3 \cdot 5H_2O$, exhibits Mössbauer spectral parameters that are both appropriate for the *light soil* and consistent with the two additional spectral techniques mentioned above. As is shown in figure 8 of reference [42], the ferristrunzite mineral exhibits two different isomer shifts and quadrupole splittings, see the two large red data points corresponding to the Paso Robles *light soil* and the two small points just below that correspond to the ferristrunzite mineral [44]. Of course this similarity does not ensure that ferristrunzite, which may be accompanied by some strengite, $FePO_4 \cdot 2H_2O$, is actually present in the light soils of Paso Robles; essentially no conclusions are presented [42] dealing with the nature of the dark soils at Paso Robles.

The paper by Adcock et al. [13] is of interest at this point even though it does not contain any Mössbauer spectral work. This paper discusses the preparation and properties of iron-bearing merrillite, $Ca_9NaMg(PO_4)_7$, and whitlockite, $Ca_9Mg(PO_3OH)(PO_4)_6$, minerals, and more specifically the stoichiometries of the new synthetic minerals are for iron(III)-merrillite, $Ca_{9.0}Fe_{0.9}Mg_{0.1}(PO_4)_{7.0}$; for iron(II)-merrillite, $Ca_{9.2}Fe_{1.0}Mg_{0.1}(PO_4)_{7.0}$; and for iron(II)-whitlockite, $Ca_{9.0}Fe_{1.0}(PO_3OH)(PO_4)_6$.

On earth the fluorapatite mineral, $Ca_5(PO_4)_3F$, is the most predominant phosphate mineral and may be considered an analog to merrillite that may be a major residence for the phosphorus found in some Martian soils [44–46]. Further, merrillite and some of its related minerals, such as whitlockite, are of interest in the study of the Mössbauer spectra measured on Mars because merrillite is often found to be the dominant phosphate mineral in Martian originated meteorites [47, 48], and thus one might conjecture that it should be found on Mars. Strangely, neither merrillite nor whitlockite are mentioned in reference [6] or [13], and it seems that there are, at this time, no published Mössbauer spectral studies of either mineral.

2.4 Analysis of Magnetically Ordered Mössbauer Spectra

The authors have noted over many years that many of the iron-57 Mössbauer-effect papers reporting fits of magnetically ordered sextet spectra *mistakenly* report the actual quadrupole shift, *QS*, as the quadrupole splitting, ΔE_Q, or more usually the magnitude of the quadrupole splitting, $|\Delta E_Q|$. Indeed, a clear example of this error may be found in the analysis of the magnetic spectra of the two phosphates containing mixed valence iron(II/III), ${}^1{}_\infty\{[C_4N_2H_{11.6}]_{1.5}[Fe^{II}Fe^{III}(PO_4)(H_{0.8}PO_4)_2]\cdot H_2O\}$ and ${}^3{}_\infty[Fe^{II}{}_5Fe^{III}{}_2(PO_4)_2(H_{0.5}PO_4)_4]$, compounds [49].

As a consequence of this widespread error, the authors believe that it is essential to bring this problem to the attention of the Mössbauer-effect scientists who are dealing with such spectra. Although this difference is discussed in several excellent papers, books, and chapters [50, 51], it seems that many authors have failed to grasp the important difference between the two parameters, *QS* and ΔE_Q, and the added highly important information that can be obtained from a full spectral analysis. Unfortunately, in reading the literature it is often difficult to determine which parameter is being reported, but the reader should not be expected to have to decipher this distinct mistake.

It seems that an excellent way to illustrate both the importance of the correct interpretation of magnetic Mössbauer spectra in the presence of a quadrupole interaction and the added information that may be obtained are well demonstrated by our study [3] of the Mössbauer spectra of α-$FePO_4$ to be discussed below.

To a first approximation, when the electric field perturbation [50, 51], i.e. the magnitude of the quadrupole splitting, $|\Delta E_Q|$, in mm/s is substantially smaller than the magnetic hyperfine field, H_{int}, splitting also in mm/s, the center of the four inner lines of an iron-57 Mössbauer spectral sextet is shifted relative to the center of the outer two lines; this shift is properly referred to as the quadrupole shift, *QS*, a quantity that *must* be carefully distinguished from the quadrupole interaction or, in the case of iron-57 Mössbauer spectra, the quadrupole splitting, ΔE_Q.

For example, the above perturbation criterion is often met in iron(III) compounds as is the case for α-$FePO_4$, in which at 4.2 K $H_{int} = 51.36$ T and the corresponding magnetic splitting of 16.54 mm/s is far larger than the quadrupole splitting of $\Delta E_Q = 0.651$ mm/s. In contrast, this criterion is often *not* met for iron(II) containing compounds, as is the case for the iron(II) in α-$Fe_2(PO_4)O$, to be discussed below [52], in which at 30 K $H_{int} = 24.1$ T and the corresponding magnetic splitting of 7.74 mm/s is only ca. three times larger than the quadrupole splitting of $\Delta E_Q = 2.706$ mm/s.

Thus, when the above perturbation criterion is met, the quadrupole shift, *QS*, is defined as:

$$QS = ½[(p_6 - p_5) - (p_2 - p_1)] \tag{2.1}$$

where, for *i* increasing from 1 to 6 from the low velocity to the high velocity, p_i is the experimental velocity of the *i*-th absorption line of the magnetic sextet; often *QS* is defined as 2ε, and

$$QS = 2\varepsilon = (1/4)eQV_{zz}(3\cos^2\theta - 1 + \eta\sin^2\theta\cos 2\phi) \tag{2.2}$$

where *e* is the proton charge, *Q* is the nuclear quadrupole moment, $V_{zz} = eq$ is the principal component of the electric field gradient tensor, θ and ϕ are the polar and azimuthal angles that the hyperfine field makes with the principal axes of the electric field gradient tensor, and η is the asymmetry parameter defined as:

$$\eta = [V_{xx} - V_{yy}]/V_{zz} \tag{2.3}$$

when $|V_{zz}| \geq |V_{yy}| \geq |V_{xx}|$. The quadrupole splitting of a paramagnetic doublet is given by:

$$\Delta E_Q = ½[eQV_{zz}(1 + \eta^2/3)^{1/2}] = ½[e^2Qq(1 + \eta^2/3)^{1/2}] \tag{2.4}$$

If $\eta = 0$, equation (2.2) yields the simpler equation:

$$QS = 2\varepsilon = ½[\Delta E_Q(3\cos^2\theta - 1)] \tag{2.5}$$

between the experimental QS and the paramagnetic quadrupole splitting, ΔE_Q.

In equation (2.4) both the eQV_{zz} and e^2Qq quantities are referred to as the nuclear quadrupole coupling constant. It is important to note that in eQV_{zz} and e^2Qq both e and Q are positive for iron-57 and thus the sign of ΔE_Q is determined by the sign of V_{zz} or q.

At this point it should be noted that when $\eta = 0$ at an iron site, its iron-57 magnetic Mössbauer spectrum will consist of a sextet. In contrast, when $\eta > 0$ at an iron site, its iron-57 magnetic spectrum will consist of an octet, where the seventh and eighth lines may typically be quite weak but are often observed especially in high-spin iron(II) magnetic spectral components; the seventh and eighth lines are forbidden when $\eta = 0$, as is often the case for high-spin iron(III) compounds that exhibit magnetic sextets.

Typically, the magnitude of the quadrupole splitting, $|\Delta E_Q|$, may be determined from the paramagnetic spectrum obtained above the magnetic ordering temperature of a compound. As a consequence, from the measurement of both QS in the magnetic sextet and $|\Delta E_Q|$ in the paramagnetic spectrum of a compound, the angle, θ, can be determined, if $\eta = 0$. Because θ in equation (2.5) increases from 0° to 90°, QS may have values from ΔE_Q for $\theta = 0°$ to $-\Delta E_Q/2$ for $\theta = 90°$ and is zero if $\theta = 54.7°$, the so-called magic angle. In this latter case, the magnetic sextet is perfectly symmetric and it is not possible to obtain any information on the electric field gradient from this symmetric spectrum. Finally, depending upon the unknown sign of ΔE_Q and the known sign and magnitude of QS, there may be only one or two possible θ angle values that satisfy equations (2.1) and (2.5).

Unfortunately, when reporting the iron-57 Mössbauer spectral hyperfine parameters obtained for magnetic sextets, many authors confuse or interchange the quadrupole shift, QS, and the quadrupole splitting, ΔE_Q, i.e. many authors refer to the actual quadrupole shift, QS, as the quadrupole splitting, ΔE_Q, see reference [49]. A reader should always check to be sure that a quadrupole splitting reported for an iron-57 magnetic sextet is not in reality the quadrupole shift – this error is often revealed as an unusually small value for this parameter.

Finally, if the first-order perturbation is valid and is used in fitting a magnetic iron-57 Mössbauer spectrum, it *cannot* yield both eQV_{zz} and η because they cannot be separated in measuring the quadrupole splitting in the corresponding paramagnetic spectrum. Finally, the four parameters eQV_{zz}, η, θ, and ϕ can never be simultaneously obtained from the measurement of the quadrupole shift given by equation (2.2). Only limiting ranges of values for these parameters compatible with the measured quadrupole shift and quadrupole splitting can be estimated [53]. Hence, the first-order perturbation method is *not* used herein to fit the Mössbauer spectra of α-$FePO_4$, but rather the fits reported below refined the three parameters, eQV_{zz}, η, and θ, in the formalism developed by Dattagupta and Blume [54]. The same code may also handle the four parameters, eQV_{zz}, η, θ, and ϕ, in complex algebra, but the user should be aware of the high correlations between these parameters when they are all adjusted. This code uses an advanced Levenberg–Marquardt fit minimization routine that both better evaluates the statistical uncertainties in the variable parameters and determines the correlation coefficients between all the pairs of fitted parameters. The code also uses the most recent values for the experimental nuclear magnetic moment of the iron-57 nuclide, values that are slightly revised relative to the earlier experimental values used. Most of the fits presented herein used the same code to fit different iron phosphate compounds, and careful attention to the correlation coefficients between adjusted parameters is given.

2.5 Structural and Thermodynamic Properties of the Polymorphs of $FePO_4$

2.5.1 Polymorphs of $FePO_4$

Any paper dealing with the magnetic, structural, and Mössbauer spectral properties of iron(III) phosphate must be careful to specify the specific polymorph of $FePO_4$ under study. Indeed, there are several iron phosphate polymorphs that are kinetically "stable" at ambient temperature and ambient pressure or high pressure. The simplest thermodynamically stable iron phosphate compound is α-$FePO_4$, and the discussion of these polymorphs will start with this important compound.

The most stable iron phosphate polymorph, herein designated as α-$FePO_4$, has a standard heat of formation, ΔH_f°, of −1404.6 kJ/mol [55], and is isostructural with α-$AlPO_4$, the mineral berlinite; both α-$FePO_4$ and α-$AlPO_4$ crystallize in the trigonal space group $P3_121$, number 152. At 980 K α-$FePO_4$ undergoes a phase transition to a high-temperature polymorph, β-$FePO_4$, which has the hexagonal space group $P6_422$, number 181; the structural changes occurring at this phase change have been studied in detail by Haines and coworkers; the iron(III) ions in both α-$FePO_4$ and β-$FePO_4$ have similar but not identical pseudotetrahedral coordination environments [56] that will be discussed in detail below.

Two metastable polymorphs of $FePO_4$ include the monoclinic and orthorhombic polymorphs that have been prepared by dehydration of the hydrothermally prepared monoclinic and orthorhombic forms of $FePO_4{\cdot}2H_2O$ compounds [57]. This dehydration occurs with no change in crystal structure upon the loss of water and yields a monoclinic polymorph γ-$FePO_4$ with the space group $P2_1/n$, number 14, and an orthorhombic polymorph ζ-$FePO_4$ in the space group *Pbca*, number 61; in both compounds the iron(III) has a pseudooctahedral coordination environment. As may be

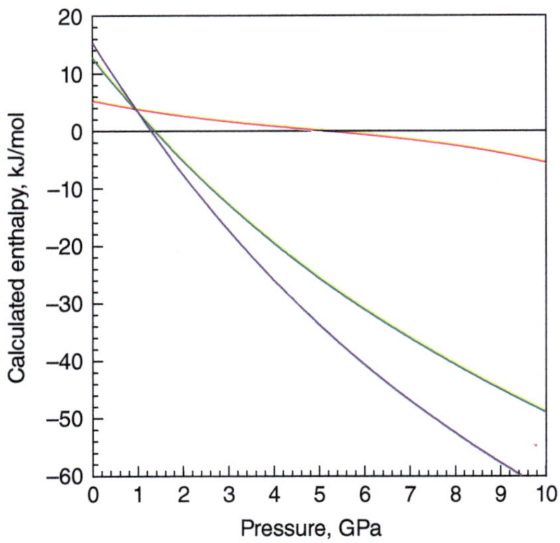

Figure 2.4 The calculated enthalpy of several polymorphs of $FePO_4$ at ambient and applied pressures shown relative to that of α-$FePO_4$, (space group $P3_121$) shown in black. The calculated value for the monoclinic polymorph, γ-$FePO_4$, ($P2_1/n$) is shown in red; the mineral heterosite, δ-$FePO_4$, (space group *Pmnb*) in green; and a high-pressure polymorph, ε-$FePO_4$, (space group *Cmcm*) that is stable only above ca. 2.5 GPa, in blue. Source: This figure is adapted from reference [62].

observed in Figure 2.4, the monoclinic form, γ-$FePO_4$, has a calculated enthalpy that is ca. 5.35 kJ/mol above that of α-$FePO_4$.

At ambient pressure a metastable polymorph of $FePO_4$ is the heterosite mineral with the $FePO_4$ orthorhombic *Pmnb*, space group, number 62, herein δ-$FePO_4$, which at ambient temperature is calculated to be ca. 12.75 kJ/mol above α-$FePO_4$; in δ-$FePO_4$ the iron(III) has a pseudooctahedral coordination environment and δ-$FePO_4$ is the iron(III) end member of the heterosite mineral [58a] that typically exists as $(Fe^{3+},Mn^{3+})PO_4$ over the entire range of cation compositions. It has also been reported [58b] that the iron(II) containing triphylite mineral, $LiFePO_4$, can be delithiated to form a polymorph of $FePO_4$; both compounds have the orthorhombic space group *Pnma*, number 62; the Mössbauer spectral parameters reported [59–61] for this polymorph are not those of $FePO_4$ but in reality correspond to the delithiated polymorph δ-$FePO_4$ and not, as might have been expected, to α-$FePO_4$.

Another polymorph of $FePO_4$ is ε-$FePO_4$, a high-pressure phase, that is only stable at and above ca. 2.5 GPa of applied pressure [62]. This reference also has reported the calculated pressure dependence of the enthalpy of several of the above $FePO_4$ polymorphs relative to that of the most stable α-$FePO_4$ compound, see Figure 2.4. The polymorph ε-$FePO_4$ is referred to either as $FePO_4$-II by Battle et al. [63] or as the crystalline high-pressure phase of α-$FePO_4$ by Pasternak et al. [64]. Both of these compounds contain pseudooctahedrally coordinated iron(III) cations and have been prepared [65] by heating to 900 °C a stoichiometric mixture of Fe_2O_3 and $(NH_4)_2HPO_4$, followed by quenching to room temperature. This metastable polymorph ε-$FePO_4$ crystallizes [64] in the orthorhombic space group *Cmcm*, number 63; it seems that the actual thermodynamic stability of this compound has not been reported.

As mentioned above one must be certain which polymorph of $FePO_4$ is under study, and this information is sometimes absent or α-$FePO_4$ is referred to variously as the berlinite-like, or the berlinite-type *b*-$FePO_4$, or α-berlinite-like, or the α-quartz-like phase, or $FePO_4$-I, or simply $FePO_4$. Herein the initial discussion will deal with α-$FePO_4$.

2.6 Mössbauer Spectra of α-$FePO_4$

The sample of α-$FePO_4$ and the corresponding magnetic and Mössbauer spectral data are the same as reported earlier [2]. The Mössbauer spectra of α-$FePO_4$ have been measured between 4.2 and 295 K on a conventional constant acceleration spectrometer that utilized a rhodium matrix cobalt-57 source at 295 K; the resulting isomer shifts are reported relative to α-iron at 295 K. The Mössbauer spectra of α-$FePO_4$ have been least-squares fit either with a symmetric Lorentzian line shape quadrupole doublet for the paramagnetic spectra or with the minimum number of Lorentzian static magnetic sextets required to fit the magnetically ordered spectra. The Mössbauer spectral fits presented herein used the code described in Section 2.4.

The Mössbauer spectra of α-$FePO_4$ at 24.5 and 24 K, shown in Figure 2.5, indicate that the Néel temperature of α-$FePO_4$ is 24.2(2) K. The hyperfine parameters of the paramagnetic symmetric quadrupole doublet observed at 24.5 K are given in Table 2.6, together with the hyperfine parameters of the paramagnetic doublets observed between 25 and 300 K [2]. The isomer shift, δ, and the magnitude of the quadrupole splitting, $|\Delta E_Q|$, at and above 24.5 K are characteristic of iron(III) in a pseudotetrahedral oxygen coordination environment. The broad blue sextet component shown in Figure 2.5 at 24 K indicates the onset of long-range antiferromagnetic ordering. The Néel temperature of 24.2(2) K agrees well with earlier reported ordering temperatures [66, 67]. In contrast, no magnetic neutron scattering was observed by Arnold [68] at 20 K; assuming that there are no thermal calibration problems in the different experiments, a possible reason for this inconsistency in the measurements by very different techniques will be discussed below.

The fully ordered magnetic Mössbauer spectra of α-$FePO_4$ observed between 4.2 and 21 K are shown in Figure 2.6. The Mössbauer spectra of α-$FePO_4$ obtained at 4.2 and 16 K have been fitted with one sextet with the spectral parameters given in Table 2.6. Rather than analyzing these two spectra with a quadrupole shift, QS, as reported earlier [2, 69], one-half of the nuclear quadrupole coupling constant,

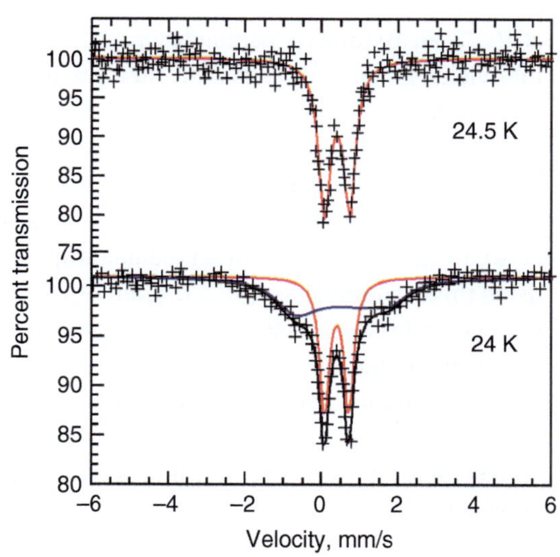

Figure 2.5 The Mössbauer spectra of α-$FePO_4$ measured at the indicated temperatures just above and just below its magnetic ordering temperature of 24.2(2) K [3].

Table 2.6 Mössbauer spectral parameters[a)] for α-$FePO_4$.

T, K	δ,[b)] mm/s	ΔE_Q,[c)] mm/s	H_{int}, T	θ, deg	η	Γ,[d)] mm/s	$\Delta\Gamma$[e)]	*Area*, %	*Area*, (%ϵ) (mm/s)
300	0.30(1)	0.62(2)	0	–	–	0.33(1)		100	
78	0.39(1)	0.63(2)	0	–	–	0.33(1)		100	
44.5	0.39(1)	0.62(2)	0	–	–	0.31(1)		100	
30.0	0.39(1)	0.62(2)	0	–	–	0.32(1)		100	
25.0	0.39(1)	0.62(2)	0	–	–	0.31(1)		100	
24.5	0.39(1)	0.649(3)	0	–	–	0.383(6)	–	100	22.75(25)
24.0	0.39	0.639(4)	0	–	–	0.30(1)	–	46	28.15(88)
		0.639(4)	8.0(3)	–	–	0.30(1)	–	54	
21.0	0.39(1)	0.636(4)	31.54(4)	85	0.55	0.288(2)	0.019(1)	50	32.67(10)
		0.636(4)	31.31(1)	25	0.55	0.288(2)	0.019(1)	50	
19.5	0.39	0.65(1)	35.42(1)	85	0.58	0.31(3)	0.017(6)	50	34.56(75)
		0.65(1)	35.47(4)	25	0.58	0.31(3)	0.017(6)	50	
18.0	0.39	0.625(5)	38.88(1)	90	0.42	0.330(3)	0.021(1)	58(1)	33.80(10)
		0.625(5)	38.43(5)	30.2(1)	0.42	0.330(3)	0.021(1)	42(1)	
16.0	0.39	0.64(1)	42.74(1)	90	0.25	0.22(1)	0.021(1)	100	28.02(50)
4.2	0.38	0.651(1)	51.36(1)	90	0.22	0.260(1)	0.014(1)	100	28.40(3)

a) Statistical uncertainties are given in parentheses; the actual uncertainties may be as much as two to three times larger. The absence of an uncertainty indicates that the value given was not refined.
b) The isomer shifts, δ, are referred to α-iron at 295 K.
c) ΔE_Q is the positive quadrupole splitting at and above 25 K and is $eQV_{zz}/2$ at and below 24.5 K.
d) Γ is the full-width at half-maximum of the Lorentzian line of a doublet and the inner two lines in a sextet.
e) The incremental line width, $\Delta\Gamma$, is given by $\Gamma(v) = \Gamma + \Delta\Gamma\ |(v - \delta)|$, as is discussed in the text.

$eQV_{zz}/2$, an asymmetry parameter, η, and an angle, θ, have been used – an approach that yields far more information than fits in terms of *QS*. Because in the absence of an extensive structural change below room temperature [68, 70], a one-half nuclear quadrupole coupling constant, $eQV_{zz}/2$, of ca. 0.63 mm/s is expected, i.e. a one-half nuclear quadrupole coupling constant that is close to the magnitude of the observed quadrupole splitting, $|\Delta E_Q|$, at 295 K, see Table 2.6 [2, 70]. Hence, between 4.2 and 16 K, the observed quadrupole shift, *QS* [2], may be reproduced by using three parameters, $eQV_{zz}/2$, η, and θ, as given in Table 2.6. Because of the high correlation between these three parameters, different fixed values of η and θ were investigated and the fit that yielded the best χ^2 and a value of $eQV_{zz}/2$ consistent with the paramagnetic quadrupole splitting was selected. In these fits only the positive value of $eQV_{zz}/2$ yielded physically acceptable fits, an observation that corresponds only to a positive value of the quadrupole splitting, ΔE_Q, as is shown in Table 2.6.

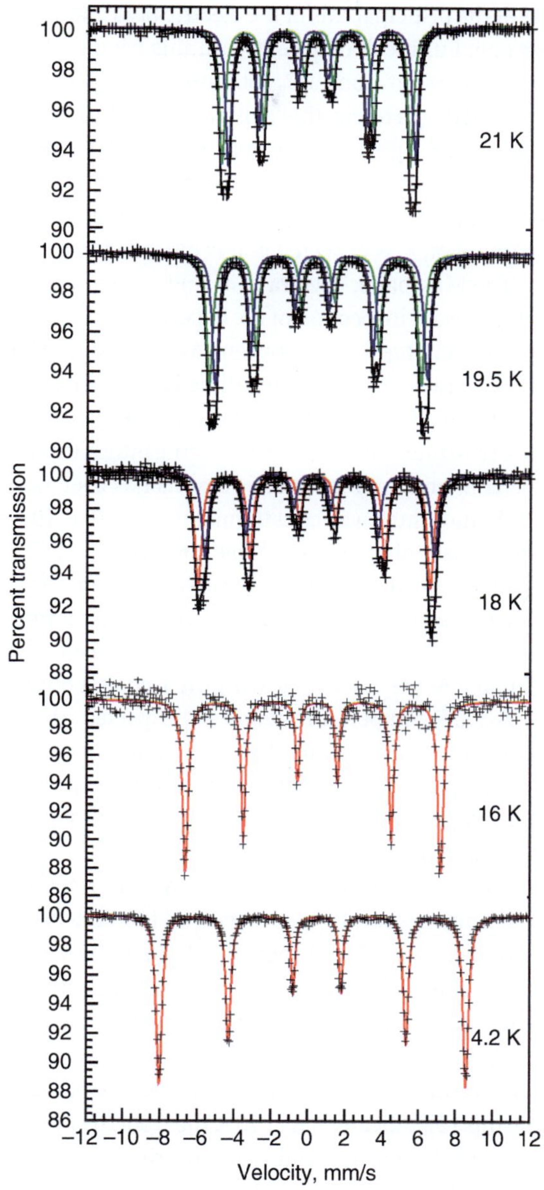

Figure 2.6 The Mössbauer spectra of α-$FePO_4$ obtained at the indicated temperatures below its critical ordering temperature of 24.2(2) K. The meaning of the red, black, blue, and green solid lines is discussed in the text [3].

2.7 Magnetic Structure of α-$FePO_4$, Obtained by Mössbauer Spectroscopy

2.7.1 Magnetic Structure of α-$FePO_4$

Neutron diffraction measurements [2] on α-$FePO_4$ reveal that the iron(III) magnetic moment, and thus the associated hyperfine field, is parallel with the crystallographic c-axis at 5 K. Thus the angle θ of 90° observed at 16 and 4.2 K, see Table 2.6, indicates that the principal axis of the electric field gradient tensor lies in the basal plane of the trigonal unit cell of α-$FePO_4$. In this trigonal $P3_121$ unit cell, the FeO_4 tetrahedron is compressed along a C_2 rotation axis that also lies in the basal plane of the unit cell of α-$FePO_4$. Herein this axis is assumed to be the most likely principal axis of the electric field gradient tensor at the iron(III) site, an axis that makes an angle, θ, of 90° with the crystallographic c-axis and the hyperfine field. Because the tetrahedral FeO_4 environment has no higher symmetry element than a C_2 axis, a non-zero asymmetry parameter, η, is expected and is indeed observed at both 4.2 and 16 K.

The Mössbauer spectra of α-$FePO_4$ obtained between 18 and 21 K are broadened, and a visual inspection of the experimental spectra reveals that the quadrupole shift, QS, has changed sign between 16 and 18 K [2]. Hence, the 18–21 K spectra have been fit with two sextets with the same isomer shift, δ, one-half of the nuclear quadrupole coupling constant, $eQV_{zz}/2$, an asymmetry parameter, η, and the same line width for the two inner lines but with differing internal hyperfine fields, H_{int}, and θ angles. The percent area ratio of the two sextets has been observed to be very close to 50 : 50 at 18 K and, thus, was constrained to 50 : 50 at 19.5 and 21 K. A unitless incremental line broadening, $\Delta\Gamma$, see Table 2.6, defined as $\Gamma(v) = \Gamma + \Delta\Gamma\ |(v - \delta)|$, where v is the source velocity, was used to account for the broadening of the four outer spectral lines relative to that of the two inner spectral lines of the sextets.

The α-$FePO_4$ Mössbauer spectral parameters observed between 18 and 21 K, see Table 2.6, are consistent with their corresponding values observed over the entire 4.2–300 K temperature range. As expected, the isomer shift, δ, is essentially independent of temperature below 78 K and decreases by a typical 0.09 mm/s between 78 and 300 K. By combining the asymmetry parameter, η, and the one-half nuclear quadrupole coupling constant, $eQV_{zz}/2$, as shown in equation (2.4), the quadrupole splitting, ΔE_Q, below the Néel temperature is ca. 0.65 mm/s, i.e. only slightly larger than the 0.62 mm/s observed above the Néel temperature. This small increase in quadrupole splitting agrees with an increase [68, 70, 71] in the distortion of the FeO_4 tetrahedra in α-$FePO_4$ between 295 and 20 K. At each temperature, the two hyperfine fields differ by no more than 0.4 T, a difference that may be associated with a small differing dipolar contribution to the two hyperfine fields.

The predominant difference between the two sextets observed between 18 and 21 K resides in the different θ angles. The powder neutron diffraction experimental results [2] indicate that at 20 K the iron(III) magnetic moments lie in the basal plane of the α-$FePO_4$ unit cell. The lowering in symmetry from trigonal expected in this case is likely to be small because it was not observed [68] at 20 K. Hence, it seems safe to assume that the C_2 axis of the FeO_4 tetrahedron is still the principal axis of the electric field gradient tensor and lies in the basal plane of the trigonal unit cell. Our analysis of the Mössbauer spectra observed between 18 and 21 K indicates that one-half of the iron(III) ions have their hyperfine field at an θ angle of 30.24 or 25° with the principal axis of the electric field gradient and the other half at an θ angle of 90° or 85°, respectively. There are two possible interpretations of these results. As suggested earlier [2], the iron(III) magnetic moments may begin to cant away from the c-axis at 18 K, a canting angle that increases with increasing temperature; the two limiting θ values may actually represent a range of θ angles. Alternatively, a very small but unobserved [68] symmetry lowering of the trigonal $P3_121$ structure may yield two slightly different but inequivalent magnetic iron(III) sites.

At this point, it is of value to discuss the absence of any magnetic scattering in α-$FePO_4$ by neutron diffraction at 20 K as has been reported by Arnold [68]. As is shown in Figure 2.6, there is a well-developed internal magnetic hyperfine field of ca. 31.4 T at 21 K and it is surprising that no magnetic neutron scattering was observed at 20 K. In order to observe magnetic neutron diffraction, there must, of course, be magnetic ordering with a sufficiently long coherence length. However, if there is a distribution of iron(III) magnetic moment canting angles between the above limiting values of θ, there may be insufficient coherence to yield observable magnetic neutron diffraction even at 20 K.

In conclusion, the Mössbauer spectra of α-$FePO_4$ measured between 4.2 and 24 K confirm the presence of a spin reorientation between 16 and 18 K. Because the magnetic anisotropy of high-spin iron(III) ions is typically very weak and stems from both dipolar interactions and high-order single-ion terms, the difference in energy between the two antiferromagnetic structures should be small. Indeed, Battle et al. [2] suggested that the spin reorientation may well be due to the differing temperature dependence of the dipolar anisotropy, an anisotropy that tends to keep the spins in the basal plane, and the weak fine-structure single-ion anisotropy, which would tend to align the spins along the *c*-axis.

The reduced hyperfine field versus reduced temperature in α-$FePO_4$ and the corresponding Brillouin curve for $S = 5/2$ is shown in Figure 2.7. Because the reduced experimental values are significantly above the Brillouin curve, the magnetic

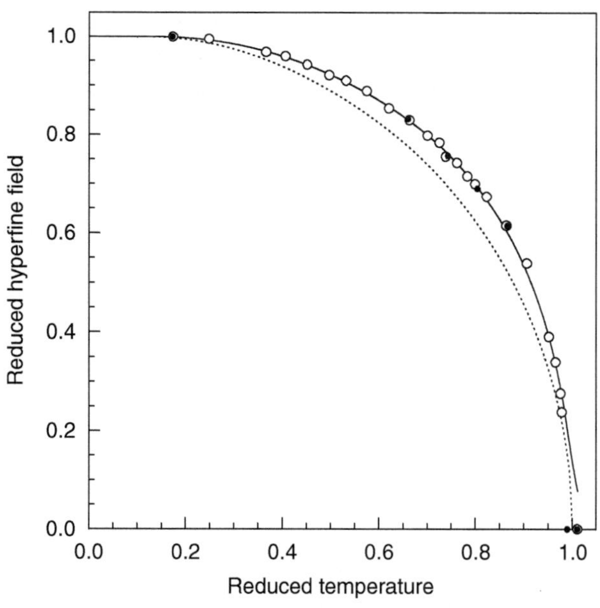

Figure 2.7 A plot of the reduced hyperfine field vs. reduced temperature for α-$FePO_4$ [3]. A Néel temperature of 24.2 K has been used to obtain the reduced temperature. The open symbols are obtained from Battle et al. [2], and a saturation hyperfine field of 51.36 T has been used. The solid symbols are obtained from the average hyperfine fields given in Table 2.6 and a saturation hyperfine field of 51.36 T. The dotted curve is the Brillouin curve for $S = 5/2$, and the solid curve is calculated from the Bean and Rodbell [73] model and corresponds to $\eta_{mag} = 0.368(4)$, a parameter that accounts for the compressibility and/or magnetostriction in the lattice at the Néel temperature.

transition at the Néel point is not a pure second-order magnetic transition. A fit [72] of the experimental points with the magnetostrictive exchange model described by Bean and Rodbell [73] yields an η_{mag} value of 0.368(4). Herein, the η_{mag} notation is used in place of the Bean and Rodbell η notation; the η_{mag} parameter should not be confused with the Mössbauer-effect asymmetry parameter, η. The η_{mag} parameter accounts for the compressibility of the lattice and the magnetostriction that may occur at the Néel temperature of 24.2 K. Because η_{mag} is less than one, the magnetic transition is not first order. Further, η_{mag} is not zero, as is the case for an $S = 5/2$ Brillouin curve, and the non-zero η_{mag} value indicates the presence of a small volume change at the Néel point of α-$FePO_4$, a change that may well be responsible for the increase in $|\Delta E_Q|$ below 24.5 K. The influence of η_{mag}, varying between zero and two, on the shape of the reduced hyperfine field versus reduced temperature curves is illustrated in the supplementary information for reference [3].

2.8 Temperature Dependence of the α-$FePO_4$ Structure Tilt Angle

Is it possible that α-$FePO_4$ structural changes upon cooling may be responsible for the changes observed in its magnetic Mössbauer spectra between 18 and 16 K? The change in the structure of α-$FePO_4$ as a function of temperature has been studied in detail [56] from 293 to 983 K; between 983 and 1003 K the trigonal space group $P3_121$ structure of α-$FePO_4$ changes to the hexagonal space group $P6_422$ of β-$FePO_4$. Because α-$FePO_4$ has many of the same properties as crystalline quartz, this phase transition has been studied by various methods including X-ray and neutron diffraction and Raman spectroscopy. A striking structural feature is the observation that although the unit-cell volume of α-$FePO_4$ increases as expected with increasing temperature, its "apparent" Fe—O bond distance, as determined from Rietveld refinements of powder neutron diffraction data, seemingly decreases with increasing temperature. This unexpected dichotomy, see the left portion of Figure 2.8, may be explained by an increasingly large underestimation of both the Fe—O and P—O bond distances in the Rietveld refinements [74] as compared to the bond distances obtained from a pair distribution functional analysis of the total neutron scattering with increasing temperature. Such an analysis indicates the presence of a significant increasing dynamic disorder [74] as the temperature approaches the phase transition from α-$FePO_4$ to β-$FePO_4$ above 980 K. Indeed, above the phase transition the discrepancy becomes even greater as the dynamic disorder increases. Concurrently with the growth in this disorder, a substantial broadening in the distribution of the Fe—O and the P—O bond distances is also observed [56b].

The dramatic differences in the temperature dependence of the Fe—O and P—O bond distances in α-$FePO_4$, as obtained either by Rietveld line profile analysis or by pair distribution functional analysis of the total neutron scattering [56b], are shown in the left portion of Figure 2.8. This figure clearly illustrates that problems can occur

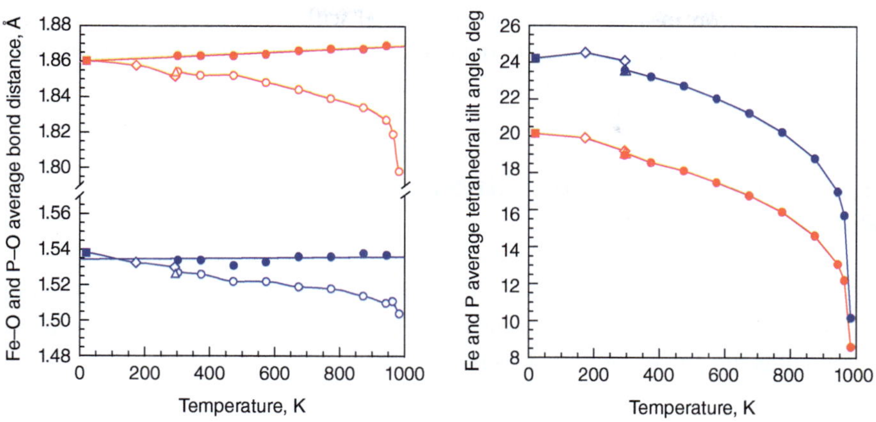

Figure 2.8 Left: The temperature dependence of the Fe—O and P—O bond distances in α-$FePO_4$ as determined by Rietveld refinement, open symbols, and from pair distribution function analysis of the total neutron scattering, closed symbols. Right: the average tilt angles as defined in reference [56a] for the FeO_4 and PO_4 tetrahedra. In both plots the red and blue symbols are for the iron and phosphorus sites, respectively. The circular points at 293–983 K are from reference [56b], the square points at 20 K are from reference [68], the diamond points at 173 and 293 K are from reference [70], and the triangle points at 295 K are from reference [75]. The diamond and triangle points are from X-ray diffraction studies, and all the remaining points are from neutron diffraction studies. Source: The figures are adapted from reference [56].

when a Rietveld refinement is used in conjunction with diffraction data obtained for an unknown dynamic compound. These problems can be especially troublesome when diffraction data is available at only one temperature. As is shown in the left portion of Figure 2.8, the Fe—O and P—O bond distances in α-$FePO_4$ obtained from a Rietveld refinement of the neutron diffraction pattern obtained [68] at 20 K, a temperature sufficiently low to avoid dynamic disorder, agree well with the distances obtained [56b] from total neutron scattering. The straight lines in the left portion of Figure 2.8 are linear fits through the solid points.

Because of the possibility of structural changes associated with the magnetic ordering of α-$FePO_4$, the temperature dependence of the tilt angles at the FeO_4 and PO_4 tetrahedra have been plotted from 20 to 983 K as is shown in the right portion of Figure 2.8. In this figure, the neutron diffraction study [56b] of Haines and his coworkers, between 293 and 983 K, has been extended down to 20 K by including additional structural results obtained from X-ray diffraction studies at 295 K [75] and 173 and 293 K [70], and a neutron diffraction study at 20 K [68]. The slightly higher than might be expected value for the two average PO_4 tetrahedral tilt angles found at 173–293 K are probably a result of the use of X-ray diffraction rather than neutron diffraction results. However, for both FeO_4 and PO_4 tetrahedra the tilt angles observed between 20 and 293 K seem to be fully consistent with those observed by Haines and his coworkers between 293 and 983 K, i.e. the tilt angles reveal no distinct changes at temperatures down to 20 K, an observation that is also consistent with the powder neutron diffraction patterns reported [2] by Battle et al. at 20 and 5 K. Thus it seems that structural changes in the tilt angle of the

FeO_4 tetrahedron in α-$FePO_4$ are not associated with the changes in the θ angle observed in the Mössbauer spectra obtained below its antiferromagnetic ordering temperature, see Figure 2.6 and Table 2.6.

2.9 Mössbauer Spectral Studies on Metastable Polymorphs of $FePO_4$

In addition to the Mössbauer spectral work on the thermodynamically stable α-$FePO_4$ compound, there have been several studies of metastable iron phosphate compounds that are found to be kinetically stable.

2.9.1 Crystallographic Structures of Two Polymorphs of $FePO_4{\cdot}2H_2O$

Perhaps the most interesting recent synthetic work on metastable $FePO_4$ compounds is that of Song et al. [57], who have reported and characterized two new metastable $FePO_4$ compounds, γ-$FePO_4$ and ζ-$FePO_4$, through the dehydration of $FePO_4{\cdot}2H_2O$. As noted above these authors have reported the crystal structures of both a monoclinic and an orthorhombic form of $FePO_4{\cdot}2H_2O$ prepared by hydrothermal syntheses, specifically a monoclinic polymorph of $FePO_4{\cdot}2H_2O$; a synthetic version of the phosphosiderite mineral, with the space group $P2_1/n$, number 14; and an orthorhombic polymorph of $FePO_4{\cdot}2H_2O$, a synthetic version of the strengite mineral, with the space group *Pbca*, number 61.

When the hydrothermal synthesis of $FePO_4{\cdot}2H_2O$ is carried out between 418 and 443 K for no more than eight hours, only the monoclinic polymorph, γ-$FePO_4$, is formed, whereas longer heating yields a mixture of both polymorphs, γ-$FePO_4$ and ζ-$FePO_4$. Fortunately, because of their differing crystal sizes and colors, these two polymorphs may be easily separated from each other with physical methods.

The monoclinic and orthorhombic $FePO_4{\cdot}2H_2O$ structures are rather similar [57] in that each has a crystallographically unique iron(III) Fe1 site in a *cis*-$FeO_4(H_2O)_2$ pseudooctahedral coordination environment involving two *cis*-coordinated water molecules and four oxygens bridging to four different phosphate trianions; there is also one unique P1 site bonded in a pseudotetrahedral coordination environment to four oxygens to form a phosphate trianion.

2.9.2 Preparation and Crystallographic Structures of the Two Polymorphs, γ-$FePO_4$ and ζ-$FePO_4$

The heating of each of the above polymorphs of $FePO_4{\cdot}2H_2O$ under vacuum for 12 hours at 353 K leads to complete dehydration, surprisingly, accompanied by a full retention of the original $FePO_4{\cdot}2H_2O$ space group [57]. More specifically, one obtains either the monoclinic polymorph, γ-$FePO_4$, with the retained space group $P2_1/n$ and an associated unit-cell volume contraction of 27% or the orthorhombic polymorph, ζ-$FePO_4$, with the retained space group *Pbca* and an associated unit-cell volume contraction of 13%. A brief description of the effect of dehydration on the structure as a contraction of the lattice resulting from the disappearance of the water molecules of crystallization is presented.

The two γ-$FePO_4$ and ζ-$FePO_4$ polymorphic structures show similarities [57] in that each has a crystallographically unique iron(III) Fe1 site and phosphorus P1 site. In monoclinic γ-$FePO_4$ the iron(III) may be considered to have a FeO_4 pseudotrigonal pyramidal or a highly distorted tetrahedral, four coordination environment, or to have a highly axially distorted FeO_5 pseudotrigonal bipyramidal five coordination environment with very different axial Fe—O bond distances of 1.809(9) and 2.234(9) Å; P1 is in a pseudotetrahedral environment. In contrast, in the orthorhombic ζ-$FePO_4$ structure both the iron(III) Fe1 site and the phosphorus P1 site have a pseudotetrahedral coordination environment. The environmental geometry of the iron(III) is very similar to that of the α-$FePO_4$ phase, but the specific molar volume of ζ-$FePO_4$ is substantially larger than that of α-$FePO_4$, see below and Table 2.7. A comparison [57] of the structures of the α-, γ-, δ-, ζ-, and ε-$FePO_4$ polymorphs shows that their backbone frameworks differ in connectivity and exhibit different shape channels.

Unfortunately, to our knowledge there have been no studies of the Mössbauer spectral properties of the two metastable γ-$FePO_4$ and ζ-$FePO_4$ phases reported by Song et al. [57]. Delacourt et al. [57] have reported the room temperature Mössbauer spectra of strengite, ζ-$FePO_4{\cdot}2H_2O$, and the related metastrengite, $FePO_4{\cdot}2.3H_2O$, of unknown structure, which have, as expected, similar isomer shifts and quadrupole splittings, isomer shifts which are smaller and quadrupole splitting which are

Table 2.7 Molar volume, V_{mol}, and density, ρ, of several $FePO_4$ polymorphs.

Compound, crystal system	Space group, number, Z	V_{mol}, $Å^3/FePO_4$	ρ, g/cm^3	References
α-$FePO_4$, trigonal	$P3_121$, 152, 3	82.19	3.05	[56a]
		82.16	3.049	[56b]
		82.4	3.04	[75]
		81.95	3.06	[70]
		81.99	3.056	[68]
Average	—	82.1(3)	3.05(1)	—
ε-$FePO_4$, orthorhombic	*Cmcm*, 63, 4	64.19	3.90	[65]
		65.19	3.84	[63]
		65.07	3.85	[64]
Average	—	64.8(6)	3.86(4)	—
ζ-$FePO_4$, orthorhombic	*Pbca*, 61, 8	94.04(2)	2.663	[57]
γ-$FePO_4$, monoclinic	$P2_1/n$, 14, 4	82.13(3)	3.049	[57]
η-$FePO_4$, orthorhombic	*Pnma*, 62, 4	67.90	3.69	[60]
η-$Fe_{0.9}Mn_{0.1}PO_4$, orthorhombic	*Pnma*, 62, 4	68.37	3.66	[60]
η-$Fe_{0.7}Mn_{0.3}PO_4$, orthorhombic	*Pnma*, 62, 4	70.59	3.54	[60]
η-$FePO_4$, orthorhombic	*Pnma*, 62, 4	66.75	3.75	[58]

somewhat larger than those of α-$FePO_4$. However, there have been several studies of the Mössbauer spectra of orthorhombic polymorphs of $FePO_4$, none of which seem to correspond to the orthorhombic phase discussed just above. Finally, it must be emphasized that the orthorhombic ζ-$FePO_4$ structure reported [57] is *not* the same as the orthorhombic high-pressure ε-$FePO_4$ structure prepared and studied by Kinomura et al. [65], Battle et al. [63], and Pasternak et al. [64]. At this point it should be noted that the very similar lattice parameters observed for the orthorhombic ε-$FePO_4$ phase with space group *Cmcm*, prepared by Kinomura et al. [65], Battle et al. [63], and Pasternak et al. [64] indicate that they are all the same metastable high-pressure phase.

For the five reported room temperature crystal structures of trigonal α-$FePO_4$, the three structures of orthorhombic ε-$FePO_4$, the orthorhombic structure of ζ-$FePO_4$, and the monoclinic structure of γ-$FePO_4$, their specific molar volume, V_{mol} per $FePO_4$, and the corresponding densities, ρ, are given in Table 2.7. For α-$FePO_4$ the average $V_{mol} = 82.13(3)\,Å^3$/mol of $FePO_4$ and $\rho = 3.05(1)\,g/cm^3$, whereas, in contrast, for orthorhombic ε-$FePO_4$, $V_{mol} =$ 64.8(6) $Å^3$/mol of $FePO_4$ and $\rho = 3.86(4)\,g/cm^3$. On average, V_{mol} for the fourfold pseudotetrahedral iron(III) coordination environment in α-$FePO_4$ is substantially larger than for the sixfold pseudooctahedral iron(III) coordination environment in the orthorhombic ε-$FePO_4$ polymorph, a difference that accounts for the substantially higher density in the latter case. It is pleasing to see the excellent internal agreement within the crystallographic results for each of the two different polymorphs given in Table 2.7 as reflected in their respective specific molar volumes, V_{mol}, and densities, ρ. It should be noted that the three orthorhombic ε-$FePO_4$ polymorphs were made by similar but not identical procedures over a period 21 years from 1976 to 1997 in three different countries on three different continents [63–65].

It is interesting to note that both V_{mol} and ρ for α-$FePO_4$ and γ-$FePO_4$, see Table 2.7, are identical within their uncertainties. Hence, it seems that the dehydration of monoclinic $FePO_4{\cdot}2H_2O$ with a 27% contraction in the lattice volume leads to V_{mol} and ρ values for γ-$FePO_4$ that are virtually identical to those observed in trigonal α-$FePO_4$ but with different backbone frameworks, as noted by Song et al. [57] and above. In contrast, the dehydration of orthorhombic $FePO_4{\cdot}2H_2O$ with a smaller 13% contraction in the lattice volume leads to V_{mol} and ρ, much larger and smaller, respectively, for orthorhombic ζ-$FePO_4$, than those observed for the orthorhombic ε-$FePO_4$ and γ-$FePO_4$ structures. This difference no doubt results from the difference in the iron(III) environment in the orthorhombic structures; a pseudotetrahedral iron(III) environment is observed in the ζ-$FePO_4$ structure, whereas a pseudooctahedral environment is observed in the ε-$FePO_4$ and γ-$FePO_4$ structures.

2.9.3 Mössbauer Spectral Studies of $FePO_4$ Metastable Polymorphs

The earliest of these studies may be the work by Kinomura et al. [65], who prepared α-$FePO_4$ by heating in air for four days at 1373 K a gel prepared from equal molar amounts of $FeCl_3$ and Na_3PO_4. The resulting α-$FePO_4$ was then sealed in a gold capsule and heated at 1273–1373 K at 4–6 GPa for one hour. The resulting metastable ε-$FePO_4$ was found to be isostructural with orthorhombic

$CrVO_4$, with the *Cmcm* space group [65]; the magnetic and Mössbauer spectral properties were reported to correspond to high-spin iron(III) in an octahedral coordination environment. No Mössbauer spectrum was shown by Kinomura et al. [65], but the reported isomer shift, $\delta = 0.27$ mm/s, and quadrupole splitting, $\Delta E_Q = 0.60$ mm/s, for the low-pressure pseudotetrahedral iron(III) seem to be typical of the 295 K parameters of α-$FePO_4$. In contrast, the $\delta = 0.40$ mm/s and $\Delta E_Q = 1.90$ mm/s doublet parameters reported for the high-pressure metastable ε-$FePO_4$ phase are consistent with a distorted octahedral iron(III) coordination environment.

A second Mössbauer study of the metastable phase ε-$FePO_4$ is the work of Battle et al. [63] who have prepared at high pressure an orthorhombic phase with space group *Cmcm*, referred to in their paper as $FePO_4$-II, by heating α-$FePO_4$ in a sealed platinum tube at 1123 K at 4 GPa for two hours. After cooling under pressure and grinding, the high-pressure orthorhombic phase, ε-$FePO_4$, was found [63] to have a density of 3.84 g/cm^3, a value that is 22.6% higher than the 3.02 g/cm^3 found for the low-pressure trigonal α-$FePO_4$ phase, see Table 2.7; these density values are in reasonable agreement with those reported by Kinomura et al. [65].

The ambient pressure Mössbauer spectra of ε-$FePO_4$ have been measured [63] from 63 to 4.2 K. At 63 K the ε-$FePO_4$ spectrum exhibits a simple quadrupole doublet with an isomer shift, $\delta = 0.54(1)$ mm/s, and quadrupole splitting, $\Delta E_Q =$ 1.81(1) mm/s, values that agree reasonably well with those reported by Kinomura et al. [65]. As is shown in figures 2 and 3 of reference [63], at 58 K and below the ambient pressure Mössbauer spectra of the high-pressure ε-$FePO_4$ phase begin to show the onset of long range magnetic order at a temperature just above 58(1) K, a temperature that is well above the 24.2(2) K Néel temperature observed [3] for α-$FePO_4$, see Figure 2.5, above.

At 4.2 K the Mössbauer spectrum of the high-pressure ε-$FePO_4$ phase is a well-developed magnetic sextet, see Figure 2.9, with an isomer shift, $\delta = 0.53(1)$ mm/s; an internal hyperfine field, H_{int}, of 51.65(1) T; and a quadrupole shift, $QS = 2\varepsilon =$ $-1.668(6)$ mm/s, where ε is $= (1/8)e^2qQ(3\cos^2\theta - 1 + \eta\sin^2\theta \cos 2\phi)$. By using the method described by Le Caër et al. [53] and assuming that the quadrupole splitting does not change at the magnetic transition temperature, Battle et al. [63] conclude that the internal hyperfine field makes an angle θ of between 0° and 13° with the principal axis, V_{zz}, of the electric field gradient tensor, with an asymmetry parameter, η, of between 0.7 and 0, respectively.

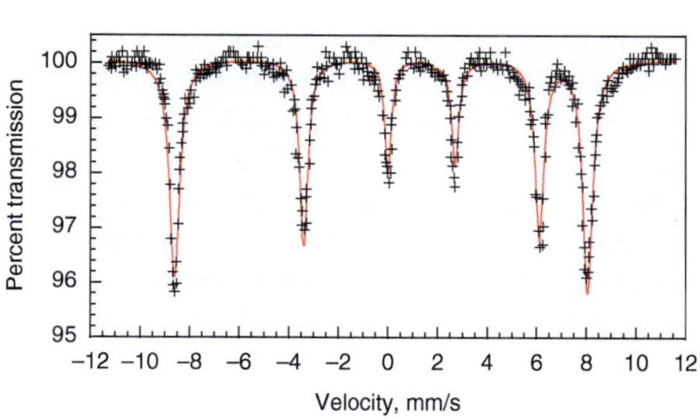

Figure 2.9 The 4.2 K Mössbauer spectrum [76] of orthorhombic ε-$FePO_4$ reported by Battle et al. [63]. The spectral hyperfine parameters corresponding to the fit shown in red are discussed in the text. Source: Adapted from reference [63].

Because Battle et al. [63] have not shown fits for any of their orthorhombic ε-$FePO_4$ Mössbauer spectra, the authors have digitized the 4.2 K experimental data shown in figure 3 of reference [63] and fit the spectrum with the two limiting sets, proposed above, for the quadrupole coupling constant, $e^2Qq/2$; the angle, θ; and the asymmetry parameter, η. A significantly smaller χ^2 is obtained [76] for the fit, shown in red in Figure 2.9, with $e^2Qq/2 = -1.811$ mm/s, $\theta = 13°$, and $\eta = 0$, constrained to the proposed values [63] and the resulting fitted parameters, $\delta = 0.529(2)$ mm/s relative to α-iron, $H_{int} = 51.45(2)$ T, $\Gamma = 0.31(1)$ mm/s, and $\Delta\Gamma = 0.021(2)$, where the incremental line width, $\Delta\Gamma$, is defined above.

Pasternak et al. [64] have reported the partially irreversible transformation of the trigonal α-$FePO_4$ polymorph with increasing pressure into the metastable high-pressure ε-$FePO_4$ orthorhombic polymorph and a secondary amorphous iron phosphate, through X-ray diffraction and Raman and Mössbauer spectral high-pressure studies; this high-pressure phase has an orthorhombic structure with the *Cmcm* space group and is the same metastable phase as that studied by Kinomura et al. [65] and Battle et al. [63].

The X-ray diffraction study indicated that α-$FePO_4$ was the stable phase from ambient pressure to 1.6 GPa and changed to the orthorhombic ε-$FePO_4$ phase at 3.9 and 13.0 GPa; the observed diffraction pattern remained essentially that of the orthorhombic phase ε-$FePO_4$ upon release of the pressure to 0 GPa. The Raman spectra exhibited the expected Raman modes for α-$FePO_4$ from ambient to 1 GPa, both the trigonal and orthorhombic modes at 3 GPa, and only the orthorhombic modes at 5 and 18 GPa with the possible presence of an amorphous phase; again the transformation was essentially irreversible upon return to ambient pressure [64].

The ambient pressure Mössbauer spectra measured at 35 K by Pasternak et al. [64] reveal a quadrupole doublet with an isomer shift relative to α-iron of $\delta = 0.36(5)$ mm/s and a quadrupole splitting $\Delta E_Q = 0.70(5)$ mm/s – parameter values that are consistent with those expected for α-$FePO_4$ at 35 K. These parameters have a somewhat higher uncertainty than usual because Pasternak et al. [64] do not report any hyperfine parameter values, and these values have been obtained from figure 3 of their paper.

The Mössbauer spectra of the high-pressure form of orthorhombic space group *Cmcm* ε-$FePO_4$, see Figure 2.10 that is adapted from figure 3 of reference [64], obtained at 35 K and 2.5, 13, and 25 GPa, consist of three components, a dominant rather broad sextet associated with the crystalline orthorhombic ε-$FePO_4$ phase, a very broad component presumably associated with an amorphous $FePO_4$ phase, and a broadened doublet that seems to have hyperfine parameters that are rather similar to the 35 K ambient pressure α-$FePO_4$ phase. The spectrum measured at 15 K and 13 GPa is different in that it seems the doublet has undergone magnetic ordering upon cooling from 35 to 15 K and 13 GPa. Once again, the transformation at high pressure is rather irreversible and the final 35 K decompressed 0 GPa spectrum obtained upon decompression does *not* resemble the initial 0 GPa paramagnetic spectrum but is quite similar to that observed at 35 K and 2.5 GPa, see the top of Figure 2.10. No spectral parameters are given in reference [64], but a determination of the positions of the outer sextet lines yields an internal hyperfine field at 35 K

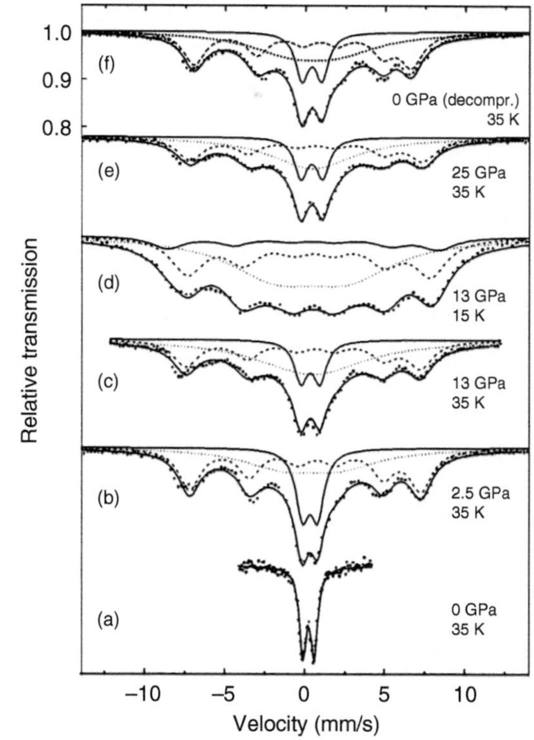

Figure 2.10 The paramagnetic Mössbauer spectrum of α-$FePO_4$, **a**, [64] is the quadrupole doublet expected at 35 K and ambient pressure. The **b–e** spectra obtained at 35 or 15 K at the indicated pressures have been fit with three components, a rather broad magnetic sextet, dashed line, corresponding to the orthorhombic ε-$FePO_4$ phase; a very broad sextet, dotted line, corresponding to an amorphous $FePO_4$ phase; and a quadrupole doublet, solid line, that is predominantly a remnant of the α-$FePO_4$ phase remaining at 35 K and the higher pressures; this phase is as expected magnetically ordered at 15 K and 13 GPa in spectrum **d**. The top spectrum, **f**, has been measured after the release of the applied pressure and indicates, as does the room temperature ambient pressure spectrum, the irreversibility of the changes produced at high pressure.

and 2.5 GPa of $H_{int} = 45.3(5)$ T; at 35 K and 5, 13, and 25 GPa a virtually constant $H_{int} = 45.8(5)$ T; and at 15 K and 13 GPa, as expected, the field is slightly larger at $H_{int} = 46.8(5)$ T.

In view of the relatively non-compact structure of trigonal α-$FePO_4$, with space group $P3_121$, which has an average specific molar volume, $V_{mol} = 82.1(3)$ Å^3/$FePO_4$ and a density of $\rho = 3.05(1)$ g/cm^3, see Table 2.7, it seems quite reasonable that it is converted at high pressure to its orthorhombic ε-$FePO_4$ polymorph with space group *Cmcm*, which has a substantially smaller $V_{mol} = 64.8(3)$ Å^3/$FePO_4$ and a much higher $\rho = 3.84(4)$ g/cm^3. In this regard it seems unexpected that a substantial portion of the α-$FePO_4$ phase has been converted to an amorphous $FePO_4$ phase at high pressure. Perhaps, α-$FePO_4$ is just on the edge of a transformation from a tetrahedral iron(III) fourfold coordination environment to sixfold environment at relatively low pressure that is driven by the two added Fe—O bonds. However, even at the low pressure there probably is a kinetic barrier that at least partially retards the crystallographic transition and leads, in part, to the partial amorphization of α-$FePO_4$.

2.9.4 Preparation and Mössbauer Spectra of Synthetic Heterosite, (Fe,Mn)PO_4

There have been more recent reports by Kwon et al. [60, 77] on the preparation, magnetic properties, and Mössbauer spectral results for η-$FeMnPO_4$, η-$Fe_{0.9}Mn_{0.1}PO_4$, and η-$Fe_{0.7}Mn_{0.3}PO_4$, all of which crystallize in the orthorhombic *Pnma* space group, number 62, of heterosite. These compounds have been prepared with a solid-state reaction starting with a stoichiometric mixture of Li_2CO_3, $FeC_2O_4 \cdot 2H_2O$, $NH_4H_2PO_4$, and $MnCO_3$, that was calcined at 350 °C for three hours in an argon atmosphere and then compressed at 5000 N/cm^2 into pellets that were then heated for 10 hours at 700 °C in a quartz tube. Next, the products, $LiFe_{1-x}Mn_xPO_4$, were delithiated by chemical oxidation through the reaction, $LiFe_{1-x}Mn_xPO_4 + NO_2BF_4 \rightarrow Fe_{1-x}Mn_xPO_4 + LiBF_4 + NO_2$. Finally, the product of this delithiation reaction was purified by filtration and vacuum evaporation of the impurities present. The product of these reactions are all crystalline orthorhombic η-$Fe_{1-x}Mn_xPO_4$ compounds with the *Pnma* space group of heterosite as opposed to the *Cmcm* or *Pbca* space groups found for the other orthorhombic compounds listed in Table 2.7. However, as might be expected, the three $Fe_{1-x}Mn_xPO_4$ compounds have specific molar volumes, V_{mol}, and densities, ρ, that are similar to orthorhombic ε-$FePO_4$ and have the *Cmcm* space group; the nature of the space group for these orthorhombic compounds seems to be very sensitive to their preparation method.

The room temperature Mössbauer spectra of some heterosite like, (Fe,Mn)PO_4, synthetic minerals, specifically, η-$FePO_4$, η-$Fe_{0.9}Mn_{0.1}PO_4$, and η-$Fe_{0.7}Mn_{0.3}PO_4$, have been measured and exhibit quadrupole splittings of $|\Delta E_Q| = 1.512$, 1.502, and 1.38 mm/s, respectively, all with an isomer shift of $\delta = 0.31$ mm/s. The ΔE_Q values are indicative of the expected pseudooctahedral high-spin iron(III) coordination environments, but the isomer shift is smaller than would be expected for this environment, but sadly the authors [77] never report a reference standard for their isomer shift values and thus these values are less useful in confirming this expected coordination environment.

The molar magnetic susceptibilities of η-$FePO_4$, η-$Fe_{0.9}Mn_{0.1}PO_4$, and η-$Fe_{0.7}Mn_{0.3}PO_4$, reported by Kwon et al. [60] yield Néel temperatures for the onset of antiferromagnetic coupling of 114, 107, and 97 K, respectively; the value of 107 K for η-$Fe_{0.9}Mn_{0.1}PO_4$ may be inconsistent with the fully paramagnetic Mössbauer spectrum observed at 107 K (see figure 2 of reference [60]), a spectrum that shows no indication of the incipient onset of magnetic exchange coupling as might be expected.

2.9.5 Fits of the Magnetic Mössbauer Spectra of η-$Fe_{0.9}Mn_{0.1}PO_4$

Of more potential interest would be the 4.2–295 K temperature dependence of the Mössbauer spectra of η-$Fe_{0.9}Mn_{0.1}PO_4$ reported by Kwon et al. [60]. In their work, the isomer shift, δ, is reported to decrease from $\delta = 0.43$ mm/s at 4.2 K to $\delta = 0.34$ mm/s at 295 K. In their related paper [77] the 295 K isomer shift was given as $\delta = 0.31$ mm/s for all three of the above compounds, including η-$Fe_{0.9}Mn_{0.1}PO_4$, but once again no reference is given for this 295 K isomer shift difference nor is any spectral line width reported in either paper. Density functional theory calculations [59] in the generalized gradient approximation, GGA, with an additional Coulomb interaction, U, predict an isomer shift of 0.47(6) mm/s relative to α-iron in the antiferromagnetic state of η-$FePO_4$ in its orthorhombic *Pnma* structure in good agreement with the observed [60] 4.2 K isomer shift value of 0.43 or the observed [61] 295 K isomer shift value 0.424(5) mm/s.

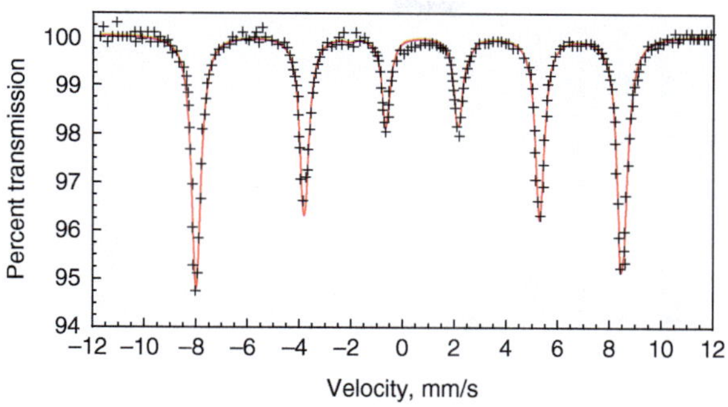

Figure 2.11 The 4.2 K Mössbauer spectrum of η-$Fe_{0.9}Mn_{0.1}PO_4$ published by Kwon et al. [60] with model 2b fit obtained by the authors of this chapter, shown in red; the model 2b fit parameters are given in the bottom row of Table 2.8. The correlation coefficients between the fitted hyperfine parameters are given in italics in Table 2.10. Source: Adapted from reference [60].

The Mössbauer spectra of η-$Fe_{0.9}Mn_{0.1}PO_4$ obtained at 77, 27, and 4.2 K are shown with fits in figure 2 of reference [60], and the corresponding six fitted parameters are given [60] in their table I; these three reported spectra are single magnetic sextets with no obvious secondary components and at most a small negative quadrupole shift, $QS = ½[(p_6 - p_5) - (p_2 - p_1)]$, of ca. −0.53 mm/s at 4.2 K, as obtained from the positions of these four lines in the digitized spectrum obtained from figure 2 in reference [60] and shown in Figure 2.11. Unfortunately, Kwon et al. [60] do not indicate how they obtained from their spectral data the corresponding fit parameters given in their table I. Hence, the reported parameters must be treated with skepticism because the observed small negative $QS = -0.53$ mm/s has led to a rather large magnitude of the quadrupole splitting, ΔE_Q, of 2.10 mm/s at 4.2 K, a value that is unexpectedly larger in magnitude than the $|\Delta E_Q| = 1.50$ mm/s exhibited at 295 and 107 K by the paramagnetic quadrupole doublets, see figure 2 and table I in reference [60]. In addition, no uncertainties in the fit parameters are reported and, of course, one would expect extensive correlation between some of the fit parameters reported in table I. It also appears likely that Kwon et al. [60] have not considered the possibility that ΔE_Q is negative.

As a consequence of the skepticism just noted above, the 4.2 K spectrum of η-$Fe_{0.9}Mn_{0.1}PO_4$, shown in figure 2 of reference [60], has been digitized with the goal of checking the reported fit in order to perhaps overcome this skepticism and check the possibility that ΔE_Q is negative.

In digitizing the 4.2 K Mössbauer spectrum of η-$Fe_{0.9}Mn_{0.1}PO_4$ in order to obtain reliable χ^2 values, one must know the percent statistical error, $100/n^{1/2}$, of the spectrum, information that is not available in reference [60]. Thus to proceed, we have estimated the number of baseline counts as ca. 0.59×10^6 counts/channel, a value that leads to a statistical error of 0.13%. The resulting digitized spectrum has been fit with the code based on the formalism developed by Dattagupta and Blume [54] described in Section 2.4. The results obtained with this code yield the model 1 and 2 results given in Table 2.8, in which the first line is the fit published in table I of reference [60].

Table 2.8 Comparison of the 4.2 K Mössbauer spectral parameter[a)] fits for η-$Fe_{0.9}Mn_{0.1}PO_4$.

δ,[b)] mm/s	H_{int}, T	ΔE_Q,[c)] mm/s	$eQV_{zz}/2$,[c)] mm/s	θ, deg	ϕ, deg	η	Γ,[d)] mm/s	$\Delta\Gamma$[e)]	χ^2	Source
0.43	49.8	2.10	1.91	66	48	0.8	f)	f)	f)	Reference [60]
0.43	50.14(4)	1.75(4)	1.55(4)	61.21(1)	62.30(4)	0.9	0.38(1)	0.006(2)	1.9302	Model 1a
0.43	50.14(2)	1.75(4)	1.55(1)	61.21(1)	62.30	0.9	0.38(1)	0.006(2)	1.9217	Model 1b
0.43	*50.15(1)*	*−1.72(1)*	*−1.72(1)*	*43.54(1)*	*19.56(7)*	*0.13*	*0.38(1)*	*0.006(1)*	*1.9296*	*Model 2a*
0.43	*50.15(2)*	*−1.72(1)*	*−1.72(1)*	*43.53(1)*	*20.0*	*0.13*	*0.38(1)*	*0.006(2)*	*1.9210*	*Model 2b*

a) Statistical uncertainties when available are given in parentheses; the actual uncertainties may be as much as two times larger. The absence of an uncertainty indicates that the value was either unknown or not refined.
b) The isomer shifts, δ, are believed to be referred to α-iron at 295 K; no isomer shift reference is given in reference [60].
c) ΔE_Q is related to $eQV_{zz}/2$, through the expression $\Delta E_Q = \frac{1}{2}[eQV_{zz}(1+\eta^2/3)^{1/2}]$.
d) Γ is the full-width at half-maximum.
e) The unitless incremental line width, $\Delta\Gamma$, is given by $\Gamma(v) = \Gamma + \Delta\Gamma\,|(v-\delta)|$, as is discussed in the text.
f) Value not given in reference [60].

Table 2.9 Correlation coefficients for model 1a and 2a, in italics, in Table 2.8.

	H_{int}	$eQV_{zz}/2$	θ	ϕ	Γ	$\Delta\Gamma$
H_{int}	1	−0.955	−0.906	0.930	0.009	−0.242
	1	*0.635*	*−0.471*	*0.333*	*−0.044*	*−0.079*
$eQV_{zz}/2$		1	0.949	−0.973	−0.021	0.236
		1	*−0.715*	*0.531*	*−0.015*	*−0.050*
θ			1	−0.991	−0.028	0.221
			1	*−0.845*	*−0.007*	*0.042*
ϕ				1	0.029	−0.232
				1	*0.005*	*−0.013*
Γ					1	−0.844
					1	*−0.743*
$\Delta\Gamma$						1
						1

Preliminary refits using the parameters found in table I of reference [60] indicated that fitting the quadrupole coupling constant, $eQV_{zz}/2$, dramatically improves the fit; fitting the angle θ further slightly improves the fit, but fitting both the angle ϕ and the asymmetry parameter, η, sometimes yields unreasonable values of η that are greater than one and always yields extremely large correlation coefficients between all the hyperfine parameters. As a consequence, η was constrained to values between 0.6 and 0.9, and the two angles, θ and ϕ, were allowed to vary for model 1a, which yields a positive $eQV_{zz}/2$, and an η of 0.9 gave the smallest χ^2 and the model 1 parameters in the second line of Table 2.8. Initially, this fit seems potentially reasonable in view of the published [60] parameters given in the first line of this table. However, this is certainly *not* the case as is shown by the correlation coefficients between the fit parameters given in Table 2.9, i.e. correlation coefficients between H_{int}, $eQV_{zz}/2$, θ, and ϕ that range in magnitude from 0.9058 to 0.9908, indicating that these parameters are almost perfectly correlated and that their values are basically meaningless.

The correlation coefficients for the line width, Γ, and the incremental line width, $\Delta\Gamma$, have been included in Tables 2.9 and 2.10 and reveal that Γ and $\Delta\Gamma$ exhibit, as expected, little correlation with the fitted hyperfine parameters but, again as expected, the fitted Γ and $\Delta\Gamma$ parameters are significantly correlated. Also the corresponding fitted quadrupole splitting is not, as published [60], essentially $\Delta E_Q = 2.10$ mm/s but is smaller at 1.75 mm/s and closer to the 1.50 mm/s splitting of the quadrupole doublet observed at 107 and 295 K. It is apparent that the 4.2 K Mössbauer spectral fit parameters reported by Kwon et al. [60] correspond to a false minimum in the applicable parameter space and the fit is unacceptable. In an attempt to avoid the large correlation coefficients between the fitted parameters, η and ϕ were constrained to 0.9 and 62.30°, and this fit yielded the resulting parameters given in line 3 of Table 2.8 and the correlation coefficients given in Table 2.10. This fit gives a similar acceptable χ^2 value and no correlation coefficient is larger in

Table 2.10 Correlation coefficients for models 1b and 2b, in italics, in Table 2.8.

	H_{int}	$eQV_{zz}/2$	θ	Γ	$\Delta\Gamma$
H_{int}	1	−0.591	0.320	−0.048	−0.072
	1	*0.660*	*−0.480*	*−0.029*	*−0.089*
$eQV_{zz}/2$		1	−0.465	0.030	0.045
		1	*−0.674*	*−0.005*	*−0.071*
θ			1	0.001	−0.070
			1	*−0.030*	*0.091*
Γ				1	−0.863
				1	*−0.864*
$\Delta\Gamma$					1
					1

absolute value than 0.6. Hence, some confidence can be placed in the hyperfine parameters given in line 3 of Table 2.8 for model 1b with a positive quadrupole splitting of +1.75 mm/s. It may be noted that with these hyperfine parameters, a quadrupole shift $QS = 2\varepsilon = (1/4)\mathrm{e}QV_{zz}(3\cos^2\theta - 1 + \eta\sin^2\theta\ \cos2\phi) = -0.54$ mm/s is obtained in good agreement with the observed value of −0.53 mm/s.

Now the above assumption that $\mathrm{e}QV_{zz}/2$ is positive must be tested. If $\mathrm{e}QV_{zz}/2$ is negative, θ must be smaller than the magic angle of 54.7° and $\eta = 0$ is possible. Thus, we tested the validity of model 2 with a negative $\mathrm{e}QV_{zz}/2$. As was done with model 1, η was constrained between 0 and 0.2, and $\eta = 0.13$ was found to give the smallest χ^2; then H_{int}, $\mathrm{e}QV_{zz}/2$, θ, and ϕ were refined; and the resulting parameters are given on line 4 of Table 2.8 for model 2a with the correlation coefficients given in italics in Table 2.9. Again large correlation coefficients, in magnitude of up to 0.845, were observed. In a final step, both η and ϕ were constrained equal to 0.13 and 20°, respectively, and the resulting parameters are given on line 5 of Table 2.8 for model 2b with the correlation coefficients given in italics in Table 2.10. None of the correlation coefficient exceeds 0.674 in magnitude. Hence, some confidence can be placed in the hyperfine parameters given in line 5 of Table 2.8 for model 2b with a negative quadrupole splitting of $\Delta E_Q = -1.72$ mm/s. It may be noted that with these hyperfine parameters, a quadrupole shift of $QS = 2\varepsilon = (1/4)eQV_{zz}(3\cos^2\theta - 1 + \eta\sin^2\theta\ \cos2\phi) = -0.54$ mm/s is obtained in good agreement with the observed value of $QS = -0.53$ mm/s.

The agreement between the calculated and observed quadrupole shifts with both models indicates that the small perturbation approach is actually valid to interpret the 4.2 K Mössbauer spectrum of η-$Fe_{0.9}Mn_{0.1}PO_4$. In contrast, the quadrupole shift, $QS = 2\varepsilon = (1/4)eQV_{zz}(3\cos^2\theta - 1 + \eta\sin^2\theta\ \cos2\phi) = -0.60$ mm/s, obtained from the published parameters in line 1 of Table 2.8 is significantly more negative than the observed value of −0.53 mm/s; a disagreement that again indicates that the published fit is wrong. Unfortunately, at this point, it is not possible to choose between model 1b and model 2b.

Finally, we are forced to conclude that the fit reported [60] for the 4.2 K Mössbauer spectrum of η-$Fe_{0.9}Mn_{0.1}PO_4$ is completely in error as are also the fit parameters reported at 77–24 K in table I of reference [60]. Further, sadly the plot of ΔE_Q versus temperature shown in figure 4 of their paper is completely wrong and must be ignored; the same may well be the case for the plot of H_{int} versus temperature results shown in their figure 3. This failure is a clear indication that any attempt to fit a Mössbauer spectrum with a complex multiparameter model, as attempted by Kwon et al. [60], must be accompanied by a careful evaluation of both the statistical uncertainties obtained from the fit and the associated correlation coefficients between all the fit parameters.

Density functional theory calculations [59] for η-$FePO_4$ in its orthorhombic *Pnma* and antiferromagnetic state yield the magnitude and orientation relative to the crystallographic axes of the three principal components of the electric field gradient tensor. In the GGA + U approximation, $V_{zz} = -9.15(1) \times 10^{21}$ V/m^2 oriented along the nearly linear O2–Fe–O1 direction at an angle, θ, of 34° with the crystallographic *a*-axis, along which a major component of the magnetic moment lies, and $\eta = 0.60(1)$. These values yield $eQV_{zz}/2 = +1.64$ mm/s and a quadrupole splitting [59] of $\Delta E_Q = 1.74(1)$ mm/s with $Q = 0.173 \times 10^{-28}$ m^2 and $eQV_{zz}/2 = +1.52$ mm/s and $\Delta E_Q = 1.61(1)$ mm/s with the experimental value of the nuclear quadrupole moment $Q = 0.16 \times 10^{-28}$ m^2. The magnitude of the calculated quadrupole splittings agrees reasonably well with the 295 K quadrupole splitting [77, 78] of 1.512 and 1.520(5) mm/s for η-$FePO_4$. In contrast, the *negative* quadrupole shift observed in Figure 2.11 is incompatible with a positive quadrupole splitting and an angle, θ, of 34° [79].

In contrast with the spectrum in Figure 2.11 [60], the 17 K Mössbauer spectrum of a natural sample of heterosite [80], with a composition close to η-$Fe_{0.65}Mn_{0.35}PO_4$, exhibits a very broad sextet with a *positive* quadrupole shift. Its fit required the use of a distribution of hyperfine field with $eQV_{zz}/2 = +1.50$ mm/s, $\eta = 0.79$, $\theta = 40°$, and $\phi = 93°$. The authors note that this spectrum is a rare occurrence of an iron(III) magnetic spectrum that could not be fitted within the first-order approximation but could well be fitted by adjusting all four parameters. Further, the fitted parameters agree reasonably well with the calculated [59] $eQV_{zz}/2 = +1.52$ mm/s, $\eta = 0.60(1)$, and $\theta = 34°$.

Unfortunately, the authors are forced to conclude that they do not understand how two samples of the same crystallographic orthorhombic structure, *Pnma*, and similar compositions, η-$Fe_{0.9}Mn_{0.1}PO_4$ and η-$Fe_{0.65}Mn_{0.35}PO_4$, give such different magnetic spectra. They hope that these compositions and the corresponding Mössbauer spectra will be carefully reinvestigated in the future.

2.10 Mössbauer Spectral Studies of Various Iron Phosphate Compounds

In addition to the Mössbauer spectral studies of the various polymorphs of $FePO_4$ discussed above, there are many papers that have dealt with related but more complex iron phosphate compounds. Several research groups have worked on the

Mössbauer spectral properties of these more complex iron phosphate compounds, and perhaps the most interesting is the work over many years of Charles Gleitzer, at the University of Nancy, France, and his several colleagues. The work of this group will be, in part, discussed below.

2.10.1 Mössbauer Spectral Properties of α-$Fe_2(PO_4)O$

The mixed-valence compound, α-$Fe_2(PO_4)O$, in powder form has been prepared [9, 81] from a mixture of $FePO_4$, Fe_2O_3, and iron placed in a silver crucible sealed inside a silica tube that was heated under vacuum at 900 °C for several days. Crystals of α-$Fe_2(PO_4)O$ were obtained after an additional heating, under vacuum, at 900 °C of the powder in the presence of $FeCl_2$. The resulting α-$Fe_2(PO_4)O$ single crystals are orthorhombic with space group *Pnma*, number 62; their corresponding specific molar volume, V_{mol}, is 91.56 $Å^3/Fe_2PO_5$ and with a rather high density, ρ, of 4.039 g/cm^3, a value that is high compared with the other orthorhombic densities given in Table 2.7.

In α-$Fe_2(PO_4)O$ the phosphorus occupies a rather isolated oxygen tetrahedra, whereas the mixed valence iron(II) and iron(III) ions are ordered in adjacent face-sharing octahedra [52]. An interesting aspect of this structure is the presence of zig-zag chains of these face-sharing octahedra running parallel to the orthorhombic *b*-axis and containing alternately the iron(II) and iron(III) ions; the resulting zig-zag chains form $[FeO_2]_n$ sheets that, because of the face-sharing octahedra, exhibit a rather short 2.97 Å iron(II)···iron(III) through-space separation. This short separation between iron(II) and iron(III) ions in face-sharing octahedra is not conducive to electron transfer between the two ions and thus α-$Fe_2(PO_4)O$ is a class 1 Robin and Day [1] mixed-valence compound with discrete iron(II) and iron(III) ions with no electron transfer observed between 4.2 and 295 K.

A powder neutron diffraction study of α-$Fe_2(PO_4)O$ between 4.2 and 280 K revealed ferromagnetic exchange within two $[FeO_2]_n$ sheets that are antiferromagnetically coupled together through a dominant 134° iron(III)–O–iron(III) intersheet-superexchange pathway. Because ferromagnetic coupling dominates within the $[FeO_2]_n$ sheets, the α-$Fe_2(PO_4)O$ magnetic structure avoids any spin frustration. Further, the neutron diffraction results [52] yield 4.2 K magnetic moments of 3.82(4) μ_B for the iron(II) sites and 4.26(5) μ_B for the iron(III) sites, moments that are oriented parallel to the orthorhombic *b*-axis of α-$Fe_2(PO_4)O$. Magnetic studies yield a Néel temperature of 218(3) K and reveal magnetostriction predominantly along the orthorhombic *c*-axis.

Mössbauer spectral studies of α-$Fe_2(PO_4)O$ have been reported [52, 82] between 220.5 and 315 K and between 423 and 873 K, revealing two paramagnetic quadrupole doublets characteristic of discrete iron(II) and iron(III) valence states, see Table 2.11. In contrast, the spectra reported [52, 83] between 4.2 and

Table 2.11 Paramagnetic Mössbauer spectral parameters for various iron phosphate compounds.a)

Compound, site	T, K	δ,b) mm/s	ΔE_Q, mm/s	Γ, mm/s	*Area*, %	Assignment	Ref.
α-$Fe_2(PO_4)O$	295	1.099	2.394	0.26	48.1	Fe(II)	[83]
		0.451	1.024	0.263	51.9	Fe(III)	
α-$Fe_2(PO_4)O$	295	1.060	2.386	0.28	47.2	Fe(II)	[52]
		0.464	0.981	0.28	52.8	Fe(III)	
	220.5	1.148	2.583	0.28	48.9	Fe(II)	
		0.487	1.029	0.28	51.1	Fe(III)	
$Fe_3(PO_4)O_3$	295	0.34	1.13	0.28	100	Fe(III)	[84]
$Fe_9(PO_4)O_8$ Fe(1)	292	1.15	0.19	0.24	11.1	~O_h Fe(II)	[81]
Model I Fe(2)		0.35	0.00	0.28	11.1	~O_h Fe(III)	
Fe(3)		1.07	0.83	0.28	22.2	~O_h Fe(II)	
Fe(4)		1.16 1.16	1.31 1.46	0.30 0.30	22.2 22.2	~O_h Fe(II)	
Fe(5)		0.97	1.88	0.25	11.1	~T_d Fe(II)	
$Fe_9(PO_4)O_8$ Fe(1)	292	1.15	0.19	0.24	11.1	~O_h Fe(II)	[81]
Model II Fe(2)		0.46	0.04	0.31	11.1	~O_h Fe(III)	
Fe(3)		1.07	0.83	0.28	22.2	~O_h Fe(II)	
Fe(4)		1.16 1.13	1.30 1.54	0.30 0.30	22.2 22.2	~O_h Fe(II)	
Fe(5)		0.96	1.87	0.26	11.1	~T_d Fe(II)	

a) No statistical uncertainties were reported.
b) The isomer shifts are referred to α-iron.

220 K reveal two magnetic spectral components again with hyperfine parameters characteristic of discrete iron(II) and iron(III) valence states, see Figure 2.12 and Table 2.12.

Since the work [52] published in 1991, the fitting software has evolved and is now capable of both evaluating the correlation coefficients between the hyperfine parameters and handling the complex form of the Hamiltonian resulting from the simultaneous presence of the asymmetry parameter, η, and azimuthal angle, ϕ, of the hyperfine field in the principal axes of the electric field gradient tensor. So, with the goal of checking all the different ranges of hyperfine parameters proposed [83] earlier, the authors carried out a complete exploration of the parameter space for α-$Fe_2(PO_4)O$ at 30 K.

The 30 K Mössbauer spectrum of α-$Fe_2(PO_4)O$ has been fitted with two equal area magnetic components. The first component, a sextet, is assigned to iron(III)

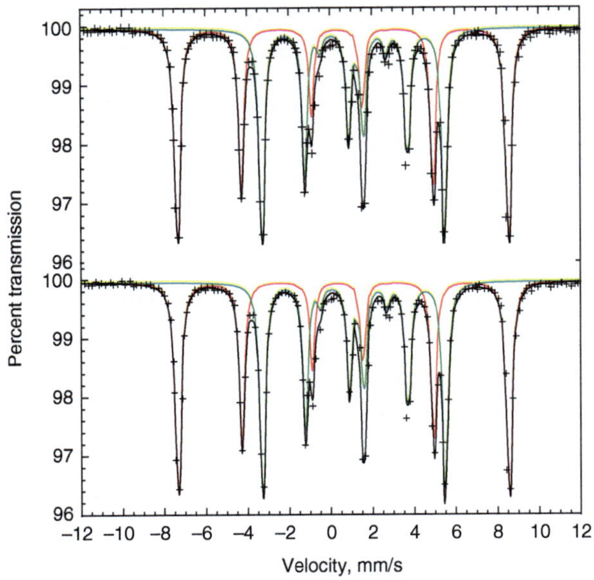

Figure 2.12 Mössbauer spectra of α-$Fe_2(PO_4)O$ at 30 K; the upper and lower plots correspond to the first two and last two lines in Table 2.13. The black, green and red solid lines correspond to the total fit and the iron(II) and iron(III) components, respectively. Note that there is some overlap between the red and black solid lines.

and may be analyzed within the first-order approximation because the magnetic perturbation, $H_{int} = 50.0$ T or 16.1 mm/s, is ca. 45 times the quadrupolar perturbation, i.e. the modulus of $|eQV_{zz}/2| = 1.12$ mm/s. For the second component, an octet, assigned to iron(II), the first-order approximation breaks down because the magnetic hyperfine field is much smaller than that observed for iron(III) and thus the magnetic perturbation, $H_{int} = 24.1$ T or 7.74 mm/s, is only ca. three times the quadrupolar perturbation, the modulus of $|eQV_{zz}/2| = 2.706$ mm/s. The resulting iron(III) and iron(II) isomer shifts, relative to α-iron at room temperature, are 0.590(5) and 1.270(5) mm/s, respectively, in good agreement with earlier published values [52]. For both components, the same line width, Γ, and the same dimensionless incremental line width, $\Delta\Gamma$, defined above, were refined to $\Gamma = 0.28(1)$ mm/s and $\Delta\Gamma = 0.006(2)$.

For the 30 K iron(III) Mössbauer spectral component in α-$Fe_2(PO_4)O$, Modaressi et al. [83] analyzed the iron(III) sextet with a quadrupole shift, *QS*. In contrast, Ijjaali et al. [52], on the basis of the temperature dependence of the iron(III) sextet, constrained both the iron(III) quadrupole interaction, $eQV_{zz}/2$, to −1.12 mm/s and the azimuthal angle, ϕ, to zero to avoid handling complex numbers, and refined the polar angle, θ, and the asymmetry parameter, η.

Further, for the 30 K iron(II) Mössbauer spectral component in α-$Fe_2(PO_4)O$, Modaressi et al. [83] analyzed the iron(II) octet on the basis of the limiting range [53] of values for $eQV_{zz}/2$, θ, ϕ, and η. For *positive* values of $eQV_{zz}/2$, they propose ranges of $2.41 < eQV_{zz}/2 < 2.70$ mm/s, $0.42 < \eta < 1$, $66° < \theta < 71°$, and $0° < \phi < 27°$. Whereas, for *negative* values of $eQV_{zz}/2$, they propose ranges of $-2.62 < eQV_{zz}/2 < -2.41$ mm/s, $0.62 < \eta < 1$, $61.7° < \theta < 64°$, and $0° < \phi < 21°$. In contrast, Ijjaali et al. [52] on the basis of the temperature dependence of the

Table 2.12 Magnetically ordered Mössbauer spectral parameters for various iron phosphate compounds.[a)]

Compound	T, K	H_{int}, T	δ,[b)] mm/s	$eQV_{zz}/2$, mm/s	θ, deg	η	ϕ, deg	Γ, mm/s	$\Delta\Gamma$,[f)] mm/s	*Area*, %	Assignment	References
α-$Fe_2(PO_4)O$	100		1.197	2.797[c)]	[d)]	[d)]	[d)]	0.268	—	50.5	Fe(II)	[83]
			0.528	0.148[e)]				0.270		49.5	Fe(III)	
α-$Fe_2(PO_4)O$	100	22.8	1.223	2.700	66	0.43	0	0.34	—	50	Fe(II)	[52]
		48.3	0.542	−1.12	80	0.35	0	0.34	—	50	Fe(III)	
	30	23.71(1)	1.270(5)	2.59(2)	66(1)	0.43(1)	0	0.28(1)	0.006(2)		Fe(II)	[85]
		49.02(1)	0.590(5)	−1.12	78(1)	0.39(1)	0	0.28(1)	0.006(2)		Fe(III)	
		23.57(5)	1.270(5)	−2.36(2)	64(1)	1	21	0.28(1)	0.006(2)		Fe(II)	[85]
		48.92(2)	0.590(5)	1.12	60(1)	1	0	0.28(1)	0.006(2)		Fe(III)	
	4.2	24.1	1.274	2.706	66	0.43	0	0.28	—	50	Fe(II)	[52]
		50.0	0.591	−1.12	82	0.39	0	0.28	—	50	Fe(III)	
$Fe_3(PO_4)O_3$	60	42.71(4)	0.32(1)	−1.13	72.8(1)	0	0	0.27(3)	0.012(5)	33.3	Fe(III)a	[85]
		43.00(3)	0.33(1)	−1.13	41.0(1)	0	0	0.46(3)	0.012(5)	66.7	Fe(III)b	
	30	44.30(4)	0.32(1)	−1.13	69.9(1)	0	0	0.22(2)	0.019(4)	33.3	Fe(III)a	[85]
		44.67(3)	0.33(1)	−1.13	41.5(1)	0	0	0.50(3)	0.013(4)	66.7	Fe(III)b	
	4.2	44.89(4)	0.311(5)	−1.13	72.7(1)	0	0	0.22(3)	0.022(5)	33.3	Fe(III)a	[85]
		45.27(4)	0.333(6)	−1.13	40.8(1)	0	0	0.50(3)	0.014(6)	66.7	Fe(III)b	

a) The statistical uncertainties, when reported, are given in parentheses.
b) The isomer shifts are referred to α-iron.
c) Value for $\Delta E_Q = (eQV_{zz}/2)(1+\eta^2/3)^{1/2}$.
d) The ranges of values are given in reference [83].
e) Value for ε.
f) The unitless incremental line width, $\Delta\Gamma$, is given by $\Gamma(v) = \Gamma + \Delta\Gamma|(v-\delta)|$, as is discussed in the text.

Table 2.13 Spectral fit parameters for the 30 K Mössbauer spectrum of $Fe(PO_4)O$.[a)]

H_{int}, T	$eQV_{zz}/2$, mm/s	θ, deg	ϕ, deg	η	χ^2	Site
49.02(1)	−1.12	78(1)	0	0.39(1)	1.5014	Fe(III)
23.71(1)	+2.59(2)	66(1)	0	0.43(1)		Fe(II)
49.02(1)	−1.12	78(1)	0	0.39(1)	1.5069	Fe(III)
23.65(2)	+2.33(2)	72(1)	27	1		Fe(II)
49.03(2)	−1.12	78(1)	0	0.39(1)	1.5236	Fe(III)
23.57(2)	−2.56(2)	62(1)	0	0.62(1)		Fe(II)
49.02(2)	−1.12	78(1)	0	0.39(1)	1.5048	Fe(III)
23.60(5)	−2.35(2)	64(1)	21	1		Fe(II)
48.92(2)	+1.12	60(1)	0	1	1.5439	Fe(III)
23.71(2)	+2.59(1)	66(1)	0	0.43(1)		Fe(II)
48.92(2)	+1.12	60(1)	0	1	1.5383	Fe(III)
23.65(2)	+2.33(1)	72(1)	27	1		Fe(II)
48.92(2)	+1.12	60(1)	0	1	1.5495	Fe(III)
23.57(2)	−2.56(1)	62(1)	0	0.62(1)		Fe(II)
48.92(2)	+1.12	60(1)	0	1	1.5384	Fe(III)
23.57(5)	−2.36(2)	64(1)	21	1		Fe(II)

a) The statistical uncertainties are given in parentheses. The absence of uncertainty indicates that the parameter was constrained to the given value.

iron(II) hyperfine parameters analyzed the octet with only a positive $eQV_{zz}/2$ value and constrained ϕ to zero to avoid complex numbers. Hence, Ijjaali et al. [52] obtained satisfactory fits between 4.2 and 220 K but, unfortunately, they did not explore the entire parameter space as suggested [83] above.

Figure 2.12 and Table 2.13 present the results of an extensive exploration of the parameter space for the 30 K spectrum of α-$Fe_2(PO_4)O$. All eight possible combinations of the two quadrupole interactions and their signs yield satisfactory fits with very similar χ^2-values ranging from 1.5014 to 1.5495. These fits reveal that the four combinations with a negative iron(III) quadrupole interaction of $eQV_{zz}/2 = -1.12$ mm/s give slightly better χ^2-values, but there is no corresponding obvious visible improvements in the fits. The upper and lower portions of Figure 2.12 show the spectral fits obtained with the first and the last combinations in Table 2.13, respectively. It is clear that the fits are identical in spite of their quite different hyperfine parameters. In the absence of additional information, such as the magnetic structure and/or calculations of the principal components of the electric field gradient tensor at both the iron(III) and iron(II) sites, it is not possible to determine the best fit by choosing between the eight combinations shown in Table 2.13 or any other combination with parameters lying within the limiting ranges given above.

2.10.2 Mössbauer Spectral Properties of $Fe_3(PO_4)O_3$

$Fe_3(PO_4)O_3$ has been prepared [84] by two different methods, either by mixing stoichiometric amounts of Fe_2O_3 and $FePO_4$ followed by calcination in an alumina crucible or by the evaporation to dryness of a solution containing stoichiometric amounts of $Fe(NO_3)_3$ and $(NH_4)_2HPO_4$, followed by successive grinding-calcination steps. In either method, the reaction yields pure $Fe_3(PO_4)O_3$, based on the X-ray diffraction patterns, after several of these steps, followed by annealing between 1050 and 1100 °C. If annealed below 1050 °C, the reaction is incomplete, whereas annealing above 1100 °C decomposes the desired product into hematite and $FePO_4$.

A single crystal X-ray diffraction study at 298 K of $Fe_3(PO_4)O_3$ indicates a trigonal unit cell with the $R3m$, space group number 160 [84]. The corresponding specific molar volume, V_{mol}, is 126.99 Å^3/$Fe_3(PO_4)O_3$ and with density, ρ, of 4.06 g/cm^3, a value that is high compared with the orthorhombic densities given in Table 2.7.

The tetrahedral PO_4 trianions are well isolated, and at 298 K there is only one crystallographic iron(III) site that has a five coordinate pseudotrigonal bipyramidal coordination environment with the iron(III) ions shifted away from the trigonal equatorial plane. The two edge-sharing trigonal bipyramids yield groups of three hexahedra; this edge-sharing environment leads to a somewhat short through-space iron(III)···iron(III) distance of 3.13 Å. As a consequence, antiferromagnetic exchange coupling with a Néel temperature of 175(5) K occurs as is observed in the magnetic susceptibility measurements [86] of $Fe_3(PO_4)O_3$.

Above the Néel temperature of 175(5) K, the Mössbauer spectra of $Fe_3(PO_4)O_3$ consist of a single quadrupole doublet [84]; at 295 K the isomer shift, δ, is 0.34 mm/s; the magnitude of the quadrupole splitting, $|\Delta E_Q|$, is 1.13 mm/s; and the line width, Γ, is 0.28 mm/s, see Table 2.11. These values would be typical of high-spin iron(III) in a pseudooctahedral coordination environment, except that the rather low isomer shift, $\delta = 0.34$ mm/s, is also consistent with the observed pseudotrigonal bipyramidal coordination found in $Fe_3(PO_4)O_3$.

The low temperature Mössbauer spectra of $Fe_3(PO_4)O_3$ have also been measured [86] at 60, 30, and 4.2 K, i.e. well below its Néel temperature and, somewhat unexpectedly, the spectra exhibit two magnetic sextets, see Figure 2.13. Clearly, the magnetic perturbation is much larger than the quadrupolar perturbation and the first-order perturbation is valid to fit these spectra as was previously reported [86] with a parameter Δ that is equal to 4ε. In order to link the quadrupole interaction between the magnetic and paramagnetic spectra in the fits shown in Figure 2.13, $eQV_{zz}/2$ was constrained to +1.13 mm/s and then to −1.13 mm/s and the angles θ_a and θ_b of the two sextets were refined with both η and ϕ constrained to zero because both cannot be obtained without extensive correlations between all the hyperfine parameters. Somewhat surprisingly, the fits with $eQV_{zz}/2$ constrained to −1.13 mm/s exhibit χ^2 values, which were significantly smaller, ca. one-half the χ^2 values obtained in the fits with $eQV_{zz}/2$ constrained to +1.13 mm/s. The resulting hyperfine parameters that correspond to high-spin iron(III) sites are given in Table 2.12. This new analysis of the Mössbauer spectra of $Fe_3(PO_4)O_3$ agrees with the proposal [86] that its antiferromagnetic structure loses a threefold

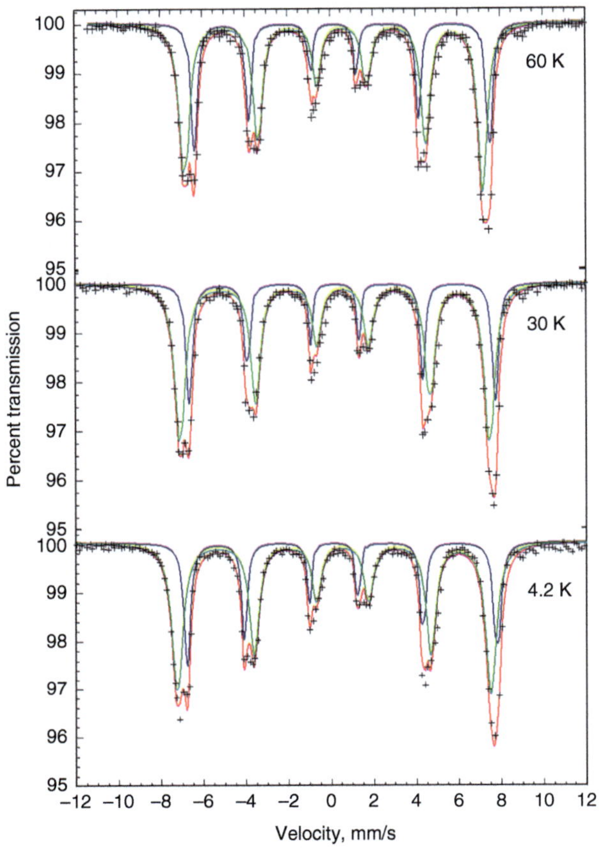

Figure 2.13 Mössbauer spectra of $Fe_3(PO_4)O_3$ at 4.2, 30, and 60 K with the spectral parameters reported in Table 2.12. The red, blue, and green solid lines represent the total fit, the iron(III)a and iron(III)b components, respectively.

symmetry axis upon cooling but retains a mirror plane, a change that yields two crystallographically equivalent but magnetically inequivalent iron(III) sites – one site iron(III)a is in the mirror plane and is thus unique, and the second consists of two equivalent sites, iron(III)b, that lie outside the mirror plane.

2.10.3 Preparation and Structural Properties of $Fe_9(PO_4)O_8$

Venturini et al. [9] have prepared the complex phosphate, $Fe_9(PO_4)O_8$, by using two different methods: first, a reaction of a stoichiometric mixture of $FePO_4 + (8/3)Fe + (8/3)Fe_2O_3$ in a sealed quartz tube under vacuum at 900 °C and, second, through reduction of a mixture of $FePO_4 + 4\ Fe_2O_3$, followed by calcination of the resulting powder at 900 °C for a limited time to avoid loss of phosphorus. Both preparations produce samples that exhibit essentially the same Mössbauer spectra.

Single crystals of $Fe_9(PO_4)O_8$ were obtained by annealing the powder with a trace of $FeCl_2$ and were found to have an orthorhombic crystal structure with the

Bmmm space group number 65. Their corresponding specific molar volume, V_{mol}, is 234.33 Å^3/$Fe_9(PO_4)O_8$, with a density, ρ, of 5.14 g/cm^3, a value that is rather high compared with the other orthorhombic densities given in Table 2.7. An electron diffraction study of a single crystal of $Fe_9(PO_4)O_8$ indicates the presence of a "pseudomonoclinic" supercell with $a' = 2a$, $b' = 2b$, $c' = c$, and $\beta' = 90°$, and with Z = 8, a supercell that was less apparent in X-ray diffraction studies because the superstructure reflections were extremely weak.

A single crystal structural study [9] of $Fe_9(PO_4)O_8$ revealed that, in the orthorhombic *Bmmm* space group number 65, there are five crystallographically different Wyckoff iron sites of 2*b*, 2*a*, 4*i*, 8*q*, and 4*j* symmetry, with the fifth, 4*j*, site randomly occupied by two iron(II) cations and two phosphorus anions. This leads to an iron site degeneracy of 1 : 1 : 2 : 4 : 1 and a total of nine iron cations per $Fe_9(PO_4)O_8$. The Fe(1), Fe(3), and Fe(4) sites contain iron(II) in somewhat different pseudooctahedral coordination environments. The Fe/P(5) site is tetrahedral and contains iron(II), whereas the pseudooctahedral Fe(2) site contains iron(III); thus charge balance is obtained with eight iron(II) cations and one iron(III) cation.

The structure of mixed-valence $Fe_9(PO_4)O_8$ consists of a stacking along the orthorhombic *c*-axis of layers of $(FeO)_6$, layers containing the Fe(1), Fe(2), and Fe(4) crystallographic sites with a composition and structure that are similar to stoichiometric wüstite. These $(FeO)_6$ layers alternate with Fe_3PO_6 layers that have a composition and structure similar to that of α-$FePO_4$ and contain the Fe(3) and Fe/P(5) crystallographic sites.

The structure of $Fe_9(PO_4)O_8$ reported [9] in the orthorhombic *Bmmm* space group is, on average, a good description of the structure; however, the observed very weak superstructure diffraction peaks strongly suggest a doubling of the unit cell along the *a* and *b* axes. These doublings have two consequences: the two Fe/P(5) site iron(II) and phosphorus(V) cations order on the 4*j* site along the *c*-axis and the Fe(3) and Fe(4) sites are split into two subsites.

The magnetic properties of $Fe_9(PO_4)O_8$ indicate that it is antiferromagnetic with a Néel temperature, T_N, of 217 K, and an exchange coupling that arises predominantly from the presence of the small wüstite-like $(FeO)_6$ blocks. Below T_N there is some indication in the magnetic scattering of a more complex pattern of unknown origin. Above room temperature, there is no indication of valence disproportionation, as is the case for wüstite. Based upon the changes in their unfit experimental Mössbauer spectra of $Fe_9(PO_4)O_8$, Venturini et al. [9] have proposed that there might be an onset of electron delocalization at or above 369 K, a delocalization that was thought to confirm the presence of electron holes that were localized within the $(FeO)_6$ blocks. Unfortunately, a subsequent study of the Mössbauer spectra discussed below has failed to confirm this proposal.

2.10.4 Mössbauer Spectral Properties of $Fe_9(PO_4)O_8$

The Mössbauer spectra of $Fe_9(PO_4)O_8$ are, as expected, very complex, see figures 3, 4, and 9 of reference [81]. An example of this complexity is shown by the experimental data for one of the many paramagnetic Mössbauer spectra of $Fe_9(PO_4)O_8$ measured

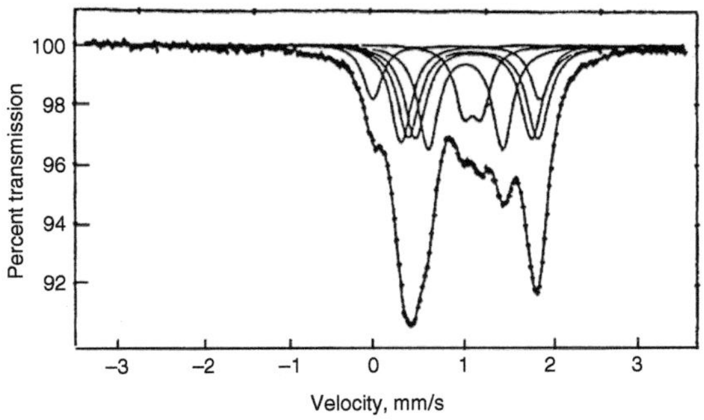

Figure 2.14 Mössbauer spectra of $Fe_9(PO_4)O_8$ measured at 292 K and fit with the model I parameters given in Table 2.11. The iron(III) component is the second left single line at ca. 0.35 mm/s. Source: Reference [81]/Springer Nature.

between 232 and 850 K, see Figure 2.14. The analysis of such a complex Mössbauer spectrum clearly poses distinct difficulties that require some special considerations.

The analysis of such complex Mössbauer spectrum, that corresponds to five crystallographically distinct iron cation sites, can be facilitated by certain steps. Perhaps the most important step is the measurement of the Mössbauer spectra over as large as possible a range of temperatures, and in the case of paramagnetic $Fe_9(PO_4)O_8$ this step included 16 spectra measured between 232 and 850 K. Such spectra should be obtained under as identical geometric conditions as possible – this sometimes proves difficult because of the need of a cryostat for low-temperature measurements and an oven for high-temperature measurements – a difficulty that may be overcome if a few spectra can be obtained in both the cryostat and oven; this has been the case for the study of the paramagnetic Mössbauer spectra of $Fe_9(PO_4)O_8$.

Once a series of spectra have been obtained over a wide range of temperatures, it is of value to obtain as much model independent information as possible from the initial unfit spectral data. This search may first be carried out by determining the temperature dependence of the median shift of the spectra. The median shift is the velocity, above and below which one-half of the spectral counts, or of the spectral absorption area, is observed. The median shift velocity is thus the average of the isomer shifts of all the crystallographic sites in the compound. The temperature dependence of the median shift may or may not reveal some differences with temperature as is the case for $Fe_9(PO_4)O_8$, see figure 2 in reference [81].

Further, the temperature dependence of the integrated counts, or the area of the spectral absorption, obtained over a fixed velocity range, will yield model free information about the lattice properties of the compound understudy. In the case of $Fe_9(PO_4)O_8$, the logarithm of the integrated area is linear between 232 and 850 K, see figure 2 in reference [81] and yields a Debye temperature of 326(26) K, a value that seems reasonable for the non-compact structure of $Fe_9(PO_4)O_8$.

Of course of equal importance in obtaining a model to fit complex Mössbauer spectra is to use all the crystallographic properties of the compound understudy. The advantage for $Fe_9(PO_4)O_8$ is that it contains iron cation sites with different oxidation states, different crystallographic site degeneracies, different coordination numbers and environments, different site average Fe—O bond distances with different distance ranges and site distortions, and different Wigner–Seitz site cell volumes, see table 1 in reference [81] for full details. By using these differences as well as their correlation with the iron-57 Mössbauer effect isomer shift presented by Ménil [87], one can construct a preliminary set of initial estimates for the hyperfine parameters at a given temperature, usually 295 K. Any preliminary fit is greatly facilitated by fixing the relative intensities of the crystallographic sites to their Wyckoff degeneracy; for $Fe_9(PO_4)O_8$ the initial relative areas were set at 11.1% for Fe(1), Fe(2), and Fe/P(5); 22.2% for Fe(3); and 44.4% for Fe(4).

This initial approach was undertaken for the 292 K Mössbauer spectrum of $Fe_9(PO_4)O_8$. Once an acceptable spectral fit was obtained, a process that took many attempts over time, extensive thought, and a careful consideration of the fit residuals, a good first approximation of a fitting model to the experimental data was obtained. The next step is to test the model by fitting the next higher and/or the next lower temperature spectrum available. Eventually, it became apparent through the fits of all the spectra obtained between 232 and 850 K, that two models, model I and model II, led to very similar quality spectral fits with hyperfine parameters that exhibited reasonable temperature dependencies, but only if the Fe(4) site spectral component was divided into two doublets with similar isomer shifts but different quadrupole splittings, see Table 2.11; the interested reader should consult the plots of the temperature dependence of these two model parameters in reference [81].

In both models I and II, the quadrupole doublet contributions of the Fe(3) and Fe(4) sites are split into two equal area contributions, in agreement with the "pseudomonoclinic" structure of $Fe_9(PO_4)O_8$. In contrast, there is no clear trace in the Mössbauer spectrum of the ordering of the Fe(5) on their 4*j* site.

It is interesting that the temperature dependence of the model independent median shift obtained from the Mössbauer spectra of $Fe_9(PO_4)O_8$ exhibits a change in slope at ca. 625 K. Below this temperature the slope is -5.35×10^{-4} (mm/s)/K, whereas above it is -11.65×10^{-4} (mm/s)/K; the decrease in the slope at higher temperatures is consistent with either an intervalent electron transfer between an iron(II) and iron(III) site or transfer of electron density into the conduction band. Virtually the same change in slope of the median shift is also observed in the temperature dependence of the weighted average isomer shifts derived from model I. Indeed, the median shift is essentially the same as the weighted average of the isomer shifts shown in the left portion of Figure 2.15 as both show a similar temperature dependence of the isomer shift as that of the Fe(4) site, a site that is found in the $(FeO)_6$ wüstite-like layers of $Fe_9(PO_4)O_8$. The Fe(4) site has been divided into two doublets of equal area, presumably because of the presence of the pseudomonoclinic supercell that partially removes the degeneracy of this site [9].

The change in both the median shift and weighted average isomer shift with increasing temperature is closely reflected in the isomer shift of one-half of the Fe(4) site spectral doublet, the portion with the higher isomer shift, whereas the isomer shift of the other-half of the Fe(4) site spectral doublet exhibits much the same

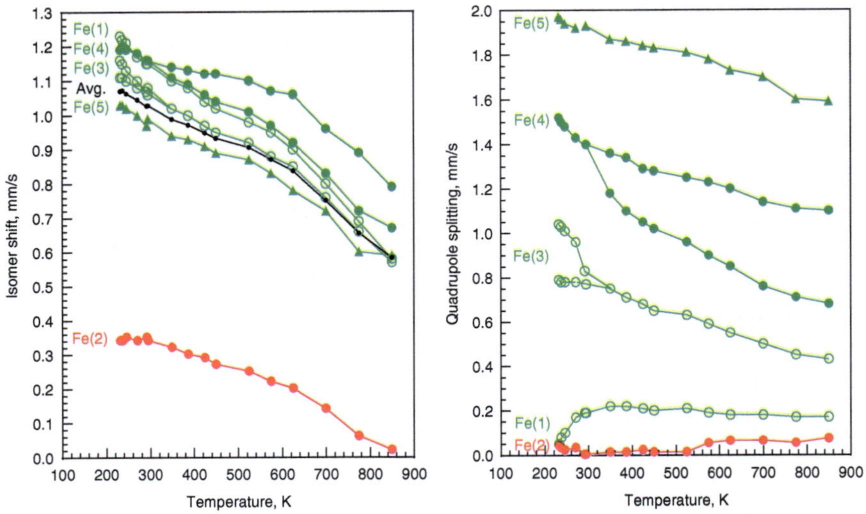

Figure 2.15 The temperature dependence of the $Fe_9(PO_4)O_8$ Mössbauer spectral isomer shifts (left) and quadrupole splittings (right) obtained with model I and reported in reference [81]. The weighted average isomer shifts are given in black, the iron(III) values are shown in red, and the iron(II) values are in green; the open symbols correspond to the Fe(1) and Fe(3) sites, the closed symbols correspond to the Fe(2) and Fe(4) sites, and the triangles correspond to the Fe(5) pseudotetrahedral site. Source: Adapted from figures 5 and 6 of reference [81].

temperature dependence as the remaining iron(II) doublets, see the left portion of Figure 2.15. This may be an indication that this half of the divalent iron(II) on the Fe(4) site participates in electron transfer to either the conduction band of $Fe_9(PO_4)O_8$ or perhaps to the iron(III) Fe(2) site that is also found in the $(FeO)_6$ wüstite-like layers.

Basically, on the basis of the quality of the model I and II fits it was difficult to distinguish which model was more valid. At this point all the relative areas of the different quadrupole doublets had been constrained to the Wyckoff degeneracy fixed values given above. As a final test of the validity of the two models, fits were carried out with the hyperfine parameters given in Table 2.11 but with only the relative areas allowed to vary. This exercise led to very small area changes and a small preference for model I.

As concluded by the authors of reference [81], it seems that it is not possible to determine unambiguously the nature of the electron transfer at higher temperatures from the paramagnetic Mössbauer spectra of $Fe_9(PO_4)O_8$. Of course this does not rule out the possible, but unlikely, electron transfer below the Néel temperature. Unfortunately, the Mössbauer spectra of $Fe_9(PO_4)O_8$ obtained just below $T_N = 217$ K consist of a broad distribution of overlapping and partially of highly broadened sextets. Further, at 4.2 K the magnetic sextets are less broad but there are ca. 40–50 partially resolved spectral absorption lines and, to date, it has not proven possible to obtain a solution to the very complex problem of sorting out the hyperfine parameters of these sextets. There are some recent papers that deal with new iron phosphate compounds, namely one that uses a pyrophosphate to form a new [88] iron phosphate compound, $Fe_7P_{11}O_{38}$, and a new [89] mixed metal mixed valence compound, $Mg_{2.88}$ $Fe(II)_{0.12}Fe(III)_{4.00}(PO_4)_6$. Finally, two additional phosphites

$Li_{1.4}$ [$Fe(II)_{4.4}Fe(III)_{0.6}(HPO_3)_6$]·1.47 H_2O and $Li_{1.43}$[$Fe(II)_{4.43}Fe(III)_{0.57}(HPO_3)_6$]·1.5 H_2O, have been studied in detail and their Mössbauer spectral fit revised [90].

Acknowledgments

The authors thank Prof. M. D. Dyar of Mount Holyoke College for providing the experimental spectral data used in the revised fits of the hagendorfite Mössbauer spectra shown in Figure 2.2 and other spectra. The authors would like to thank Drs. Raphaël Hermann, Dimitri Hautot, and Margaret L. Buhl and Profs. Peter Battle, Charles Gleitzer, Anthony K. Cheetham, and Charles E. Johnson for many enlightening discussions over the course of their work. Also, we thank Drs. Dimitri Hautot and Margaret L. Buhl for their excellent help in obtaining some of the data reported herein.

References and Notes

1 Robin, M.B. and Day, P. (1967). Mixed valence chemistry. A survey and classification. *Adv. Inorg. Chem. Radiochem.* 10: 247–422.

2 Battle, P.D., Cheetham, A.K., Gleitzer, C., Harrison, W.T.A., Long, G.J., and Longworth, G.A. (1982). A novel magnetic phase transition in anhydrous iron(III) phosphate, $FePO_4$. *J. Phys. C Solid State Phys.* 15: L919–L924.

3 Grandjean, F. and Long, G.J. (2019). Mössbauer spectral study of the low temperature magnetic properties of $FePO_4$ and the mixed valence iron(II/III) phosphate, $SrFe_3(PO_4)_3$. *Inorg. Chem.* 58: 13314–13322. https://doi.org/10.1021/acs.inorgchem.9b02208. See also, Song, Z., Liu, X., Ochani, A., Shen, S., Li, Q., Sun, Y., and Ruggiero, M.T. (2021). Low-frequency vibrational spectroscopy: a new tool for revealing crystalline magnetic structures in iron phosphate crystals. *Phys. Chem. Chem. Phys.* 23: 22241–22245.

4 La Iglesia, A. (2009). Estimating the thermodynamic properties of phosphate minerals at high and low temperature from the sum of constituent units. *Estud. Geol.* 65: 109–119.

5 Nriagu, J.O. (1984) Chapter 1. Phosphate minerals: their properties and general modes of occurrence. In: *Phosphate Minerals* (ed. J.O. Nriagu and P.B. Moore), 1–136. Berlin: Springer-Verlag.

6 Dyar, M.D., Jawin, E.R., Breves, E., Marchand, G., Nelms, M., Lane, M.D., Mertzman, S.A., Bish, D.L., and Bishop, J.L. (2014). Mössbauer parameters of iron in phosphate minerals: implications for interpretation of Martian data. *Am. Mineral.* 99: 914–942.

7 Vieillard, P. and Tardy, Y. (1984) Chapter 4, Table 6). Thermochemical properties of phosphates. In: *Phosphate Minerals* (ed. J.O. Nriagu and P.B. Moore), 171–198. Berlin: Springer-Verlag.

8 Iyer, R.G., Delacourt, C., Masquelier, C., Tarascon, J.M., and Navrotsky, A. (2006). Energetics of $LiFePO_4$ and polymorphs of its delithiated form, $FePO_4$. *Electrochem. Solid-State Lett.* 9: A46–A48.

9 Venturini, G., Courtois, A., Steinmetz, J., Gérardin, R., and Gleitzer, C. (1984). Préparation et étude d'un oxyphosphate de fer à valence mixte $Fe_9(PO_4)O_8$.

J. Solid St. Chem. 53: 1–12. See also Blois, M. (2023). Lithium iron phosphate comes to America. *Chem. Eng. News.* 101: 23–27. Baker, P.J., Franke, I., Pratt, F.L., Lancaster, T., Prabhakaran, D., Hayes, W., and Blundell, S.J. (2011). Probing magnetic order in $LiMPO_4$ (M=Ni, Co, Fe) and lithium diffusion in Li_xFePO_4. *Phys. Rev. B* 84: 174403 (8p.).

10 (a) In attempting to reproduce the calculations presented for ammonium taranakite, $(NH_4)_3Al_5(PO_4)_3H_6{\cdot}18H_2O$, presented on page 110 of reference [4], we obtain $\Delta H_f^{\circ} = -18\,482(38)$ kJ/mol that agrees within the uncertainty with the −18 485(38) kJ/mol reported on page 110, values that are both close to the experimental value of −18 532.60 kJ/mol reported by Vieillard and Tardy [10b]; similar, small differences occur for the calculation of $\Delta G_f^{\circ} = -16\,108(42)$ kJ/mol, which agrees within the uncertainty with the −16 106(42) kJ/mol reported on page 110. The reason for the small differences in these values is not clear but may result from a differing number of significant figures in table 2 of reference [4] and those actually used by the author of reference [4]. Here, the uncertainties are those calculated by equation (3) of reference [4] and use an error ± value of 1.75 that, for some reason, is missing from table 2 of reference [4]. (b) Vieillard, P. and Tardy, Y. (1984). Thermochemical properties of phosphates. In: *Phosphate Minerals* (ed. J.O. Nriagu and P.B. Moore). Berlin: Springer-Verlag, 442 pp.

11 Wagman, D.D., Evans, W.H., Parker, V.B., Schumm, R.H., Halow, I., Bailey, S.M., Churney, I.C., and Nuttall, R.L. (1982). The NBS tables of chemical thermodynamic properties: selected values for inorganic and C_1 and C_2 organic substances in SI units. *J. Phys. Chem. Ref. Data* 11 (Suppl. 2), 392 pp.

12 http://chemistry-reference.com/tables/Standard%20Thermodynamic%20Values.pdf derived from Dean, J. A. *Lange's Handbook of Chemistry*, 11th ed.; McGraw-Hill: New York, 1979, pp 9:4–9:128. Lide, D. R. *CRC Handbook*, 84th ed.; CRC Press: Boca Raton, FL, **2003**, pp 5:5–5:60, 5:85–5:86.

13 Adcock, C.T., Hausrth, E.M., Forster, P.M., Tschauner, O., and Sefein, K. (2014). Synthesis and characterization of the Mars-relevant phosphate minerals Fe- and Mg-whitlockite and merrillite and a possible mechanism that maintains charge balance during whitlockite to merrillite transformation. *Am. Mineral.* 99: 1221–1232. Dyar, M.D. and Skulte, E.C. (2020). Mössbauer spectroscopy: theory and laboratory spectra of geologic materials. In: *Remote Compositional Analysis* (ed. J.C. Bishop, J.F. Bell, and J.E. Moersch), 147–167. Cambridge: Cambridge University Press. Morris, R.V., Schröder, C., Klingelhöfer, G., and Agresti, D.G. (2020). Mössbauer spectroscopy at Gusev Crater and Meridiani Planum: iron mineralogy, oxidation state, and alteration on Mars. In: *Remote Compositional Analysis* (ed. J.C. Bishop, J.F. Bell, and J.E. Moersch), 538–554. Cambridge: Cambridge University Press.

14 Moore, P.B. (1971). Crystal chemistry of the alluaudite structure type contribution to the paragenesis of pegmatite phosphate giant crystals. *Am. Mineral.* 56: 1955–1974.

15 Corbin, D.R., Whitney, J.F., Fultz, W.C., Stucky, G.D., Eddy, M.M., and Cheetham, A.K. (1986). Synthesis of open-framework transition metal phosphates using organometallic precursors in acidic media preparation and structural characterization of $Fe_5P_4O_{20}H_{10}$ and $NaFe_3P_3O_{12}$. *Inorg. Chem.* 25: 2279–2280.

16 Yakubovich, O.V., Matvienko, E.N., Simonov, M.A., and Mel'nikov, O.K. (1986). Crystal structure of synthetic Fe^{3+}-arrojadite with ideal formula of $K_2Na_5Fe^{2+}{}_{14}Fe^{3+}(PO_4)_{12}(OH)_2$. *Moscow Univ. Geol. Bull.* 35–47.

17 Blount, A.M. (1974). The crystal structure of crandallite. *Am. Mineral.* 59: 41–47.

18 Redhammer, G.J., Tippelt, G., Roth, G., Lottermoser, W., and Amthauer, G. (2000). Structure and Mössbauer spectroscopy of barbosalite $Fe^{2+}Fe_2{}^{3+}(PO_4)_2(OH)_2$ between 80 K and 300 K. *Phys. Chem. Minerals* 27: 419–429.

19 Giuseppetti, G. and Tadini, C. (1984). The crystal structure of childrenite from Tavistock (SW England) $Ch_{89}Eo_{11}$ term of children-eosphorite series. *Neues Jahrb. für Mineral. Monatshefte* 6: 263–271.

20 Vochten, R. and De Grave, E. (1990). Mössbauer and infrared spectroscopic characterization of ferristrunzite from Blaton, Belgium. *Neues Jahrb. für Mineral. Monatshefte* 176–190.

21 Segeler, C.G., Moore, P.B., Dyar, M.D., Leans, F., and Ferraiolo, J.A. (2012). Ferrolaueite, a new mineral from Monmouth County, New Jersey, U.S.A. *Aust. J. Mineral.* 16: 69–76.

22 Redhammer, G.J., Roth, G., Tippelt, G., Bernroider, M., Lottermoser, W., Amthauer, G., and Hochleitner, R. (2006). Manganoan rockbridgeite $Fe_{4.32}Mn_{0.62}Zn_{0.06}(PO_4)_3(OH)_5$: structure analysis and ^{57}Fe Mössbauer spectroscopy. *Acta Cryst. C* 62: i24–i28.

23 Le Bail, A., Stephens, P.W., and Hubert, F. (2003). A crystal structure for the souzalite/gormanite series from synchrotron powder diffraction data. *Eur. J. Mineral.* 15: 719–723.

24 Nord, A.G. and Ericsson, T. (1982). Cation distributions in $(Fe_{1-z}Me_z)_3(PO_4)_2$ graftonite-type solid solutions determined by Mössbauer spectroscopy. *Z. Krist.* 161: 209–224.

25 Redhammer, G.J., Tippelt, G., Bernroider, M., Lottermoser, W., Amthauer, G., and Roth, G. (2005). Hagendorfite $(Na,Ca)MnFe_2(PO_4)_3$ from type locality hagendorf (Bavaria, Germany): crystal structure determination and ^{57}Fe Mössbauer spectroscopy. *Eur. J. Mineral.* 17: 915–932.

26 Moore, P.B. and Araki, T. (1973). Hureaulite, $Mn_5{}^{2+}(H_2O)_4[PO_3(OH)]_2[PO_4]_2$: its atomic arrangement. *Am. Mineral.* 58: 302–207.

27 Lindberg, M.L. and Christ, C.L. (1959). Crystal structures of the isostructural minerals lazulite, scorzalite, and barbosalite. *Acta Cryst.* 12: 695–697.

28 Vencato, I., Mattievich, E., and Mascarebhas, Y.P. (1989). Crystal structure of synthetic lipscombite: a redetermination. *Am. Mineral.* 74: 456–460.

29 Abrahams, S.C. and Bernstein, J.L. (1966). Crystal structure of paramagnetic ludlamite, $Fe_3(PO_4)_2{\cdot}4H_2O$, at 298K. *J. Chem. Phys.* 44: 2223–2229.

30 Léone, P., Palvadeau, P., and Moëlo, Y. (2000). Structure cristalline d'un nouvel hydroxyphosphate naturel de strontium, fer et aluminium (lulzacite), $Sr_2Fe(Fe_{0.63}Mg_{0.37})_2Al_4(PO_4)_4(OH)_{10}$. *Chimie de l'Etat Solide et Cristallochimie* 3: 301–308.

31 Le Page, Y. and Donnay, G. (1977). The crystal structure of the new mineral maricite $NaFePO_4$. *Can. Mineral.* 15: 518–521.

32 Fleck, M. and Kolitsch, U. (2003). Natural and synthetic compounds with kröhnkite-type chains. An update. *Z. Krist.* 218: 553–567.

33 Chukanov, N.V., Scholz, R., Aksenov, S.M., Rastsvetaeva, R.K., Pekov, I.V., Belakovskiy, D.I., Krambrock, K., Paniago, R.M., Righi, A., Martins, R.F., Belotti, F.M., and Bermanec, V. (2012). Metavivianite, $Fe^{2+}Fe_2^{3+}(PO_4)_2(OH)_2{\cdot}6H_2O$: new data and formula revision. *Mineral. Mag.* 76: 725–741.

34 Losey, A., Rakovan, J.F., Hughes, J.M., Francis, C.A., and Dyar, M.D. (2004). Structural variation in the lithiophilite-triphylite series and other olivine-group structures. *Can. Mineral.* 42: 1105–1115.

35 Tadini, C. (1981). Magniotriplite: its crystal structure and relation to the triplite-triploidite group. *Bull. Mineral.* 104: 677–680.

36 Grey, I., Mumme, W.G., Neville, S.M., Wilson, N.C., and Birch, W.D. (2010). Jahnsite-whiteite solid
solutions and associated minerals in the phosphate pegmatite at Hagendorf-Süd, Bavaria, Germany. *Mineral. Mag.* 74: 969–978.

37 Dyar, M.D., Agresti, D.G., Schaefer, M.W., Grant, C.A., and Sklute, E.C. (2006). Mössbauer spectroscopy of Earth and planetary materials. *Ann. Rev. Earth Planet. Sci.* 34: 83–125. Unfortunately, it was not possible to distinguish which points in figure 9A in reference [6] came from "this study" or from other literature. Further, the reader should note that there are significant differences between figure 9A in reference [6], figure 6 in this reference, and the left portion in Figure 2.1 in this chapter.

38 From table 1 in reference [6], the entries with χ^2 between 8.05 and 76.795 have been deleted.

39 Morris, R.V., Klingelhöfer, G., Schröder, C., Rodionov, D.S., Yen, A., Ming, D.W., de Souza, P.A. Jr., Wdowiak, T., Fleischer, I., Gellert, R. et al. (2006). Mössbauer mineralogy of rock, soil, and dust at Meridiani Planum, Mars: opportunity's journey across sulfate-rich outcrop, basaltic sand and dust, and hematite lag deposits. *J. Geophys. Res.* 111: E12S15.

40 Morris, R.V., Klingelhöfer, G., Schröder, C., Rodionov, D.S., Yen, A., Ming, D.W., de Souza, P.A. Jr., Wdowiak, T., Fleischer, I., Gellert, R. et al. (2006). Mössbauer mineralogy of rock, soil, and dust at Gusev Crater, Mars: spirit's journey through weakly altered olivine basalt on the plains and pervasively altered basalt in the Columbia Hills. *J. Geophys. Res.* 111, E02S13.

41 Klingelhöfer, G., Morris, R.V., Bernhardt, B., Schröder, C., Rodionov, D.S., de Souza, P.A. Jr., Yen, A., Gellert, R., Evlanov, E.N., Zubkov, B., Foh, J., Bonnes, U., Kankeleit, E., Gütlich, P., Ming, D.W., Renz, F., Wdowiak, T., Squyres, S.W., and Arvidson, R.E. (2004). Jarosite and hematite at Meridiani Planum from Opportunity's Mössbauer spectrometer. *Science* 306: 1740–1745.

42 Lane, M.D., Bishop, J.L., Dyar, M.D., King, P.L., Parente, M., and Hyde, B.C. (2008). Mineralogy of the Paso Robles soils on Mars. *Am. Mineral.* 93: 728–739.

43 The reader should note that the spectra shown in figure 7 of reference [42] have been manipulated and are not the backscattered Mössbauer spectra actually measured on Mars. This manipulation may also account for the absorption (%) labels in figure 7, labels that are incorrect for the backscattered emission

spectra but are probably correct for the manipulated spectra that are shown as transmission spectra.

44 The two large red points shown in figure 8 in reference [42] correspond to a two doublet fit and not to the *one* iron(III) doublet fit reported by Morris et al. in reference [40].

45 Greenwood, J.P. and Blake, R.E. (2006). Evidence for an acidic ocean on Mars from phosphorus geochemistry of Martian soils and rocks. *Geology* 34: 953–956.

46 Brückner, J., Dreibus, G., Gellert, R., Squyres, S.W., Wänke, S.W., Yen, A., and Zipfel, J. (2008). Mars exploration rovers: chemical composition by the APXS. In: *The Martian Surface-Composition, Mineralogy, and Physical Properties* (ed. J. Bell), 58–100. U.K.: Cambridge University Press.

47 McSween, H.Y., Eisenhour, D.D., Taylor, L.A., Wadhwa, M., and Crozaz, G. (1998). QUE94201 Shergottite: crystallization of a Martian basaltic magma. *Geochim. Cosmochim. Acta* 60: 4563–4569.

48 McSween, H. and Treiman, A.H. (1998). Martian mereorites. In: *Planetary Materials*, Reviews in Mineralogy, vol. 36 (ed. J. Papike), F1–F53. Chantilly, Virginia: Mineralogical Society of America.

49 Abu-Shandi, K., Winkler, H., Gerdan, M., Emmerling, F., Wu, B., and Janiak, C. (2003). Mixed-valence phosphato–hydrogenphosphato–iron network compounds $^{1}_{\infty}\{[C_4N_2H_{11.6}]_{1.5}\ [Fe^{II}Fe^{III}(PO_4)(H_{0.8}PO_4)_2]\cdot H_2O\}$ and $^{3}_{\infty}[Fe^{II}{}_5Fe^{III}{}_2(PO_4)_2(H_{0.5}PO_4)_4]$: structure elucidation with the help of Mössbauer spectroscopy and a caveat on X-ray diffraction. *Dalton Trans.* 2815–2823. The reader should note that the claim for electron tunneling at low temperature is not strongly substantiated in this reference.

50 Greenwood, N. and Gibb, T.M. (1971). *Mössbauer Spectroscopy*. London: Chapman & Hall.

51 Gütlich, P., Bill, E., and Trautwein, A.X. (2011). *Mössbauer Spectroscopy and Transition Metal Chemistry, Fundamentals and Applications*. Heidelberg: Springer.

52 Ijjaali, M., Malaman, B., Venturini, G., Gletizer, C., Long, G.J., and Grandjean F. (1991). A neutron diffraction and Mössbauer effect study of the ferromagnetic sheets in antiferromagnetic α-$Fe_2(PO_4)O$. *J. Phys. Condens. Matter* 3: 9597–9612.

53 Le Caër, G., Dubois, J.M., Häggström, L., and Ericcson, T. (1978). The determination of $\theta = (V_{zz},H)$ from $\frac{1}{2} \rightarrow 3/2$ Mössbauer spectra. *Nucl. Inst. Methods* 157: 127–129.

54 Dattagupta, S. and Blume, M. (1974). Stochastic theory of line shape. I. Nonsecular effects in the strong-collision model. *Phys. Rev. B* 10: 4540–4550.

55 This value has been obtained from the standard heat of formation of the mineral strengite, $FePO_4\cdot 2H_2O$, and water vapor, and the uncertainty in the quoted value is ca. ± 4 kJ/mol.

56 (a) Haines, J., Cambon, O., and Hull, S. (2003). A neutron diffraction study of quartz-type $FePO_4$: high-temperature behavior and α–β phase transition. *Z. Krist.* 218: 193–200. On p. 196 of this paper, equation (3) is in error and should read: $\tan(45° - \delta_{1(PO4)}) = (x_{O1} - y_{O1})\ a \sin \gamma/(\frac{1}{2} - z_{O1})\ c$. (b) Bhalerao, G.M., Hermet, P., Haines, J., Cambon, O., Keen, D.A., Tucker, M.G., Buixaderas, E., and Simon, P. (2012). Dynamic disorder and the α-β phase transition in quartz-type $FePO_4$ at high temperature investigated by total neutron scattering, Raman spectroscopy, and density functional theory. *Phys. Rev. B* 86: 134104(12p.).

57 Song, Y., Zavalij, P.Y., Suzuki, M., and Whittingham, M.S. (2002). New iron(III) phosphate phases: crystal structure and electrochemical and magnetic properties. *Inorg. Chem.* 41: 5778–5786. Delacourt, C., Poizot, P., Bonnin, D., Masquelier, C. (2009). Lithium-Insertion Mechanism in Crystalline and Amorphous $FePO_4 \cdot nH_2O$. *J. Electrochem. Soc.* 156: A595–A605.

58 (a) Evenstoff, W. and Martin, R. (1972). The crystal structure of heterosite. *Am. Mineral.* 57: 45–51. (b) Rousse, G., Rodriguez-Carvajal, J., Patoux, S., and Masquelier, C. (2003). Magnetic structures of the triphylite $LiFePO_4$ and of its delithiated form $FePO_4$. *Chem. Mater.* 15: 4082–4090.

59 El Khalifi, M. and Lippens, P.E. (2016). First-principles investigation of the ^{57}Fe Mössbauer parameters of $LiFePO_4$ and $FePO_4$. *J. Phys. Chem. C* 120: 28375–28389.

60 Kwon, W.J., Lee, B.W., and Kim, C.S. (2013). Investigation of spin ordering in antiferromagnetic Fe_{1-x} Mn_xPO_4 with Mössbauer spectroscopy. *J. Appl. Phys.* 113: 17E121 (3p.).

61 Aldon, L., Perea, A., Womes, M., Ionia-Bousquet, C.M., and Jumas, J.C. (2010). Determination of the Lamb-Mössbauer factors of $LiFePO_4$ and $FePO_4$ for electrochemical *in situ* and *operando* measurements in Li-ion batteries. *J. Solid State Chem.* 183: 218–222.

62 Arroyo y de Dompablo, M.E., Biskup, N., Gallardo-Amores, J.M., Moran, E., Ehrenberg, H., and Amador, U. (2010). Gaining insights into the energetics of $FePO_4$ polymorphs. *Chem. Mater.* 22: 994–1001.

63 Battle, P.D., Gibb, T.C., Hu, G., and Munro, D.C. (1986). The magnetic properties of the high pressure phase of ferric phosphate, $FePO_4$-II. *J. Solid State Chem.* 65: 343–350. The density value of 3.90 g/cm^3 given in this reference is in error and should read 3.84 g/cm^3.

64 Pasternak, M.P., Rozenberg, G.K., Milner, A.P., Amanowicz, M., Zhou, T., Schwarz, U., Syassen, K., Taylor, R.D., Hanfland, M., and Brister, K. (1997). Pressure-induced concurrent transformation to an amorphous and crystalline phase in berlinite-type $FePO_4$. *Phys. Rev. Lett.* 79: 4409–4412. In reference [15] in this paper, the hyperfine parameters quoted are wrong and are not found in this reference [15] that is reference [2] herein. In reference [16] in this paper, the cited hyperfine parameters are measured at 4.2 K and not 300 K as indicated. See also, Bull, C.L., Ridley, C.J., Funnell, N.P., Wilson, C.W., and MacLeod, S.G. (2021). The distortion of two $FePO_4$ polymorphs with high pressure. *Mater. Adv.* 2: 5096–5104.

65 Kinomura, N., Shimada, M., Koizumi, M., and Kume, S. (1976). Synthesis of a high pressure phase of $FePO_4$. *Mater. Res. Bull.* 11: 457–460.

66 Bruckner, W., Fuchs, W., and Ritter, G. (1967). Mössbauer effect in calcinated $FePO_4$ below 26 K. *Phys. Lett.* 26A: 32–33.

67 Beckmann, V., Bruckner, W., Fuchs, W., Ritter, G., and Wegener, W. (1968). The measurement of the anisotropy constant of antiferromagnetic $FePO_4$ by means of the Mössbauer effect. *Phys. Stat. Sol.* 29: 781–791.

68 Arnold, H. (1986). Crystal structure of $FePO_4$ at 294 and 20 K. *Z. Krist.* 177: 139–142. In table 2 in this reference, the O(2) 6(c) *y* positional parameter is in error and should read 0.264(2) as is found in the crystallographic information file.

69 Note that the quadrupole shift in reference [2] was defined as the negative of the quadrupole shift given in equation (1).

70 Goiffon, A., Jumas, J.C., and Philippot, E. (1986). Phases de type quartz α: structure de $FePO_4$ et spectrométrie Mössbauer du fer-57. *Rev. Chim. Mineral.* 23: 99–110.

71 Goiffon, A., Bayle, G., Astier, R., Jumas, J.C., Maurin, M., and Philippot, E. (1983). Cristallochimie des phases $GaPO_4$, $AlAsO_4$, et $GaAsO_4$, étude comparée des structures de type quartz α. *Rev. Chim. Mineral.* 20: 338–350.

72 Hermann, R.P., Tegus, O., Brück, E., Bushow, K.H.J., de Boer, F.R., Long, G.J., and Grandjean, F. (2004). Mössbauer spectral study of the magnetocaloric $FeMnP_{1-x}As_x$ compounds. *Phys. Rev. B* 70: 214425(9p.).

73 Bean, C.P. and Rodbell, D.S. (1962). Magnetic disorder as a first-order phase transformation. *Phys. Rev.* 126: 104–115.

74 Mittal, R., Chaplot, S.L., Kolesnikov, A.I., Loong, C.K., Jayakumar, O.D., and Kulshreshtha, S.K. (2002). Inelastic neutron scattering, lattice dynamics, and synchrotron X-ray diffraction study of $FePO_4$. *Phys. Rev. B* 66: 174304(7p.).

75 Ng, H.N. and Calvo, C. (1975). Refinement of the crystal structure of the low-quartz modification of ferric phosphate. *Can. J. Chem.* 53: 2064–2065.

76 Because the statistical error on the percent transmission of the 4.2 K Mössbauer spectrum in figure 3 of reference [63], also shown in Figure 2.9 herein, defined as $100/N^{1/2}$, where N is the number of counts in the spectral baseline, is unknown, the statistical error was assumed to be 0.14; with this assumption $\chi^2 = 2.749$ for the fit shown in Figure 2.9. Further, a $\chi^2 = 3.751$ is obtained for the fit (not shown) with $e^2Qq/2 = -1.679$ mm/s, $\theta = 0°$, and $\eta = 0.7$, a fit that is substantially poorer and also yields a larger line width of 0.33 mm/s.

77 Kwon, W.J., Shim, I.-B., and Kim, C.S. (2013). Mössbauer spectra of olivine $Fe_{1-y}Mn_yPO_4$. *J. Korean Phys. Soc.* 62: 1922–1924. See also Kyu, L., Moon, S.J., Shim, I.-B., and Chul Sung Kim, C.H. (2009). The structural transition and magnetic properties of lithium deintercalation in $LiFePO_4$. *IEEE Trans. Magn.* 45: 4268–4270; references [2] and [3] in this paper are in error, the two journal titles should both read *Physical Review B.*

78 Andersson, A., Kalska, B., Häggström, L., and Thomas, J.O. (2000). Lithium extraction/insertion in $LiFePO_4$: an X-ray diffraction and Mössbauer spectroscopy study. *Solid State Ionics* 130: 41–52.

79 A fit, only slightly poorer than that shown in Figure 2.11, could only be achieved with the fixed following parameters, $eQV_{zz}/2 = -1.502$ mm/s, $\theta = 38°$, $\eta = 0.6$,

and $\phi = 65°$; parameters that are similar to the calculated values. However, the negative sign of $eQV_{zz}/2$ does not agree with the calculations [59].

80 De Grave, E., Magela da Costa, G., Van Alboom, A., and Vandenberghe, R.E. (2013). Low temperature Mössbauer spectra of heterosite, (Fe,Mn)PO_4. *Spectro. Chim. Acta A* 100: 104–108.

81 Long, G.J. and Gleitzer, C. (1990). A Mössbauer effect study of the electronic and structural properties of $Fe_9(PO_4)O_8$. *Hyperfine Interact.* 62: 147–166.

82 Iraldi, G., Le Caër, G., and Gleitzer, C. (1981). Electron localization in Fe_2PO_5. *Solid State Commun.* 40: 145–147.

83 Modaressi, A., Courtois, A., Gérardin, R., Malaman, B., and Gleitzer, C. (1981). Fe_2PO_5, un phosphate de fer de valence mixte. Préparation et études structurale, Mössbauer et magnétique. *J. Solid State Chem.* 40: 301–311.

84 Modaressi, A., Courtois, A., Gérardin, R., Malaman, B., and Gleitzer, C. (1983). Fe_3PO_7, un cas de coordinence 5 du fer trivalent, étude structurale et magnétique. *J. Solid State Chem.* 47: 245–255.

85 Grandjean, F. and Long, G.J. (2023). Unpublished results.

86 Gavoille, G., Gleitzer, C., and Long, G.J. (1987). Etude magnétique de $Fe_3(PO_4)O_3$ avec notamment des effets de covalence et de frustration. *Rev. Chim. Mineral.* 24: 42–54.

87 Ménil, F. (1985). Systematic trends of the ^{57}Fe Mössbauer isomer shifts in (FeO_n) And (FeF_n) polyhedra. Evidence of a new correlation between the isomer shift and the inductive effect of the competing bond *T*-*X* (→Fe) (where *X* is O or F and *T* any element with a formal positive charge). *J. Phys. Chem. Sol.* 46: 763–789.

88 Goj, P., Wajda, A., Blachowski, A., and Stoch, P. (2021). A new iron-phosphate compound ($Fe_7P_{11}O_{38}$) obtained by pyrophosphate stoichiometric glass devitrification. *Sci. Rep.* 11: 22957 (12p.).

89 Saad, Y., Hidouri, M., Alvarez-Serrano, I., Veiga, M.L., Wattiaux, A., and Amara, M.B. (2014). Crystal structure and Mössbauer spectroscopy of a new iron phosphate $Mg_{2.88}Fe_{4.12}(PO_4)_6$. *J. All. Compds.* 584: 625–630.

90 Grandjean, F., Long, G.J., Evrard, R., Chung, U-C., Larrea, E.S., de Larramendi, I.R., Arriortua, M.I., and Rojo, T. (2020). Impact of Lithium and Potassium Cations on the Mössbauer Spectral and Electrical Properties of Two Mixed-Valence Iron(II/III) Phosphites. *Chem. Mater.* 32: 5534–5540.

3

Mössbauer Spectroscopic Investigation of Fe-Based Silicides

Xiao Chen[1], Junhu Wang[2], and Changhai Liang[1]

[1] *Dalian University of Technology, School of Chemical Engineering, No. 2 Linggong Road, Dalian 116024, China*
[2] *Chinese Academy of Sciences, Dalian Institute of Chemical Physics, No. 457 Zhongshan Road, Dalian 116023, China*

3.1 Introduction

Transition metal silicides represent a rich variety of refractory materials and have specific crystal and electronic structures different from those of their component metals due to the strong interaction between metals and silicon [1–3]. Transition metal silicides have received much attention because of their unique physical and chemical properties, such as high thermal stability, low electrical resistivity, low work function, and low density [4]. Such metal silicide materials could meet the requirements of modern technological applications, including nanoelectronics, field emitters, spintronics, thermoelectric, solar energy conversion, and heterogeneous catalysis [5–22]. Especially, iron silicide can be used as a potential candidate material in new silicon-compatible devices for photovoltaic [23], spintronic [24], and thermoelectric [25] applications and the Kondo insulator FeSi, the semiconductor β-$FeSi_2$, and the intermetallic compound α-$FeSi_2$ have been found among them.

Generally, the multiple stoichiometries and complex phase behavior exhibit on iron silicide compounds. As shown in Figure 3.1, there are six known iron silicide intermetallic compounds (Fe_3Si, Fe_2Si, Fe_5Si_3, FeSi, α-$FeSi_2$, and β-$FeSi_2$), and only three of which are stable at room temperature [26]. The crystal structure of iron silicides also presents great difference, including cubic (C1, $L1_2$, DO_3, B20), hexagonal, tetragonal, and complex construction. For example, the Fe_3Si crystallizes in the DO_3 structure, which is composed of two sublattices in unit cell. One of the sublattices contains eight Fe atoms, and another sublattice contains four Fe atoms and four Si atoms, which alternately occupy the cube vertices, as shown in Figure 3.2. However, the FeSi crystallizes in the B20 structure, which is formed by the simple cubic structure composed of Fe atoms and Si atoms, respectively. There are two stable phases for $FeSi_2$ – the tetragonal metallic α-$FeSi_2$ that is stable at high temperature and the low-temperature β-$FeSi_2$, which is orthorhombic phase.

Mössbauer Spectroscopy: Applications in Chemistry and Materials Science,
First Edition. Edited by Yann Garcia, Junhu Wang, and Tao Zhang.

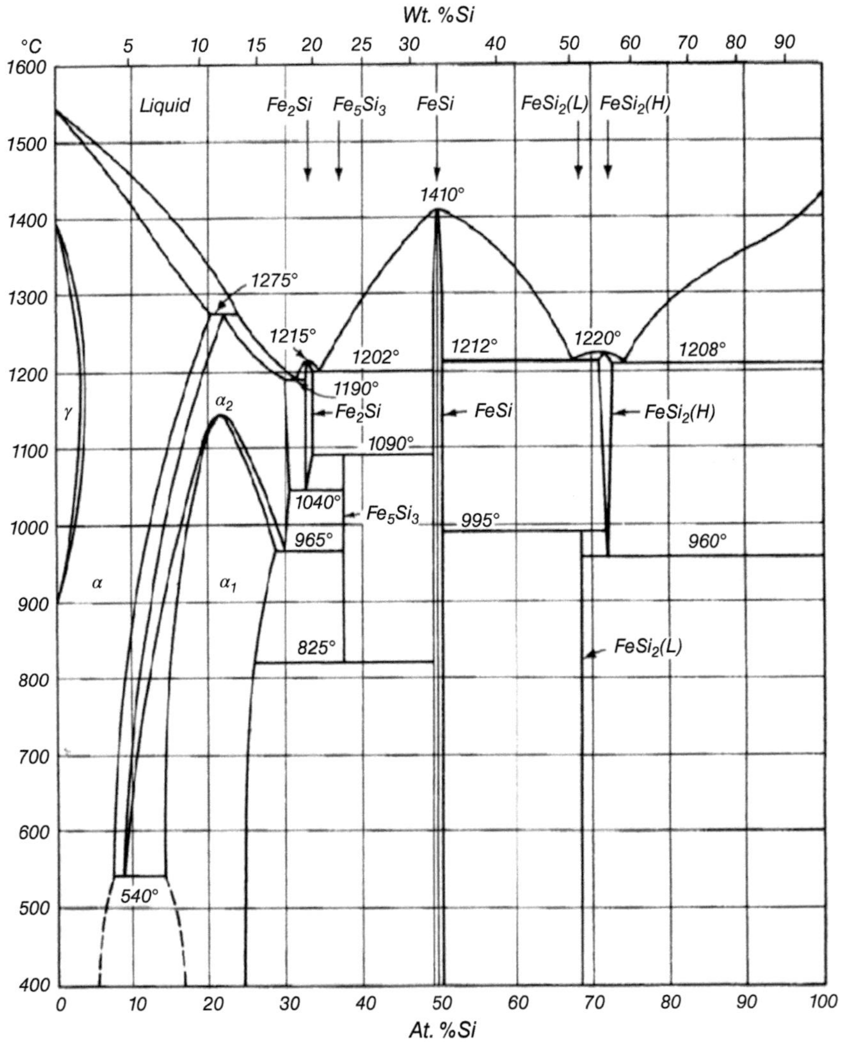

Figure 3.1 Fe–Si binary phase diagram. Source: Ref. [26]/MDPI/Public Domain.

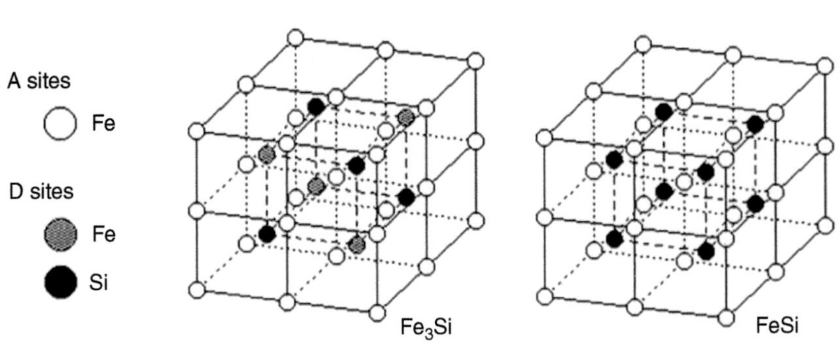

Figure 3.2 Unit cells of Fe_3Si with the DO_3 structure and FeSi with the B20 structure.

In such complex material systems, it can be difficult to accurately control and characterize Fe-based synthesized silicides. Over the past year, multiple characterization technologies have been developed to analyze the resulting phases, such as X-ray diffraction (XRD), high-resolution transmission electron microscopy (HRTEM), scanning tunneling microscopy (STM), and high-energy electron diffraction (HEED). In some cases, ^{57}Fe Mössbauer spectrometry has been also successfully applied to clarify the chemical state and relative amount of the iron species in Fe-based silicides [27].

Aiming to characterize many kinds of iron-bearing environmental and chemical state during the formation of iron silicides, this chapter summarizes the Mössbauer spectroscopic investigation of the structure of iron-based silicides produced by different techniques. In addition, the Mössbauer effect characterization of the substituted Fe-based silicides also will have been reviewed. ^{57}Fe Mössbauer spectroscopic results provide strong evidences of the relationship between the structure and properties.

3.2 Mössbauer Spectroscopic Investigation of Iron Silicides Prepared By Mechanical Alloying and Heat Treatment

Mechanical alloying as a viable solid-state processing route has been widely applied to the synthesis of a variety of equilibrium and non-equilibrium phases of iron silicides [28]. Since mechanical alloying can reach to the atomic level alloying, the resulting material has the advantages of uniform composition and simple processing. By controlling the milling time, the atomic ratio of the raw materials and the input of mechanical energy, multiple phases, and various structures of iron silicides could be synthesized. Mössbauer spectroscopy is a very sensitive method to observe the changes of the short range structure of iron containing materials and recognize the different phases formed.

Dézsi et al. investigated the phases formed after long time mechanical alloying of powders with the Fe/Si atomic ratio of 1 : 2 by Mössbauer spectroscopy [29]. As shown in Figure 3.3 and Table 3.1, the relative intensity of the resulting phases between Fe and Si is strongly dependent on the milling time. Mössbauer spectrum showed significant changes after two days of mechanical alloying. The large quadrupolar component can be described by three doublets attributed to ε-FeSi, α-$Fe_{1-x}Si_2$, and disordered $FeSi_2$. At longer milling period, the relative intensity of α-$FeSi_2$ with an asymmetric doublet increased. β-$FeSi_2$ phase with two doublets was formed after annealing the gridded powder at 850 °C for 10 days, which may be a promising material for thermoelectric devices.

Recently, the aluminothermic reaction has been introduced in the synthesis of iron silicide by mechanical alloying of Fe_3O_4, Al, and Si powder mixtures and by heat treatment [30]. During this processing, the atomic ratio of the raw materials was studied and the phases of the nanocomposite as-prepared were detected by XRD and

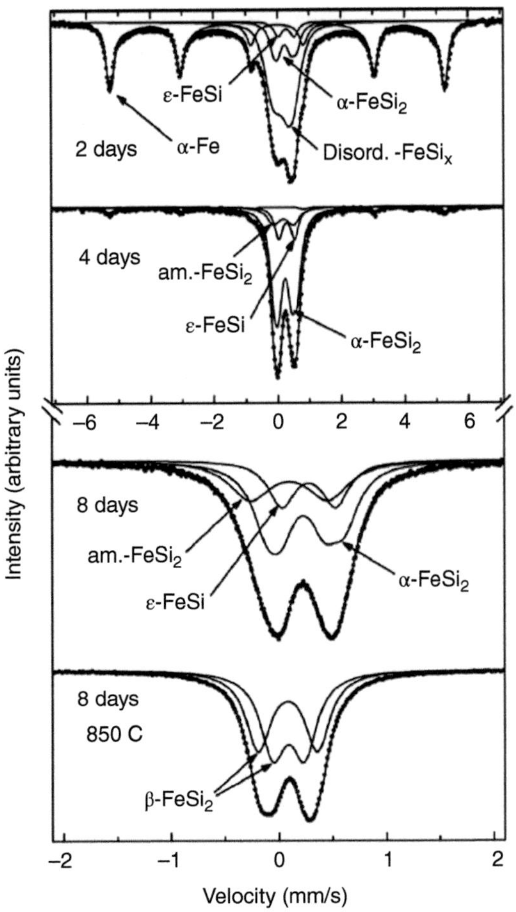

Figure 3.3 Mössbauer spectra of the alloyed and annealed sample with the Fe/Si atomic ratio of 1 : 2 measured at 300 K. Source: Reproduced from Ref. [29]/with permission from Elsevier.

Mössbauer spectra. The XRD patterns confirmed that crystalline transformations occur from the raw materials to the $Fe_3Si–Al_2O_3$ nanocomposite during mechanical alloying and annealing and from Fe to Fe_3Si to a mixture of Fe_3Si, Fe_5Si_3, and FeSi with the increase in the proportion of Si in the raw materials. The reaction is given in Eq. (3.1):

$$3Fe_3O_4 + 3Si + 8Al \rightarrow 3Fe_3Si + 4Al_2O_3 \tag{3.1}$$

^{57}Fe Mössbauer spectra of the as-prepared samples measured at room temperature are shown in Figure 3.4, and corresponding Mössbauer parameters are listed in Table 3.2. In this chapter, δ is referred to α-Fe foil at room temperature and the uncertainty is 5% of all the reported values. The Mössbauer spectrum of sample 1Si-4h-900 could be fitted into four sextets and one doublet. One sextet with $\delta = -0.11\ \text{mm s}^{-1}$ and $B_{hf} = 33.8\ \text{T}$ was indexed to Fe^0, representing 36.3%. The two sextets with $\delta = 0.12\ \text{mm s}^{-1}$ and $B_{hf} = 19.1\ \text{T}$ and $\delta = -0.08\ \text{mm s}^{-1}$, $\Delta E_Q = 0.02\ \text{mm s}^{-1}$, and $B_{hf} = 30.9\ \text{T}$ were, respectively, assigned to the Fe atoms at

Table 3.1 Mössbauer parameters of the milled sample with the Fe/Si atomic ratio of 1 : 2 measured at 300 K.

Milling days	Component	δ/ mm s^{-1}	ΔE_Q/ mm s^{-1}	B_{hf}/T	W/ mm s^{-1}	RI/%
2	α-Fe	0.01	0.00	32.0	0.37	47.4
	ε-FeSi	0.28	0.49		0.30	3.7
	Disord.-$FeSi_x$	0.15	0.54		0.26	37.4
	α-$Fe_{0.8}Si_2$	0.23	0.58		0.26	11.5
4	α-Fe	0.01	0.00	32.0	0.37	7.7
	ε-FeSi	0.28	0.49		0.30	14.3
	Am.-$FeSi_2$	0.18	0.56		0.26	12.4
	α-$Fe_{0.8}Si_2$	0.23	0.58		0.26	65.6
8	ε-FeSi	0.28	0.49		0.30	22.6
	Am.-$FeSi_2$	0.18	0.56		0.26	58.4
	α-$Fe_{0.8}Si_2$	0.23	0.58		0.26	19.0
8 Annealed 1120 K	β-$FeSi_2$ (a)	0.08	0.53		0.28	50.0
Implanted	β-$FeSi_2$ (b)	0.09	−0.32		0.28	50.0
	Am.-$FeSi_2$	0.18	0.56		0.26	100.0

Isomer shift (δ), quadrupole splitting (ΔE_Q), hyperfine field (B_{hf}), width (W) in mm s^{-1}, and relative intensities (RI) in percent.
Source: Reproduced from Ref. [29].

nonequivalent lattice sites A and B of Fe_3Si with the DO_3 cubic structure [31–33]. However, no diffraction peaks due to Fe_3Si were observed in the XRD pattern of sample 1Si-4h-900, which indicated that Fe_3Si might exist in an amorphous state. The doublet with $\delta = -0.14$ mm s^{-1} and $\Delta E_Q = 0$ mm s^{-1} can probably be identified as amorphous metallic iron, representing 4.3%. The Mössbauer spectrum of sample 1.5Si-4h-900 was composed of five sextets. With the exception of two sextets corresponding to Fe_3Si, the other two sextets with $\delta = 0.18$ mm s^{-1} and $B_{hf} = 17.6$ T and $\delta = 0.21$ mm s^{-1}, $\Delta E_Q = -0.19$ mm s^{-1}, and $B_{hf} = 11.9$ T were, respectively, indexed to the two types of Fe atoms in Fe_5Si_3 [31, 34], representing 11.5%. Similarly, the above Fe_5Si_3 could be amorphous, as no diffraction peaks due to Fe_5Si_3 were observed in the XRD pattern of the 1.5Si-4h-900 sample. The Mössbauer spectrum of sample 3Si-4h-900 could be fitted into three sextets and a doublet. The doublet with $\delta = 0.15$ mm s^{-1} and $\Delta E_Q = 0.48$ mm s^{-1} was assigned to FeSi [32, 35], representing 24.4%. In addition, Fe_3Si and Fe_5Si_3 also existed in the 3Si-4h-900 sample. Especially for samples 1Si-4h-900 and 1.5Si-4h-900, the sextet with $\delta = 0.06$ (0.08) mm s^{-1} and $B_{hf} = 25.0$ (25.2) T had not obtained a reasonable attribution and required further confirmation. In a word, ^{57}Fe Mössbauer spectra revealed the existence of amorphous phases Fe_3Si and Fe_5Si_3, which effectively complemented XRD result.

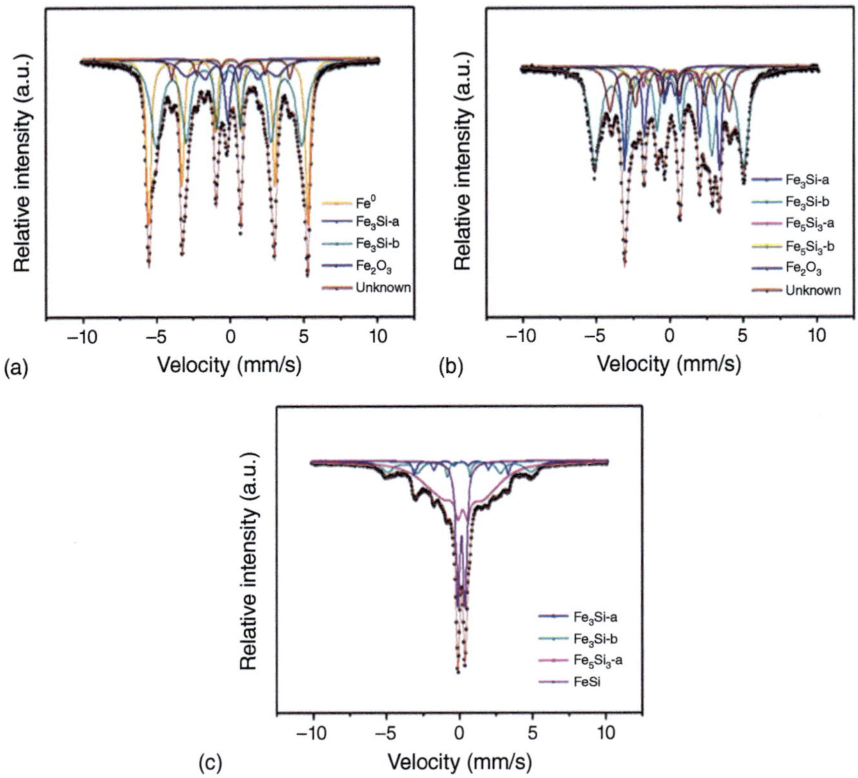

Figure 3.4 ^{57}Fe Mössbauer spectra of samples (a) 1Si-4h-900, (b) 1.5Si-4h-900, and (c) 3Si-4h-900. Source: Adapted from Ref. [30].

Table 3.2 ^{57}Fe Mössbauer parameters of samples 1Si-4h-900, 1.5Si-4h-900, and 3Si-4h-900.

Sample	Assignment	δ/mm s^{-1}	ΔE_Q/mm s^{-1}	B_{hf}/T	RI/%	LW/mm s^{-1}
1Si-4h-900	Fe^0	−0.11	—	33.8	36.3	0.28
	Fe_3Si-a	0.12	—	19.1	11.1	0.34
	Fe_3Si-b	−0.08	0.02	30.9	44.7	0.39
	Fe_2O_3	0.14	—	—	4.3	0.46
	Unknown	0.06	—	25.0	3.6	0.22
1.5Si-4h-900	Fe_3Si-a	0.15	—	20.2	24.7	0.34
	Fe_3Si-b	−0.05	0.04	31.4	48.0	0.44
	Fe_5Si_3-c	0.18	—	17.6	7.4	0.46
	Fe_5Si_3-d	0.21	−0.19	11.9	4.1	0.33
	Unknown	0.08	−0.01	25.2	15.8	0.31
3Si-4h-900	Fe_3Si-a	0.14	0.01	20.0	4.7	0.29
	Fe_3Si-b	−0.02	0.03	30.5	12.6	0.22
	Fe_5Si_3-d	0.19	−0.10	12.6	58.3	0.53
	FeSi	0.15	0.48	—	24.4	0.30

Source: Reproduced from Ref. [30].

3.3 Mössbauer Spectra of Iron Silicide on Silica Prepared by Pyrolysis of Ferrocene-Polydimethylsilane Composites

Metal containing polysilane as a single-source precursor is also a promising material for the synthesis of metal silicides and offers a number of advantages over the conventional method. In our previous research, iron silicide nanoparticles of c. 6–9 nm well dispersed on silica were prepared by pyrolysis of ferrocene-polydimethylsilane composites [31]. Typically, the $Fe(Si(CH_3)_2R)_2$ composites obtained with a mass ratio of ferrocene and PDMS 1 : 1 were, respectively, heated to 600 °C, 700 °C, and 800 °C at a rate of 10 °C min^{-1} in Ar, and then maintained at these temperatures for 120 min, which were designated as Ar-600, Ar-700, and Ar-800, respectively. TEM images confirmed that highly dispersed iron silicide (Fe_3Si and Fe_5Si_3) nanoparticles on silica had been prepared by pyrolysis of ferrocene-polydimethylsilane composites.

To study the effect of thermolysis temperature on the composition of iron silicides, ^{57}Fe Mössbauer spectra of the samples Ar-600, Ar-700, and Ar-800 were also measured at room temperature (as shown in Figure 3.5). The refined values of the corresponding Mössbauer parameters are listed in Table 3.3. The Mössbauer spectrum of the sample Ar-600 could be fitted into two sextets and an asymmetric doublet. The two sextets with $\delta = 0.21$ mm s^{-1} and $B_{hf} = 19.8$ T and $\delta = 0.06$ mm s^{-1} and $B_{hf} = 31.2$ T were attributed to the Fe atoms at sites A and B of the nonequivalent

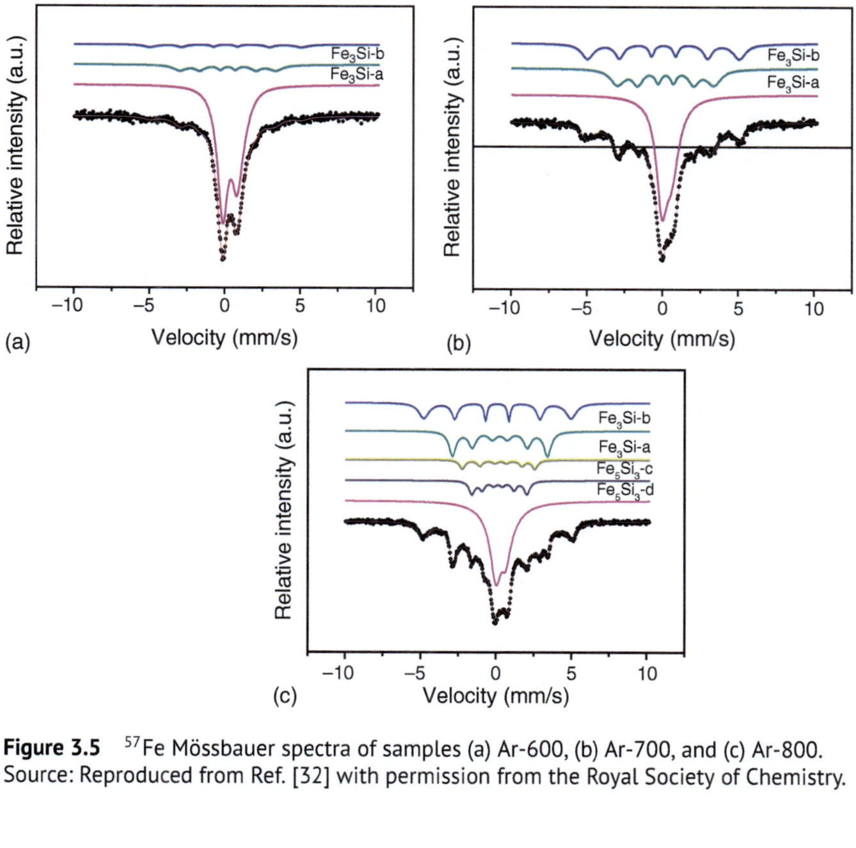

Figure 3.5 ^{57}Fe Mössbauer spectra of samples (a) Ar-600, (b) Ar-700, and (c) Ar-800. Source: Reproduced from Ref. [32] with permission from the Royal Society of Chemistry.

Table 3.3 ^{57}Fe Mössbauer parameters of samples Ar-600, Ar-700, and Ar-800.

Sample	Assignment	δ/mm s^{-1}	ΔE_Q/mm s^{-1}	B_{hf}/T	RI/%	LW/mm s^{-1}
Ar-600	Fe_3Si-a	0.21	—	19.8	11.7	0.50
	Fe_3Si-b	0.06	—	31.2	5.2	0.43
Magnetic relaxation		0.35	—	5.0	83.1	0.53
Ar-700	Fe_3Si-a	0.21	—	19.8	24.6	0.50
	Fe_3Si-b	0.06	—	31.2	21.5	0.43
Magnetic relaxation		0.27	—	6.3	53.9	0.74
Ar-800	Fe_3Si-a	0.25	—	19.6	21.5	0.65
	Fe_3Si-b	0.07	—	30.3	19.3	0.24
	Fe_5Si_3-c	0.24	−0.16	14.9	6.1	0.47
	Fe_5Si_3-d	0.17	0.09	11.3	9.2	0.47
Magnetic relaxation		0.29	—	15.0	43.8	0.38

Source: Based on Ref. [32].

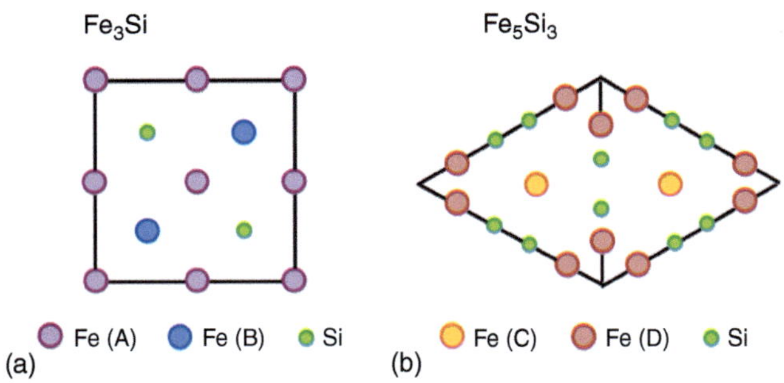

Figure 3.6 The structure of (a) Fe_3Si and (b) Fe_5Si_3 projected onto the basal plane. Source: Reproduced from Ref. [31] with permission from the Royal Society of Chemistry.

Fe_3Si lattice having the DO_3 cubic structure, respectively [32, 33]. As shown in Figure 3.6a, the Fe atoms at the B sites have eight nearest-neighbor Fe atoms, while the Fe atoms at the A sites have four nearest-neighbor Fe atoms and four Si atoms. The asymmetric doublet was speculated to be the magnetically relaxed subspectrum possibly arising from low-size Fe_3Si particles, representing 83.1% [36]. Similarly, for the Ar-700 sample, the Mössbauer spectrum was fitted into two sextets for Fe_3Si and an asymmetric doublet for its magnetic relaxation. And the proportions of small Fe_3Si particles in the Ar-700 sample decreased down to 53.9% due to the partial aggregation particles with the increasing temperature. The size histograms of iron silicide nanoparticles derived from TEM images further confirmed that average particle sizes of samples increased from 7.6 nm to 16.2 nm with increasing pyrolysis temperature due to the aggregation effect at higher temperatures (as shown in Figure 3.7). The Mössbauer spectrum of the sample Ar-800 was composed

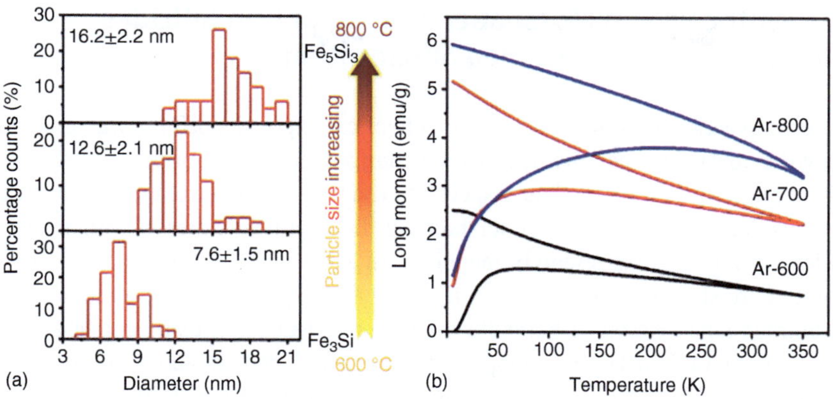

Figure 3.7 (a) The size histograms and (b) the FC and ZFC curves of samples Ar-600, Ar-700, and Ar-800. Source: Reproduced from Ref. [31] with permission from the Royal Society of Chemistry.

of four sextets and an asymmetric doublet. The two sextets with $\delta = 0.25\ \mathrm{mm\,s^{-1}}$ and $B_{hf} = 19.6$ T and $\delta = 0.07\ \mathrm{mm\,s^{-1}}$ and $B_{hf} = 30.3$ T were assigned to Fe_3Si, while the other two sextets with $\delta = 0.24\ \mathrm{mm\,s^{-1}}$, $\Delta E_Q = -0.16\ \mathrm{mm\,s^{-1}}$, and $B_{hf} = 14.9$ T and $\delta = 0.17\ \mathrm{mm\,s^{-1}}$, $\Delta E_Q = 0.09\ \mathrm{mm\,s^{-1}}$, and $B_{hf} = 11.3$ T were assigned to the two types of Fe atoms in Fe_5Si_3, respectively. Figure 3.6b shows its projection on the basal plane. The Mössbauer parameters of Fe_5Si_3 obtained here were somewhat different from those reported in the literature [34, 37], which were also not consistent and further confirmation is needed. And the proportions of small Fe_3Si (and Fe_5Si_3) particles in the Ar-800 sample decreased sequentially to 43.8%. Therefore, the ^{57}Fe Mössbauer spectra indicated the chemical state and relative amount of the iron species of all the three samples. They displayed superparamagnetic behavior at room temperature.

For understanding the magnetic properties of the samples, the ^{57}Fe Mössbauer spectra measured at −196 °C have been tested. They were not dominated by the central paramagnetic component anymore, which may attribute that the superparamagnetic blocking temperature of the majority of iron silicides nanoparticles are lying between −196 °C and room temperature so that the central component became magnetically split. Field cooling (FC) and zero field cooling (ZFC) curves were measured for the three samples to further reveal their magnetic properties. As shown in Figure 3.7, the intersections of FC and ZFC curves indicated that the Curie temperature (TC) of sample Ar-600 was c. 27 °C and Curie temperature of both samples Ar-700 and Ar-800 was close to 77 °C. It has been reported that the Curie temperature of Fe_3Si and Fe_5Si_3 was 530 °C and 112 °C, respectively [38, 39]. The decrease in Curie temperatures of samples was supposed to be a consequence of the reduced iron silicide nanoparticle size [40]. In summary, all the three samples showed superparamagnetic behavior at 27 °C and ferromagnetic behavior at −268 °C. The particle sizes had a great impact on their magnetic properties. The as-prepared iron silicide nanoparticles may apply in magnetically recording materials, magnetically separable catalysts, and electrode materials for spintronic devices.

3.4 Synthesis and Mössbauer Spectra of Iron Silicides by Temperature-Programmed Silicification

The decomposition of silicon on iron by chemical vapor deposition was used to modify the surface of metallic iron, which changed the nature of the coating and catalytic properties. The Mössbauer spectrum is a very effective means of detecting the deposition processing and understanding the chemical states of iron in iron silicides. In our previous research, magnetic Fe@SiO_2 and $FeSi_2$@SiO_2 nanoparticles with core-shell structure have been successfully synthesized by temperature-programmed silicification of hematite α-Fe_2O_3 nanoparticles, using silane as the silicon source [41]. α-Fe_2O_3 nanoparticles were reduced at 400 °C in flowing hydrogen (30 mL min^{-1}) for 3 h in a quartz microreactor and then silicified at different temperatures (350 °C, 450 °C, and 550 °C) for 15 min in a SiH_4/H_2 mixture of 110 mL min^{-1} ($V_{SiH4:H2} = 1 : 10$). To avoid further oxidation, the samples were passivated in 1% O_2/Ar overnight. The solids obtained were designated as Fe@SiO_2, Fe&$FeSi_2$@SiO_2, and $FeSi_2$@SiO_2, respectively. XRD patterns evidenced that $FeSi_2$ was formed by the reaction between iron and Si, according to the following reaction described in Eqs. (3.2) and (3.3):

$$Fe_{(s)} + Si_{(s)} \rightarrow FeSi_{2(s)} \tag{3.2}$$

$$Fe_{(s)} + SiH_{4(g)} \rightarrow FeSi_{2(s)} + H_{2(g)} \tag{3.3}$$

In order to determine the chemical state and coordination of the iron species in the samples, the ^{57}Fe Mössbauer spectra were measured at room temperature [42]. The results are shown in Figure 3.8, while the refined values of the Mössbauer parameters are summarized in Table 3.4. The Mössbauer spectrum of Fe@Fe_2O_3 (Figure 3.8a) could be fitted into a Fe^{3+} doublet and a sextet for Fe^0 [43]. From the spectral area, this doublet is accounted for about 38% while the sextet is accounted for the remaining 62% of the total spectral area. After reduction at 400 °C and silicification at 350 °C, there was only a sextet for Fe^0 in the sample Fe@SiO_2, with no other component detected, indicating that there was only one chemical state for iron. For the Fe&$FeSi_2$@SiO_2 sample, the Mössbauer spectrum (Figure 3.8c) could be fitted with one doublet and one sextet. The sextet associated with Fe^0 accounted for 66% of the total spectral area, while the doublet having ^{57}Fe Mössbauer parameters of $\delta = 0.14$ mm s^{-1} and $\Delta E_Q = 0.47$ mm s^{-1} accounted for about 34% and is attributed to $FeSi_2$ [44]. When the silicification temperature was raised to 550 °C to form $FeSi_2$@SiO_2, the Mössbauer spectrum (Figure 3.8d) could also be fitted into one doublet ($FeSi_2$) and one sextet (Fe^0). Compared to the silicification temperature of 450 °C, the $FeSi_2$@SiO_2 had a higher $FeSi_2$ intensity (79%) and a lower Fe^0 intensity (21%). Therefore, the ^{57}Fe Mössbauer spectra further evidenced the structures of the sample changed from cubic Fe to orthorhombic $FeSi_2$ with increasing silicification temperature, due to the dissolution of Si atoms into the iron lattices [41]. The SiO_2-coated iron silicide nanoparticles also can find applications in magnetically separable catalysts, biomedicines, and magnetically recording materials.

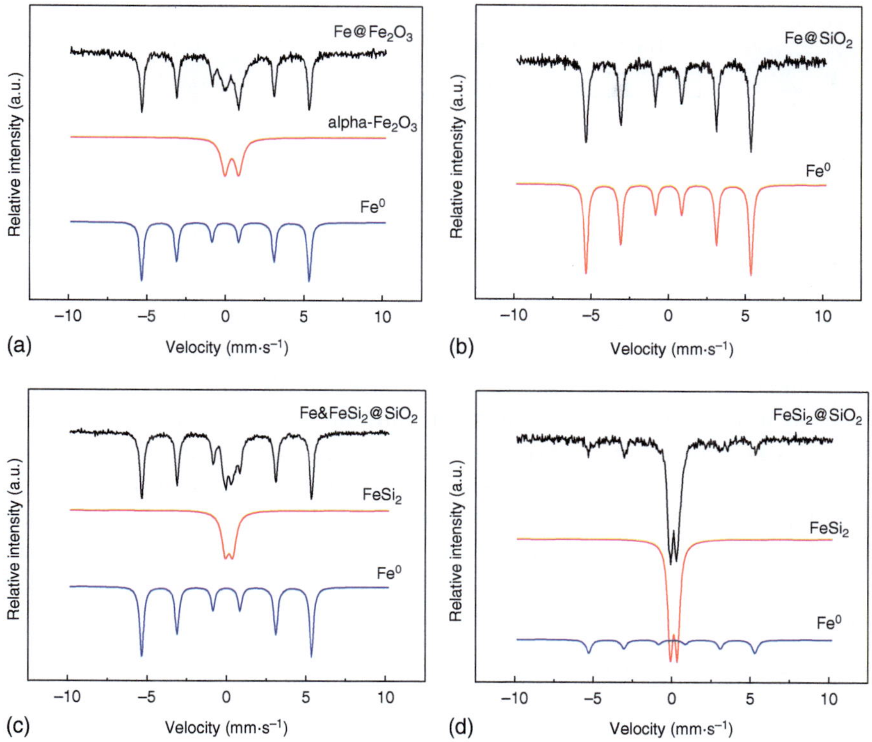

Figure 3.8 Mössbauer spectra of $Fe@Fe_2O_3$ (a), $Fe@SiO_2$ (b), $Fe\&FeSi_2@SiO_2$ (c), and $FeSi_2@SiO_2$ (d). Source: Reproduced from Ref. [41] with permission from the Royal Society of Chemistry.

Table 3.4 ^{57}Fe Mössbauer parameters of $Fe@Fe_2O_3$, $Fe@SiO_2$, $Fe\&FeSi_2@SiO_2$, and $FeSi_2@SiO_2$ samples.

Sample	State	δ/mm s^{-1}	ΔE_Q/mm s^{-1}	B_{hf}/T	RI/%	LW/mm s^{-1}
$Fe@Fe_2O_3$	Fe^{3+}	0.39	0.88		38	0.58
	Fe^0	0	0	33.2	62	0.28
$Fe@SiO_2$	Fe^0	0	0	33.2	100	0.29
$Fe\&FeSi_2@SiO_2$	$FeSi_2$	0.14	0.47		34	0.52
	Fe^0	0	0	33.2	66	0.29
$FeSi_2@SiO_2$	$FeSi_2$	0.12	0.44		79	0.4
	Fe^0	0.01	−0.01	32.9	21	0.40

Source: Reproduced from Ref. [41] with permission from the Royal Society of Chemistry.

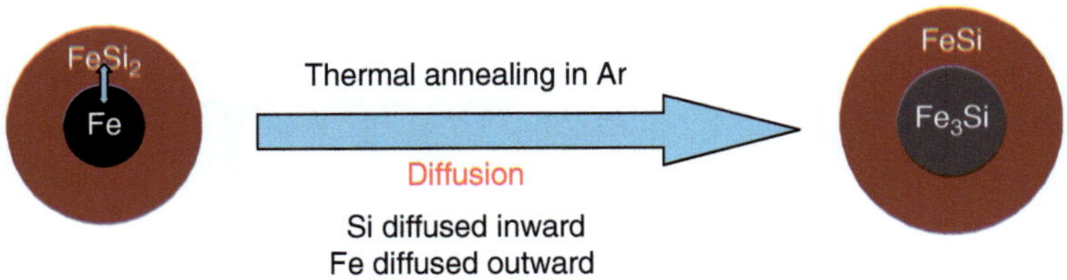

Figure 3.9 Schematic illustration of the crystal transformation after annealing Fe@$FeSi_2$ sample in Ar atmosphere.

Additionally, the ^{57}Fe Mössbauer spectra were measured to confirm the crystal structure transformation of $FeSi_2$ after annealing at high temperature in Ar atmosphere. The results indicated that the crystal transformation proceeded from Fe–$FeSi_2$ to FeSi and Fe_3Si with increasing annealing time and temperature [32]. A brief schematic crystal transformation of Fe–$FeSi_2$ after thermal annealing is shown in Figure 3.9. The mechanism of this process could be attributed to the diffusion of Fe and Si in opposite directions. It has been reported that FeSi is the first silicide phase among the possible iron silicides due to its favorable ΔGv, and Fe_3Si phase is then generated subsequent to FeSi [45, 46]. The possible reaction pathways have been proposed as follows in Eqs. (3.4)–(3.6):

$$\mathrm{Fe + FeSi_2 \rightarrow 2FeSi} \tag{3.4}$$

$$\mathrm{5Fe + FeSi_2 \rightarrow 2Fe_3Si} \tag{3.5}$$

$$\mathrm{FeSi + 2Fe \rightarrow Fe_3Si} \tag{3.6}$$

FeSi and Fe_3Si were formed while Fe and $FeSi_2$ disappeared when annealing the Fe@$FeSi_2$ sample (not exposed to the air before annealing) at 700 °C for 2 h in the Ar atmosphere, which was due to the thermal interdiffusion of the Fe and Si atoms. This phenomenon was similar to that of chemical vapor deposition of silicon and silicides on iron [47].

3.5 Mössbauer Spectroscopic Investigation of Doped Iron Silicides

Although iron silicides, in particular β-$FeSi_2$, have attracted the attention in the field of optoelectronic, thermoelectric, and catalytic conversion, the control of the electrical conductivity regime of the system is still critical. The doping of V, Cr, Mn, Al, Co, Ni, or Pt in iron silicides could tune the geometric and electronic structure, which would make it produce special performances [48–55]. ^{57}Fe Mössbauer spectroscopy is also a well-known tool for the identification of phases, chemical states, structures, particle sizes, and transformation kinetics of doped iron silicides.

Desimoni et al. investigated the phase formation in the M doped iron silicides (M = Pt, Mn, and Al) obtained by ball milling and annealing using XRD and Mössbauer spectroscopy [48–52]. The second metal dopant incorporation into iron silicides influenced the Mössbauer parameters, lattice parameters, electronic density, and dopant location. For the Mn doped iron silicides, the preferential substitution of Fe by Mn atoms at the Fe_{II} site in the β-$FeSi_2$ structure has been confirmed [48–50]. For the Al-doped samples, the β-FeSi fraction was reduced, which indicated that the doping with Al atoms enhanced the stabilization of the β-$FeSi_2$ phase due to the deviations of the Fe + 2Si (Al) stoichiometry [51]. However, there was no noticeable incorporation of Pt into the β-$FeSi_2$ lattice for the Pt-doped iron silicides due to the segregation of Pt silicides [52]. Bender et al. investigated the influence on the structure of Co or Ni doped β-$FeSi_2$ by conversion electron Mössbauer spectroscopy [53, 54]. The hyperfine interaction parameters of ^{57}Fe were slightly modified by the dissolved Co and Ni atoms in the β-$FeSi_2$ lattice. Both lattice sites of Fe in the orthorhombic β-$FeSi_2$ lattice were substituted by Co and Ni. Co was preferred to occupy Fe_{II} site and the lattice parameter of $Fe_{1-x}Co_xSi_2$ with base-centered orthorhombic increased with increasing the Co content up to 0.2.

Recently, Fe substituted Ni–Si intermetallics with different stoichiometric proportion have been prepared by the temperature-programmed silicification of the corresponding oxides [55]. Figure 3.10 displays the room temperature ^{57}Fe Mössbauer spectra of samples $Ni_{0.75}Fe_{0.25}Si_2$, $Ni_{0.50}Fe_{0.50}Si_2$, $Ni_{0.25}Fe_{0.75}Si_2$, and Fe–Si. The corresponding refined values of the Mössbauer parameters were summarized in Table 3.5. In the Mössbauer spectrum of $Ni_{0.75}Fe_{0.25}Si_2$, one resolved doublet,

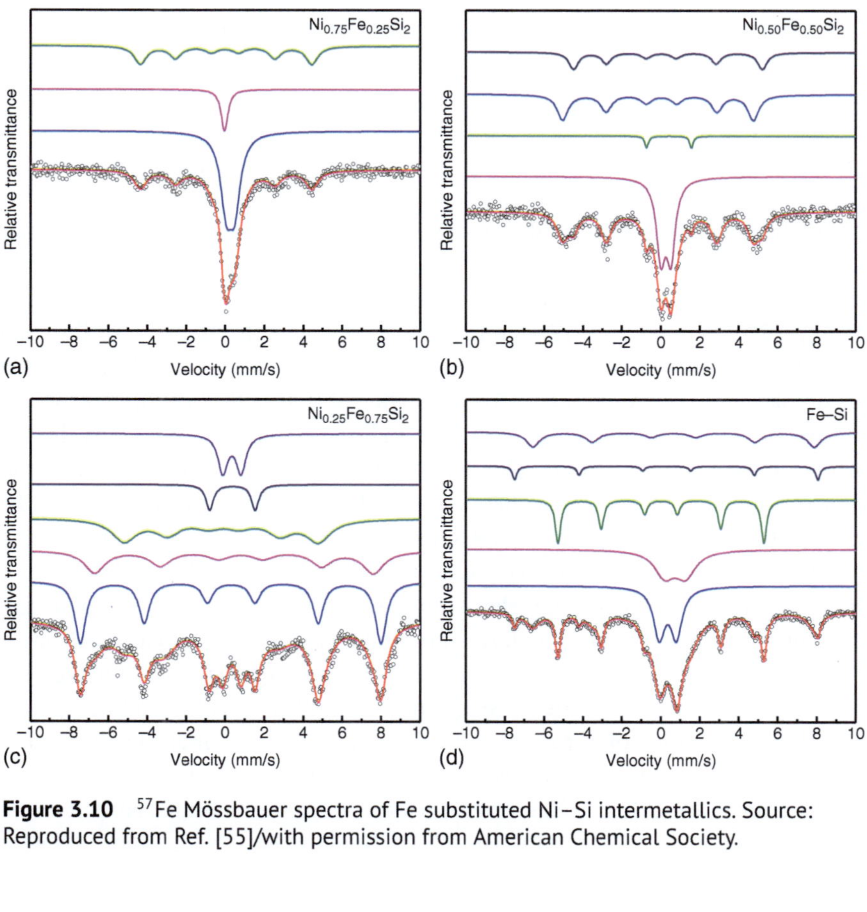

Figure 3.10 ^{57}Fe Mössbauer spectra of Fe substituted Ni–Si intermetallics. Source: Reproduced from Ref. [55]/with permission from American Chemical Society.

Table 3.5 ^{57}Fe Mössbauer parameters of the Fe substituted Ni–Si intermetallics.

Sample	Iron sites	δ/mm s^{-1}	ΔE_Q/ mm s^{-1}	B_{hf}/T	RI/%	LW/ mm s^{-1}
$Ni_{0.75}Fe_{0.25}Si_2$	FeSi	0.27	0.43	—	56.4	0.70
	$FeSi_2$, cubic	−0.03	—	—	10.5	0.42
	$FeNi_3$	0.02	0.05	27.3	33.1	0.77
$Ni_{0.50}Fe_{0.50}Si_2$	$Fe_{0.64}Ni_{0.36}$	−0.06	−0.18	30.4	35.7	0.68
	FeSi	0.26	0.52	—	41.7	0.55
	$FeSiO_3$	0.42	2.29	—	2.6	0.22
	$FeNi_3$	0.19	0.34	30.1	20.0	0.59
$Ni_{0.25}Fe_{0.75}Si_2$	Fe–Ni, taenite	0.29	−0.03	47.8	33.6	0.66
	Fe–Ni, kamacite	0.63	−0.35	44.4	23.6	1.30
	$Fe_{0.64}Ni_{0.36}$	−0.14	−0.12	30.9	28.4	1.56
	$FeSiO_3$	0.37	2.33	—	4.9	0.46
	γ-Fe_2O_3	0.35	0.93	—	9.5	0.61
Fe–Si	γ-Fe_2O_3	0.36	0.88	—	27.5	0.71
	Si doped γ-Fe_2O_3	0.73	1.08	—	24.5	1.25
	Fe^0	0.01	−0.01	32.8	22.4	0.33
	Fe^{3+}, Fe_3O_4	0.29	−0.02	48.3	5.8	0.27
	(Fe^{3+}, Fe^{2+}), Fe_3O_4	0.65	0	44.8	19.9	0.85

Source: Reproduced from Ref. [55].

one singlet, and one sextet could be easily identified. The doublet with Mössbauer parameters of $\delta = 0.27$ mm s^{-1} and $\Delta E_Q = 0.43$ mm s^{-1} accounting for 56.4% was attributed to FeSi [56], while the sextet with $\delta = 0.02$ mm s^{-1}, $\Delta E_Q = 0.05$ mm s^{-1}, and $B_{hf} = 27.3$ T could be attributed to $FeNi_3$ [57], representing 33.1% of the total spectral area. The single with the $\delta = -0.03$ mm s^{-1} could be ascribed to cubic silicide like $FeSi_2$, representing 10.5% [58]. The cubic structure could be expected because of zero quadrupole splitting that indicated spherical symmetry in the surroundings of the iron atoms. The theoretical calculations based on an electron transport model confirmed that the structure of nickel silicide changed when Fe atoms with higher *d*-electron counts substituted Ni atoms, which distorted the six rings in the structure and was traced to the formation of Ni—Fe bonds [59]. Therefore, the Fe atoms occupy the relative sites in the $Ni_{0.75}Fe_{0.25}Si_2$ material, which preferentially bind to Ni to form alloy and combine with Si to form silicide.

Similarly, for the $Ni_{0.50}Fe_{0.50}Si_2$ sample, the Mössbauer spectrum was fitted in two doublets for FeSi and ferrosilite ($FeSiO_3$) and two sextets for $FeNi_3$ and $Fe_{0.64}Ni_{0.36}$ [60, 61]. The formation of ferrosilite in the Fe substituted Ni–Si intermetallics could be attributed to incomplete silicification of the iron oxides or oxidation of the iron silicide formed. Due to the Fe:Ni molar ratio increasing, the amount of Ni-rich Fe–Ni alloy decreased to 20.0%, while the Fe-rich Fe–Ni alloy ($Fe_{0.64}Ni_{0.36}$) appeared with a content of 35.7%. By further increasing the iron content, the components assigned to two types of Fe–Ni alloys (taenite and kamacite) increased significantly while that of γ-Fe_2O_3 phase appeared, resulting in the fact that the zero-valent Fe was hardly silicificated due to the content of Ni as catalyst decreasing but oxidized by air during the passivation. When the bulk Fe was treated in SiH_4/H_2 at 450 °C to form Fe–Si, the Mössbauer spectrum could also be fitted into two doublets (γ-Fe_2O_3 and Si doped γ-Fe_2O_3) and three sextets (Fe^0, one attributed to Fe^{3+} in the tetrahedral site of spinel magnetite and the other to Fe ions located in the octahedral site) [62, 63]. Based on the refined values of the Mössbauer parameters of the Fe substituted Ni–Si intermetallics, those structures were confirmed clearly, which provided the useful information to understand the positive synergistic effect on hydrodesulfurization activity. The formation of Ni–Fe and Fe–Si bonds causes the metal site to have a high *d*-electron density, which promotes hydrogenation activity.

3.6 Conclusion and Perspective

This chapter deals with Mössbauer spectroscopic studies on Fe-based silicides. Using various synthesis methods, such as mechanical alloying, pyrolysis of silicon-containing metallic precursor, and temperature-programmed silicification, the bulk and supported Fe-based silicides could be synthesized in a controlled manner from the macroscopic to the microscopic scale. These methods can also be derived for the synthesis of doped iron silicides. ^{57}Fe Mössbauer spectrometry has become one of the classical techniques for the characterization of Fe-based silicides. The technique is used to identify phases and provide useful information on the structure of silicides. ^{57}Fe Mössbauer spectroscopic results provided strong evidence of the relationship between the structure of iron silicides and their magnetic/catalytic properties. However, the following points for the characterization of iron silicides by ^{57}Fe Mössbauer spectroscopy still need to be improved in the future.

(a) In general, iron silicides have complex phase behaviors and complex stoichiometries and structures, which leads to different chemical states of the iron-silicon system. Given their ability to detect all iron centers, identify their electronic structures, and quantify the ratios of the different iron forms present in a Fe-based silicide, many researchers turn to Mössbauer spectroscopy when they want to address structural and mechanistic issues related to iron silicides. However, this technique applied to materials chemistry is provided by only a few specialized teams worldwide. Technical difficulties ranging from sample preparation to data analysis and interpretation make it necessary for the material scientists and Mössbauer spectroscopists to work together.
(b) As mentioned above, the particle size of iron silicides plays an important role in the catalytic and magnetic properties of iron silicides. The related hyperfine interaction parameters detected at room temperature ^{57}Fe Mössbauer spectroscopy are also influenced by the particle sizes. The ^{57}Fe Mössbauer spectra measured at low temperature may not be dominated by the central

paramagnetic component, which may provide some additional analysis results for the iron silicide nanoparticles. However, there is relatively little information on the low temperature Mössbauer spectroscopy of iron silicides. Research in this area should be strengthening.

(c) The surface oxidation layer (Fe_2O_3 and SiO_2) around the iron silicide nanoparticles forms easily when exposed to air, which influences the intrinsic hyperfine interaction parameters detected by the ^{57}Fe Mössbauer spectroscopy. The quasi in situ Mössbauer spectroscopy for the iron silicides should be developed, which would provide the important information on the distribution of different Fe species on the surface and in the bulk of iron silicides under reaction conditions.

References

1 Schlesinger, M.E. (1990). Thermodynamics of solid transition-metal silicides. *Chem. Rev.* 90 (4): 607–628.

2 Schmitt, A.L., Higgins, J.M., Szczech, J.R. et al. (2010). Synthesis and applications of metal silicide nanowires. *J. Mater. Chem.* 20 (2): 223–235.

3 Chen, X. and Liang, C. (2019). Transition metal silicides: fundamentals, preparation and catalytic applications. *Catal. Sci. Technol.* 9: 4785–4820.

4 Maugh, T.H. (1984). A new route to intermetallics. *Science* 225 (4660): 403.

5 Maszara, W.P. (2005). Fully silicided metal gates for high-performance CMOS technology: a review. *J. Electrochem. Soc.* 152 (7): G550–G555.

6 Wu, Y., Xiang, J., Yang, C. et al. (2004). Single-crystal metallic nanowires and metal/semiconductor nanowire heterostructures. *Nature* 430: 61–65.

7 Liu, Z.H., Zhang, H., Wang, L. et al. (2008). Controlling the growth and field emission properties of silicide nanowire arrays by direct silicification of Ni foil. *Nanotechnology* 19 (37): 375602.

8 Manyala, N., Sidis, Y., Ditusa, J.F. et al. (2004). Large anomalous Hall effect in a silicon-based magnetic semiconductor. *Nat. Mater.* 3 (4): 255–262.

9 Zhou, F., Szczech, J., Pettes, M.T. et al. (2007). Determination of transport properties in chromium disilicide nanowires via combined thermoelectric and structural characterizations. *Nano Lett.* 7 (6): 1649–1654.

10 Wang, H.T., Wu, J.C., Shen, Y.Q. et al. (2010). $CrSi_2$ hexagonal nanowebs. *J. Am. Chem. Soc.* 132 (45): 15875–15877.

11 Lin, Y.J., Zhou, S., Liu, X.H. et al. (2009). $TiO_2/TiSi_2$ heterostructures for high-efficiency photoelectrochemical H_2O splitting. *J. Am. Chem. Soc.* 131 (8): 2772–2773.

12 Zhao, A., Zhang, X., Chen, X. et al. (2010). Cobalt silicide nanoparticles in mesoporous silica as efficient naphthalene hydrogenation catalysts by chemical vapor deposition of $Co(SiCl_3)(CO)_4$ precursor. *J. Phys. Chem. C* 114 (9): 3962–3967.

13 Chen, X., Zhao, A., Shao, Z. et al. (2010). Synthesis and catalytic properties for phenylacetylene hydrogenation of silicide modified nickel catalysts. *J. Phys. Chem. C* 114 (39): 16525–16533.

14 Chen, X., Li, M., Guan, J. et al. (2012). Nickel-silicon intermetallics with enhanced selectivity in hydrogenation reactions of cinnamaldehyde and phenylacetylene. *Ind. Eng. Chem. Res.* 51 (9): 3604–3611.

15 Chen, X., Jin, J., Sha, G. et al. (2014). Silicon-nickel intermetallic compounds supported on silica as a highly efficient catalyst for CO methanation. *Catal. Sci. Technol.* 4 (1): 53–61.

16 Chen, X., Zhang, M., Yang, K. et al. (2014). Raney Ni-Si catalyst for selective hydrogenation of highly concentrated 2-butyne-1,4-diol to 2-butene-1,4-diol in aqueous phase. *Catal. Lett.* 144 (7): 1118–1126.

17 Zhang, L., Chen, X., Jin, S. et al. (2014). Rapid microwaves synthesis of $CoSi_x$/CNTs as novel catalytic materials for hydrogenation of phthalic anhydride. *J. Solid State Chem.* 217: 105–112.

18 Chen, X., Liu, X., Wang, L. et al. (2013). High sulfur tolerance of Ni-Si intermetallics as hydrodesulfurization catalysts. *RSC Adv.* 3 (6): 1728–1731.

19 Chen, X., Wang, X., Xiu, J. et al. (2012). Synthesis and characterization of ferromagnetic nickel-cobalt silicide catalysts with good sulfur tolerance in hydrodesulfurization of dibenzothiophene. *J. Phys. Chem. C* 116 (47): 24968–24976.

20 Zhang, L., Chen, X., Peng, Z. et al. (2019). Acid-tolerant intermetallic cobalt-nickel silicides as noble metal-like catalysts for selective hydrogenation of phthalic anhydride to phthalide. *Catal. Sci. Technol.* 9 (5): 1108–1116.

21 Yang, K., Chen, X., Qi, J. et al. (2019). A highly efficient and sulfur-tolerant Pd_2Si/catalyst for desulfurization of dibenzothiophenes. *J. Catal.* 369: 363–371.

22 Yang, K., Chen, X., Wang, L. et al. (2017). SBA-15 supported metal silicides as efficient catalysts towards semihydrogenation of phenylacetylene by chemical vapor deposition. *Chemcatchem* 9 (7): 1337–1342.

23 Senthilarasu, S., Sathyamoorthy, R., and Lalitha, S. (2004). Synthesis and characterization of β-$FeSi_2$ grown by thermal annealing of Fe/Si bilayers for photovoltaic applications. *Sol. Energy Mater. Sol. Cells* 82 (1–2): 299–305.

24 Varadwaj, K.S.K., Seo, K., In, J. et al. (2007). Phase-controlled growth of metastable Fe_5Si_3 nanowires by a vapor transport method. *J. Am. Chem. Soc.* 129 (27): 8594–8599.

25 Du, X., Hu, P., Mao, T. et al. (2019). Ru alloying induced enhanced thermoelectric performance in $FeSi_2$-based compounds. *ACS Appl. Mater. Interfaces* 11 (35): 32151–32158.

26 Kolel-Veetil, M.K. and Keller, T.M. (2010). Organometallic routes into the nanorealms of binary Fe-Si phases. *Materials* 3 (2): 1049–1088.

27 Berry, F.J. (1984). *Mössbauer Spectroscopy Applied to Inorganic Chemistry.* New York: Plenum Press.

28 Murty, B.S. and Ranganathan, S. (1998). Novel materials synthesis by mechanical alloying/milling. *Int. Mater. Rev.* 43 (3): 101–141.

29 Dézsi, I., Fetzer, C., Bujdosó, L. et al. (2010). Mechanical alloying of Fe-Si and milling of α- and β-$FeSi_2$ bulk phases. *J. Alloys Compd.* 508 (1): 51–54.

30 Guan, J., Chen, X., Zhang, L. et al. (2016). Rapid preparation and magnetic properties of Fe_3Si–Al_2O_3 nanocomposite by mechanical alloying and heat treatment. *Phys. Status Solidi A* 213 (6): 1585–1591.

31 Guan, J., Chen, X., Yang, K. et al. (2014). Preparation and size dependent magnetism of highly dispersed iron silicide nanoparticles on silica. *J. Mater. Chem. C* 2 (27): 5292–5298.

32 Li, M., Chen, X., Guan, J. et al. (2013). Thermally induced phase transition and magnetic properties of Fe-$FeSi_2$ with core-shell structure. *Phys. Status Solidi A* 210 (12): 2710–2715.

33 Krumme, B., Weis, C., Herper, H.C. et al. (2009). Local atomic order and element-specific magnetic moments of Fe_3Si thin films on MgO(001) and GaAs(001) substrates. *Phys. Rev. B* 80 (14): 144403.

34 Shinjo, T. and Nakamura, Y. (1963). Magnetic study of Fe_3Si and Fe_5Si_3 by mössbauer effect. *J. Phys. Soc. Jpn.* 18 (6): 797–801.

35 Wertheim, G.K., Wernick, J.H., and Buchanan, D.N. (1966). Mössbauer Effect in $Co_{1-x}Fe_xSi$. *J. Appl. Phys.* 37 (9): 3333–3337.

36 Chuev, M.A. (2011). Multi-level relaxation model for describing the Mössbauer spectra of single-domain particles in the presence of quadrupolar hyperfine interaction. *J. Phys. Condens. Matter* 23 (42): 426003.

37 Li, T., Li, Y.Z., and Zhang, Y.H. (1997). Phases in ball-milled $Fe_{0.6}Si_{0.4}$. *J. Phys. Condens. Matter* 9: 1381.

38 Ionescu, A., Vaz, C.A.F., Trypiniotis, T. et al. (2005). Structural, magnetic, electronic, and spin transport properties of epitaxial Fe_3Si/GaAs(001). *Phys. Rev. B Condens.Matter Mater. Phys.* 71 (9): 094401.

39 Kolel-Veetil, M.K., Qadri, S.B., Osofsky, M. et al. (2009). Carbon nanocapsule-mediated formation of ferromagnetic Fe_5Si_3 nanoparticles. *J. Phys. Chem. C* 113 (33): 14663–14671.

40 Jiang, Q., Cui, X.F., and Zhao, M. (2004). Size effects on Curie temperature of ferroelectric particles. *Appl. Phys. A Mater. Sci. Process.* 78: 703–704.

41 Li, M., Chen, X., Guan, J. et al. (2012). A facile and novel approach to magnetic Fe@SiO_2 and $FeSi_2$@SiO_2 nanoparticles. *J. Mater. Chem.* 22 (2): 609–616.

42 Murad, E. and Cashion, J. (1984). *Mössbauer Spectroscopy of Environmental Materials and Their Industrial Utilization*. Boston, MA: Kluwer Academic Publishers.

43 Liu, K., Wang, A.Q., Zheng, W.S. et al. (2010). Quasi *in situ* ^{57}Fe mössbauer spectroscopic study: quantitative correlation between Fe^{2+} and H_2 concentration for PROX over Ir–Fe/SiO_2 catalyst. *J. Phys. Chem. C* 114 (18): 8533–8541.

44 Schaaf, P., Milosavljevic, M., Dhar, S. et al. (2002). Mössbauer optimization of the direct synthesis of β-$FeSi_2$ by ion beam mixing of Fe/Si bilayers. *Hyperfine Interact.* 139 (1–4): 615–621.

45 Walser, R.M. and Bene, R.W. (1976). First phase nucleation in silicon-transition metal planar interfaces. *Appl. Phys. Lett.* 28 (10): 624–625.

46 Du, V.A., Sidorenko, A., Bethge, O. et al. (2011). Iron silicide nanoparticles in a SiC/C matrix from organometallic polymers: characterization and magnetic properties. *J. Mater. Chem.* 21 (33): 12232–12238.

47 Rebhan, M., Rohwerder, M., and Stratmann, M. (1999). CVD of silicon and silicides on iron. *Appl. Surf. Sci.* 140 (1–2): 99–105.

48 Desimoni, J., Cotes, S.M., Taylor, M.A. et al. (2006). Formation of Mn-doped iron silicides by ball milling. *Hyperfine Interact.* 168 (1–3): 959–964.
49 Desimoni, J., Cotes, S.M., Taylor, M.A. et al. (2007). Phase stabilization in the $Fe_{1-x}Mn_xSi_2$ system induced by ball milling. *Intermetallics* 15 (10): 1297–1302.
50 Martínez, J., Runco, J., Cotes, S.M. et al. (2007). Evolution under thermal annealing of Mn-doped iron disilicides obtained by ball milling. *Hyperfine Interact.* 178 (1–3): 13–17.
51 Desimoni, J., Martínez, J., Cotes, S.M. et al. (2009). The effect of Mn and Al dopants on the β-$FeSi_2$ stabilization by a two-step route: mechanical alloying and annealing. *J. Alloys Compd.* 477 (1–2): 789–794.
52 Desimoni, J. and Sánchez, F.H. (2001). On the formation of pure and Pt-doped iron silicides using ball milling. *J. Phys. Condens. Matter* 13 (11): 2737–2747.
53 Vantomme, A., Wu, M.F., Langouche, G. et al. (1995). The formation and thermal stability of ion-beam-synthesized ternary $Me_xFe_{1-x}Si_2$ (Me = Co, Ni) in Si (111). *Nucl. Instrum. Meth. B* 106 (1–4): 404–408.
54 Fetzer, C., Dézsi, I., Vantomme, A. et al. (2002). Ternary $Co_xFe_{1-x}Si_2$ and $Ni_xFe_{1-x}Si_2$ formed by ion implantation in silicon. *J. Appl. Phys.* 92 (7): 3688–2693.
55 Chen, X., Wang, J., Yang, K. et al. (2015). Structure investigation and dibenzothiophene hydrodesulfurization properties of Fe-substituted Ni–Si intermetallics. *J. Phys. Chem. C* 119 (52): 29052–29061.
56 Cranshaw, T.E., Johnson, C.E., and Ridout, M.S. (1966). The Mössbauer and NMR spectra of FeSi alloys. *Phys. Lett.* 21 (5): 481–483.
57 Zeifert, B.H., Salmones, J., Hernández, J.A. et al. (2000). X-Ray diffraction and Mössbauer characterization of Raney Fe-Ni catalysts. *J. Radioanal. Nucl. Chem.* 245 (3): 637–639.
58 Vondráček, M., Dudr, V., Tsud, N. et al. (2006). Surface segregation in FeSi alloys. *Surf. Sci.* 600 (18): 4108–4112.
59 Landrum, G.A., Hoffmann, R., Evers, J. et al. (1998). The TiNiSi family of compounds: structure and bonding. *Inorg. Chem.* 37 (22): 5754–5763.
60 Abdu, Y.A. and Ericsson, T. (1997). Mössbauer spectroscopy, X-ray diffraction, and electron microprobe analysis of the new haifa meteorite. *Meteoritics Planet. Sci.* 32 (3): 373–375.
61 Cuttler, A.H. (1980). The behaviour of a synthetic ^{57}Fe doped kaolin: Mössbauer and electron paramagnetic resonance studies. *Clay Miner.* 15 (4): 429–444.
62 Mitra, A., Mohapatra, J., Meena, S.S. et al. (2014). Verwey transition in ultrasmall-sized octahedral Fe_3O_4 nanoparticles. *J. Phys. Chem. C* 118 (33): 19356–19362.
63 Hafner, S.S. and Huckenholz, H.G. (1971). Mössbauer spectrum of synthetic ferri-diopside. *Nat. Phys. Sci.* 233 (36): 9–11.

4

Mössbauer Spectroscopy of Catalysts

Károly Lázár

Centre for Energy Research, ELKH, Department for Nuclear Analysis, Konkoly Thege Miklós Street 29-33, 1121 Budapest, Hungary

4.1 Introduction

Catalytic processes play important roles in the production of various fuels, chemicals, plastics, medicines, etc. The primary focus is usually studying the reactions themselves to understand and describe the elementary steps in the sequence of transformations. An important stage in these studies is the characterization of catalysts. In common practice, the publications on connected studies are composed of two basic parts. First, the characterization of catalysts (with X-ray diffraction [XRD], transmission electron microscopy [TEM], specific surface area measurements performed with the Brunauer-Emmet-Teller [BET] method, Fourier transform infrared spectroscopy [FTIR], ultraviolet-visible [UV-Vis] spectroscopy, etc.) is conventionally described. These studies are usually performed under conditions that are, to a small or large extent, far from the working conditions of the real catalysts. The performance observed under the working conditions (e.g. the yield, activity, and selectivity) is usually described in the second part, and the combination of the two parts with the interpretation follows afterwards. Special cells can be constructed to provide experimental conditions for in situ and even for operando measurements to narrow the gap between the relevant working conditions and those existing during the ex situ characterizations.

Mössbauer spectroscopy is one of the techniques suitable for the sensitive characterization of catalysts. The principal information lays in the determination of the slight differences between the energies observed during recoilless emission and the absorption of gamma rays in appropriate nuclear transitions. By detecting these small energy differences, hyperfine interactions can be traced. From the aspect of the characterization of catalysts, two of them play important roles: (i) the electronic one, correlated to the densities of s-electrons at the nucleus, and the (ii) electric quadrupole interaction, correlated to the symmetry of the electric field around the nucleus. Due to the nuclear origin of the basic phenomenon, certain limitations also arise. First, the effect is practically observable only in the solid state. Another

Mössbauer Spectroscopy: Applications in Chemistry and Materials Science,
First Edition. Edited by Yann Garcia, Junhu Wang, and Tao Zhang.

particular feature is related to the nuclear origin, namely, the occurrence of the effect is connected to selected elements or, more precisely, to isotopes of given elements.

For further discussion, it is worth separating the various catalytic processes to a certain extent in the first approach by distinguishing between heterogeneous, bioenzymatic, and homogeneous processes. Heterogeneous catalysts are suitable primarily for Mössbauer studies since the catalyst particles are usually in the solid state. However, the obtained information usually is a mixture originated from various locations of catalytic relevance. Namely, in catalytic processes, the atoms located on the surface of particles play a primary role. In contrast, the proportion of surface atoms is small compared to the total amount of atoms in the particle. Thus, the contribution of surface atoms to the recorded spectra is generally also small. The situation will be illustrated in more detail later in Section 4.3. In addition, collective particle effects (e.g. magnetism), without any relevance to catalytic processes, may complicate the further evaluation of the spectra. Regardless, despite these limitations, most Mössbauer studies are performed on heterogeneous catalysts. The types of heterogeneous catalysts can be subdivided further depending on which kind of activation is promoted by them: conventional processes (by thermal activation), liquid–solid-state charge transfer (electrocatalysts), or photochemical activation (photocatalysts).

Enzyme catalysis is of primary importance for biological processes. The single catalytic centers are usually spread and wrapped into sophisticated flexible helices and globuli of proteins. These aggregates in a strict sense are not in the solid state, but their studies can be performed at low temperatures. The centers are usually separated, and thus, bonding conditions, covalencies, etc. can be advantageously studied.

Homogeneous catalytic processes are also widely applied in several processes and, in most cases, are applied in solutions. Solutions can easily be converted to the solid state, namely, by quick freezing. The structure of solutions can even be preserved, preferably by quick freezing. Thus, separate catalytic entities can be observed in suitable circumstances. Under appropriate conditions, even the transient substrate–catalyst complexes can be fixed and analyzed with time-resolved quick freezing technique. In instances when the concentration of the catalyst is low for detection, lyophilization can also be applied.

The previous considerations are summarized in Figure 4.1. The predominant part of the studies performed on catalysts with the help of Mössbauer spectroscopy can be described as the structural characterization of static catalysts (central part). These studies may supply important and, in certain cases, even unique information on the structure in proximity to the catalytically active atoms and ions. A possible further subdivision, determined from the point of view of the various types of catalytic processes, is shown on the right side. Another aspect is shown at the bottom, in which the conversion of the liquid state to the solid state occurs by freezing, which is suitable primarily for studying homogeneous catalysts. A further approach is represented on the left, namely, when efforts are made to approach the working conditions by applying in situ or operando conditions during the Mössbauer measurements.

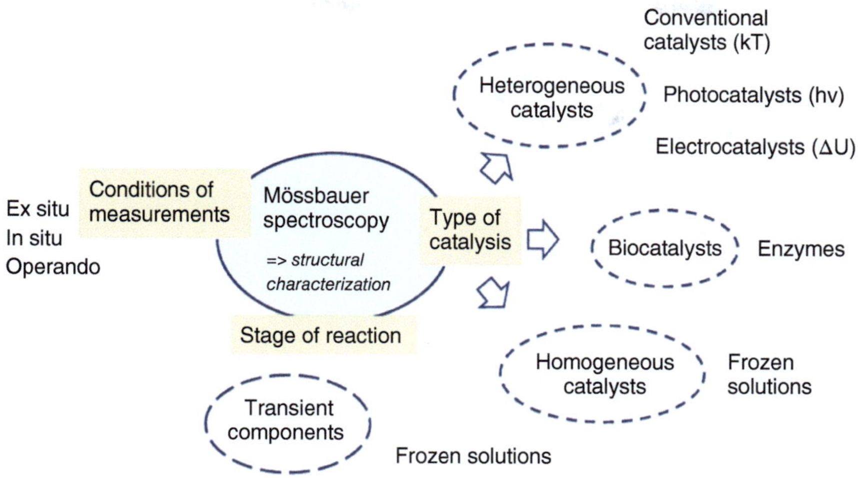

Figure 4.1 Various aspects of the characterization of catalysts by Mössbauer spectroscopy.

The subsequent part of the chapter is composed of the following sections. First, the principles of the Mössbauer effect are considered from the aspects that have expressed importance for catalysts studies. Variants of the method (conversion electron Mössbauer spectroscopy [CEMS], nuclear forward scattering [NFS], emission spectroscopy, etc.) and the conditions under which the measurements are performed (ex situ, in situ, and operando) are shortly discussed in the subsequent section. In the following sections, the particular features of studying heterogeneous, bioenzymatic, and homogeneous catalysts are discussed with selected case studies.

To provide a certain impression of the acceptance and the extent of the spread of Mössbauer spectroscopy in studying catalysts, a number of Science Citation Index (SCI) publications can be referred. Namely, a search with "Mössbauer AND *cataly*" keywords in the Web of Science database returns 3627 results in the full available time span (from 1975 until recently) [1]. This simple approach is certainly a strong underestimation, since the study of catalysts has already commenced shortly after the discovery of the effect in 1958. Further, the pool of the relevant journals is probably not complete in the Web of Science database, and publications describing results of other emerging techniques that are very close to the methodology of the Mössbauer effect (e.g. nuclear forward scattering) have probably also not been retrieved by this search. Thus, the total number of publications related to catalyst studies obtained with this method is probably larger than 4000. Another query can also be carried out for the number of yearly publications, e.g. for the period starting at the turn of the new millennium (Figure 4.2), and a still growing tendency is reflected in this representation [1].

Many relevant reviews have already been published on the characterization of catalysts by Mössbauer spectroscopy since the early days (e.g. [2–11]). Further, a *Catalysts Handbook* was also published in the series of the Mössbauer Effect Data Center's topical collections [12].

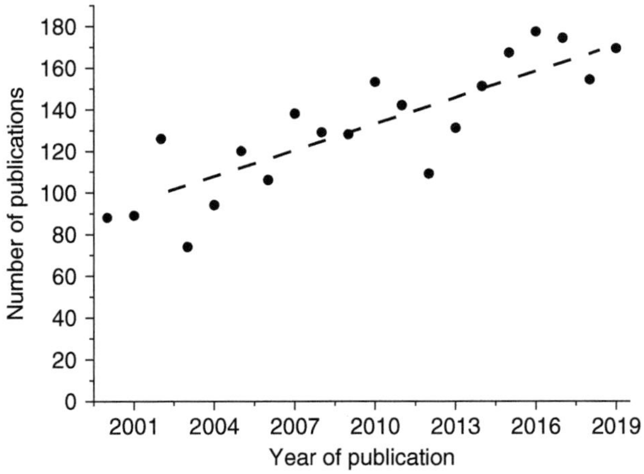

Figure 4.2 Number of yearly publications retrievable from the Web of Science database by the search with "Mössbauer AND *cataly*" keywords [1].

4.2 Principles of the Mössbauer Effect and Outlook of Its Application for Catalyst Studies

4.2.1 Brief Overview of the Basics of Mössbauer Spectroscopy

Principally, the method is based on a very sensitive comparison of the difference in the energies of γ-rays emitted or absorbed in a pair of events, namely, the recoilless emission from a source nucleus and the subsequent recoilless absorption of the emitted γ-ray in another probed nucleus. The very small change (on the 10^{-11} order of relative magnitude) carries information on the differences in the states of the emitting and absorbing nuclei. Usually, the energies of the γ-rays emitted and absorbed in nuclear transitions are rather high (several keV-s). Thus, these events are accompanied with recoils of the emitting/absorbing nuclei due to the principle of the conservation of momentum. Due to the loss in the recoil energy (on the order of 10^{-5} relative magnitude), resonance absorption cannot be observed. In certain cases, the event of recoil can be avoided, and as a consequence, the subsequent absorption process may also proceed. This conservation of the whole energy in the emission/absorption processes can be completed in certain solid structures due to restrictions connected to the quantization in energies, which can be absorbed in the excitation of lattice vibrations. See the discussion in further detail in Section 4.2.3. The tiny energy changes (a relative intensity with an order of magnitude of 10^{-11}) necessary to restore the conditions of resonance absorption can be amended by Doppler shift (i.e. by moving either the source or absorber). Thereby, spectroscopy, i.e. scanning the occasional appearance of resonance absorption with respect to the energy modulation, can be carried out.

These minuscule differences between the emitted and absorbed γ-rays can be originated from differences in the (i) electron densities in the locus of nuclei and (ii) symmetry of the electron structure around the nuclei (symmetry of ligands, which occasionally may generate an electric quadrupole moment), and (iii) occasionally, magnetic interactions may also appear, which split the accessible nuclear energy levels of the transitions. The result in case (i) is the so-called isomer shift (δ), i.e. the maximum resonance absorption is shifted in the spectrum of energy modulation. In case (ii), the quadrupolar interactions split the lines, and doublets appear in the spectra (quadrupole splitting, Δ). The occasional magnetic interaction may result in the Zeeman splitting of the nuclear levels, as reflected by the appearance of six-line sextets. The isomer shift and quadrupole splitting carry information on the very proximity of the emitting/absorbing nucleus. The third parameter, the magnetic splitting, usually develops in an interaction with several atoms (collective phenomenon), and thus, the information is connected to larger assemblies of atoms.

An important feature of the Mössbauer effect is that the basic phenomenon is not common for each atomic nuclei. The related appropriate nuclear transitions are specifically connected only to different isotopes of particular elements. The most common isotope used for Mössbauer studies is ^{57}Fe. This isotope has only a 2.12% natural abundance, i.e. only this small percentage of iron atoms present in a conventional sample could respond to Mössbauer excitation, and the remaining ~98% residing in other isotopes of Fe is silent. For illustration, some principal data obtained for this transition are listed as follows. The energy of the utilized nuclear transition is 14.4 keV. The excited ^{57}Fe nuclear level is generated by the K electron capture of ^{57}Co, and the lifetime of the emitting excited state is c. 140 ns. The half-life of the parent ^{57}Co is 270 days, and due to its spontaneous decay, it is a radionuclide. In common practice, sources of 1–2 GBq activities are applied with corresponding radiation protection.

A more detailed description of the effect can be found online on Wikipedia, for example [13]. A broad choice of excellent textbooks (e.g. [14–16]) and shorter introductory reviews (e.g. [17, 18]) are also available. A longer list of the corresponding literature can be found at a website devoted exclusively to the Mössbauer effect and spectroscopy [19]. The Mössbauer Effect Data Center maintains a Mössbauer spectroscopy database since the dawn of the method; the number of records related to publications is close to 100 000 at present [20].

4.2.2 Mössbauer Spectroscopy from the Point of View of Catalyst Studies – Particular Features

In the characterization of catalysts, the information on electron density (correlated to valency) plays an important role. As mentioned before, Mössbauer spectroscopy provides a sensitive tool, namely, the isomer shift (δ) in the spectra, which is correlated in particular to the density of s electrons at the nucleus. Further, the symmetry of the electric field around the nucleus is also important information correlated directly to the arrangement of various ligands around the given atom or ion. This information can also be extracted from the spectra, namely, from quadrupole splitting (Δ).

As referenced earlier, an essential point is that the occurrence of the effect is strongly connected to isotopes of various elements. In conventional Mössbauer spectroscopy, the predominantly used isotope is ^{57}Fe. ^{119}Sn is the second most frequently used source nucleus, and all the remaining Mössbauer active nuclei share only c. 10% of the total applications of the method. In catalytic processes, primarily noble metals (Pt, Pd, Ru, Rh, Au, etc.) play a role, and only a few of their isotopes possess appropriate Mössbauer transitions. Further, the appropriate nuclei of noble metals (^{197}Au, ^{193}Ir, and ^{99}Ru) have resonance γ-transitions with rather high energies; thus, a very low temperature is necessary for Mössbauer measurements, and in situ measurements practically cannot be performed. Another unfavorable condition is that the half-lives of these noble metal parent source nuclei are rather short (a few days, at most). Another outcome of the isotope dependence of the effect is that the sensitivity of the method with various elements is determined by the abundance of the particular Mössbauer active isotope in the sample. Namely, in the most frequently used element, iron, the natural abundance of the ^{57}Fe isotope is low (only 2.12%), ^{119}Sn possesses 8.58%, and in certain favorable cases, the natural abundance even reaches 100% (e.g. for ^{197}Au). The sensitivity of measurements performed with elements having low natural abundances can be improved by using samples prepared from enriched isotope contents of the Mössbauer active nuclei.

Transition metals can also be used to catalyze various processes, and among them, iron is also used frequently. Due to its beneficial properties, ^{57}Fe is an appropriate nucleus to perform in situ studies with various catalysts. Properties of elemental metallic catalysts can advantageously be modified in several cases, so alloy catalysts are used in many processes. This widely opens the possibilities of applications of ^{57}Fe for catalyst studies, and the changes taking place due to alloy formation can be identified and correlated with the improvements in the catalytic properties.

4.2.3 The Probability of the Mössbauer Effect – f-Factor and Size Effects

In the assessment and comparison of spectroscopic methods, usually an important point is to what extent can the components be quantified from the spectra, i.e. to what extent are the intensities of the components in the spectra proportional to their true presences in the analyzed samples. The situation arising in Mössbauer spectroscopy from this aspect is somewhat complex, so is worth discussing the conditions in further detail. It will be demonstrated that the obtained spectra are composed of components contributing to different intensities and, further on, that this contribution depends strongly on the temperature of the measurement.

An important aspect in catalyst studies is that the phenomenon of recoilless nuclear γ-ray absorption is strongly connected to the vibration states in a solid crystal lattice. The probability of the Mössbauer effect, f_A, can be described by a first approximation in the framework of the Debye model of lattice vibrations [15]:

$$f_A = \exp\left[\frac{-6E_R}{k_B\theta_D}\left[\frac{1}{4} + \left(\frac{T}{\theta_D}\right)^2 \int_0^{\frac{\theta_D}{T}} \frac{y\,dy}{e^y - 1}\right]\right]$$

where E_R is the energy of recoil, k_B is the Boltzmann constant, θ_D is the Debye temperature, and T is the absolute temperature. The energy of recoil is proportional to the energy of the original nuclear transition, $E_{\gamma,n}$, and inversely proportional to the mass of the nucleus or to the assembly of the strongly bonded nuclei that absorb the energy of recoil, m.

$$E_R = \frac{E_{\gamma,n}}{2mc^2}$$

From these two equations, important consequences can be derived. First, the greater the energy of the nuclear transition, i.e. the emitted γ-ray, the smaller is the probability of the effect. Second, the temperature of the measurement, or more exactly, the ratio of T/θ_D, is also important. The lower the temperature, the better is the efficiency. In this model, all the lattice vibrations exited at the Debye temperature, i.e. recoil may take place with any energy; thus, above θ_D, the probability of the Mössbauer effect is practically zero. For illustration, the solutions for the previous f_A function in case of ^{57}Fe with different θ_D values are represented graphically in Figure 4.3.

Different probabilities of the Mössbauer effect may exist for the same nucleus in different valence states. The character of bonds (i.e. the extent of the ionicity or the covalency) may strongly depend on valency. A good example for this effect is the Sn^{4+}/Sn^{2+} pair. The tin with four valence electrons usually has a high $\Theta_{D,}$ whereas for Sn^{2+}, the value is much smaller. Thus, in room temperature spectra of samples containing both Sn^{2+} and Sn^{4+} components in comparable amounts, Sn^{2+} is hardly noticeable, and lower temperature measurements (<80 K) are necessary to reveal the reliable presence of Sn^{2+} components.

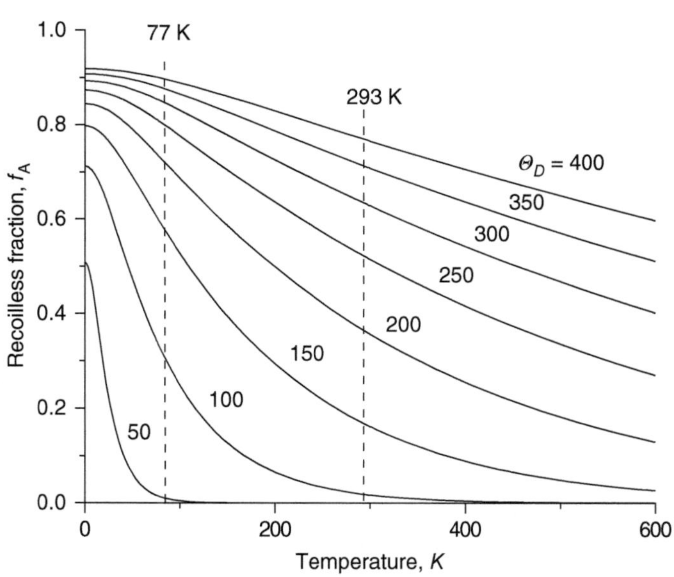

Figure 4.3 Dependence of the recoil-free fraction (f_A) on the temperature in the case of ^{57}Fe at various θ_D values.

Θ_D may even be different in various parts of a small particle. The coordination number of the atoms located on the surface of particles is smaller than that of atoms located inside the particle with fully complete coordination. Correspondingly, the strength of bonding is also different, which results in a decrease in the Θ_D temperature of surface atoms. An excellent demonstration of this effect is described in [21]. Gold nanoparticles with different sizes in the 0.5–50 nm range were deposited on Mylar. The same amount of gold resulted in 4 K spectra with very different relative spectral areas, which depend monotonously on the size (dispersion): if the contribution of the 50 nm particles is 100%, then the relative spectral area is only 20% for the smallest 0.5 nm particles [21].

Thus, the data obtained from Mössbauer spectra can only be used with strong precaution for quantitative analysis. In the case of similar Θ_D values, the corresponding signal can be proportional to the amount of the given component in the sample. In the case when the sample is composed of components with different Θ_D values, it is favorable to perform measurements that depend on temperature as well to obtain an approximation of Θ_D from the temperature dependence. Note that for components with low Θ_D values, the estimation of their amount from spectral areas, even from very low temperature measurements, can still be misleading; for example, in the case of a hypothetic component with $\Theta_D = 50\,K$, the value of f_A is only 0.5, even at the lowest 0 K temperature (see Figure 4.3).

An impressive demonstration of the drastic changes of the f-factors depending on the treatments can be found in [22], for example. A carbon-supported Fe–N–C oxygen reduction reaction (ORR) catalyst was prepared by the pyrolysis of aniline (900 °C), and the formed precursor was treated with chlorine (600 °C) and was reduced in the subsequent step with hydrogen (600 °C) afterwards. The evolution of the catalyst is demonstrated in the corresponding Mössbauer spectra. Pyrolysis resulted mostly in the formation of FeN_4 and Fe_3C (Figure 4.4a). The treatment with chlorine altered the state of iron nanoparticles due to the formation of loosely bound $FeCl_3$ complexes that significantly weakened the bonding with the carbon matrix, as reflected in the striking decrease in the intensity of the corresponding Mössbauer signal (Figure 4.4b). The subsequent treatment in H_2 restored the higher f-factor, and the presence of α-iron and Fe_3C was observed again (Figure 4.4c).

4.2.4 Variants of the Technique

4.2.4.1 ^{57}Co Emission Spectroscopy

The conventional arrangement for Mössbauer spectroscopy is the transmission mode, where the source ⟹ sample ⟹ detector sequence is used. For instance, in the case of ^{57}Fe, radioactive ^{57}Co parent nuclei are used, in which K electron capture takes place, and the 14.4 keV radiation is generated in the last step of the de-excitation of the formed ^{57}Fe nucleus. This source is attached to the drive unit, which provides the necessary modulation of the energy of gamma rays with the Doppler effect. This technique can be reverted by exchanging the roles of the source and absorber. Namely, the studied sample can be prepared by using the ^{57}Co isotope, and the velocity modulation can be performed via the velocity scanning of an iron

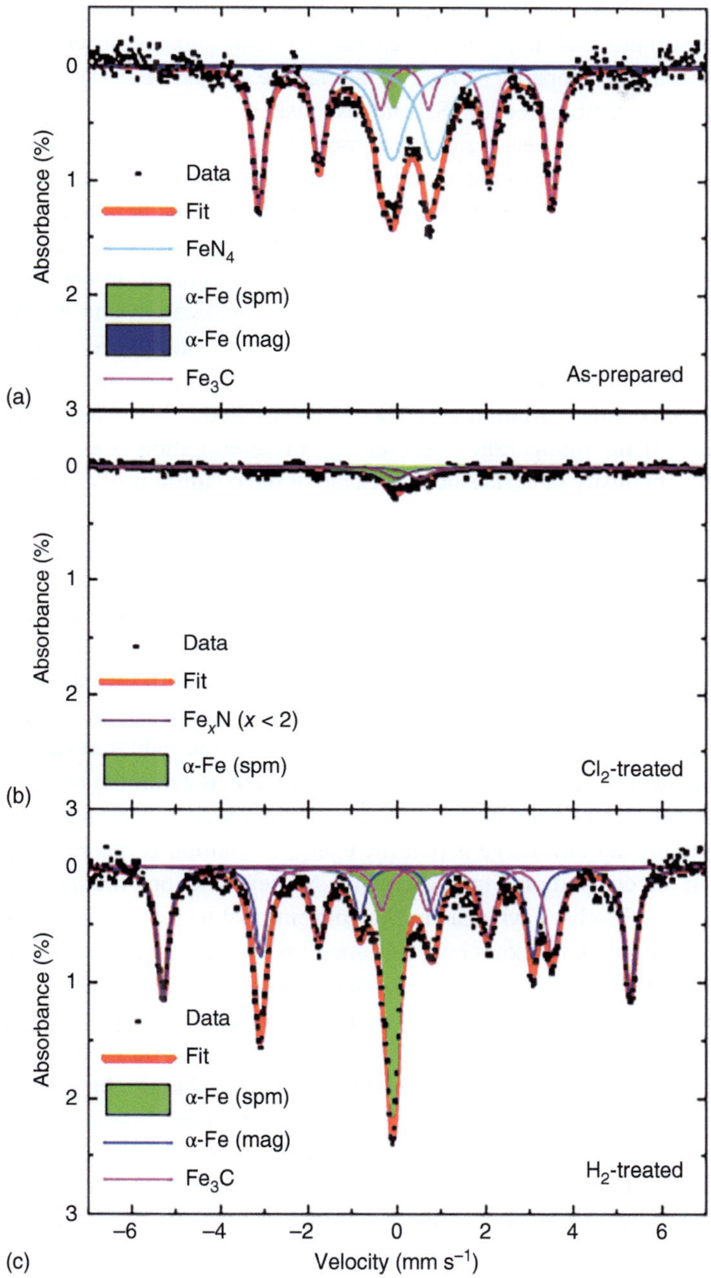

Figure 4.4 Illustration of the drastic change in the probability of the Mössbauer effect depending on the conditions in a carbon-supported Fe–N–C catalyst. 300 K spectra of samples: as-prepared with the pyrolysis of aniline (900 °C) (a), treated with chlorine gas (600 °C) (b), and reduced with hydrogen (600 °C) (c). Ref. [22]/Springer Nature/CC BY 4.0.

absorber. In this case, information on the environment of a primary Co nucleus is obtained. Four important conditions should be considered for these measurements. First, in the case that the prepared sample is radioactive, preparation should be carried out under strict radioprotection. Second, expensive ^{57}Co is used to prepare the sample and used for a few measurements or used for a series of measurements performed on the very same sample under different experimental conditions. Correspondingly, the number of measurements is usually far smaller than that of the conventional arrangement, where one source is used for one or two years, providing a means to study several various samples. The third point, in contrast, is a strong advantage; namely, if ^{57}Co is used for sample preparation and all the Co is "active," the nominal isotope enrichment is 100% (As mentioned earlier, for conventional ^{57}Fe, the natural abundance is only 2.12%.). The fourth point is related to the rate of the de-excitation processes of the excited ^{57}Fe nucleus after K electron capture. If the dissipation of the energy of de-excitation is slow, the so-called hot-atom chemistry of Co is studied instead of the conditions of a regular stabilized state.

4.2.4.2 Synchrotron-Based NFS (Nuclear Forward Scattering)

This method possesses several important advantages over the conventional technique. A principal advantage is that the intensity of the synchrotron-generated γ-rays considerably exceeds that of the conventional Mössbauer sources with several order of magnitudes. Further, there is no need to obtain different Mössbauer sources for each corresponding nuclei, as the applied energy window of the radiation can be tuned by selecting the appropriate range. For the same reason, the number of available nuclei is increased, since the half-life of the parent source isotope does not matter anymore. A further advantage is that the energy resolution is theoretically better than that in the conventional measurements. Namely, only one natural line width matters (in the absorber), whereas in the conventional technique, both the line widths of the source and absorber are summed up in the energy resolution. Additionally, the time of the measurement is much shorter due to the high intensity of the synchrotron source, and only a few tens of minutes is usually sufficient instead of several hours. Another important advantage is that the amount of sample can be much smaller, since the radiation can be focused more sharply than in the usual conventional arrangement. In contrast, the experiment is very sophisticated, and since the energy density of the probing γ-radiation is high, certain damage to the sample may take place [23]. In a novel experiment, Mössbauer spectra with a conventional shape can also be obtained with synchrotron radiation by vibrating the sample. Recently, spectra were obtained with a ^{61}Ni isotope, which has a short-life excited state (5.3 ns) [24].

4.2.4.3 Conversion Electron Mössbauer Spectroscopy

In a further variant of the technique, secondary electrons are detected instead of detecting the resonance absorption of the direct stimulating γ-radiation. A certain part of the emitted resonance γ-rays is immediately absorbed, primarily by the K and L electrons on the orbitals around the ^{57}Fe nucleus, and the corresponding electrons are kicked out with 7.3 and 13.6 keV energies. These high energy electrons may depart from the sample and can be detected. Since electrons interact strongly with the solid matrix, their escape depth is only 300–350 nm, and hence, this type of measurement is "surface sensitive" at this depth range. Thus, it is an advantage that the surface of bulk samples can be studied directly, and the thickness of samples is not limited as in the case of the transmission arrangement. This condition is favorable, for example, for corrosion studies or for studying thin layers on a substrate. For in situ catalyst studies, this method is hardly applicable, since the atmosphere in the measuring chamber is fixed, and the electrons are detected by the ionization principle, usually in a CH_4/Ar gas mixture. Although compacted powders of supported catalysts have been studied to an extent, primarily sorption processes have been analyzed [25].

4.2.5 Technical Implementations – Experimental Conditions

In most cases, the conventional *transmission arrangement* is used. Namely, the Mössbauer source is mounted on the drive unit, which provides the Doppler modulation for the energy of the resonant nuclear transition. The γ-radiation passes through the sample where the various possible resonant absorption processes take place, the detector counts the pulses, and the count number vs. modulating energy spectrum develops after a sufficient collection time had elapsed. With this conventional arrangement, usually the catalyst precursors and the spent catalyst samples can be studied, since the locations of the catalytic process and the Mössbauer measurement are separated.

A significant improvement can be achieved by applying *in situ cells*, in which the treatments and Mössbauer measurements can be performed in the very same place (or within the same apparatus). In conventional catalytic processes, the reactions are performed at elevated temperatures. The elevated temperatures are unfavorable for Mössbauer measurements due to the strong inverse dependence of the effect on the temperature. In common practice, the catalytic reactions are carried out in an in situ cell at the higher activation/reaction temperatures, and then, the spectra are recorded after cooling. In the next step, the catalyst can be exposed to the subsequent treatment/reaction, and after cooling, the catalyst can be characterized again. This is the most common practice of monitoring the changes of catalysts in the sequence of various stages of catalytic processes in in situ cells. This is an advantageous possibility for studying supported catalysts, since the extent of attenuation (Al_2O_3, SiO_2, etc.) is different than that of higher atomic number catalyst substances. Several examples are presented for operando and in situ cell designs in [8, 10, 26–29].

Further important progress in the application of the method is the usage of *operando cells*, in which spectra are collected during the catalytic reaction, providing a means for studying the catalysts directly under the actual catalytic conditions. Thereby, the chance of capturing the catalytically active species can be increased significantly. However, there is a trade-off between the length of treatments and the simultaneous measurements. Namely, the catalytic processes are very short, and in contrast, at least several minutes are necessary to collect the corresponding spectrum. In addition, note that the increase in the temperature decreases the probability of the Mössbauer effect, and a few hundred centigrade increase in temperature practically deteriorates the probability (see Figure 4.3). Thus, the study of catalysts under steady conditions can be preferred in the in operando cells. The occasional aging of catalysts can be checked, namely, the spectra recorded before and after the treatments can be compared [30, 31].

An excellent demonstration of the successful usage of operando cells is their application in studying electrochemical processes. Their usage for studying the electrochemical reduction of oxygen in fuel cells was proposed already in the 2000s [32, 33]. Most recently, Fe and NiFe layered double oxyhydroxide catalysts were studied in water oxidation [34], and a review has also been published on the in situ/operando characterization of electrocatalysts [35].

As was shortly mentioned in the Introduction, the potentials of catalyst studies can significantly be enhanced by studying frozen solutions. Provided the freezing of a sample is sufficiently fast, the conditions of the reaction taking place in a solution can be preserved, and, for instance, even transient molecular complexes can be identified by the selection of appropriate reaction conditions. Enzymes and other biocatalysts are usually also studied in the frozen state.

4.3 Heterogeneous Catalysts

By definition, in heterogeneous catalysis, the catalyst and the reactants are in different phases. In most cases, the catalyst is in the solid state, and the reactants can be either in the liquid or, more frequently, in the gaseous phase. Further, if the catalyst is composed of certain precious metals, supported catalysts can be used to provide a high dispersion on cheaper neutral oxides (e.g. silica and alumina magnesia). The heterogeneous catalytic processes are of primary importance in the chemical industry.

4.3.1 Sites on Supported Particles with Different Participation in Catalytic Processes

It is worth examining some further details of the heterogeneous catalytic processes that take place on small supported particles. These small crystallites are connected to the support with an interface layer. On the surfaces of the crystalline particles, various crystal faces can be stabilized, and the densities of the atoms (i.e. the distances between the nearest neighbors) may be different depending on the orientation of the corresponding surface plane. In a simple approach, the process of catalytic importance proceeds, usually on the surface of the small crystallites. The surface atoms contribute to catalysis to different extents depending on their coordination. In most cases, the atoms that have a low coordination and that are located in the vertexes play the primary role, whereas the atoms embedded in the surface planes are usually less active in the processes. To obtain certain insight, the most interesting part of the catalytic processes is the study of the interaction of the reactants with the catalyst

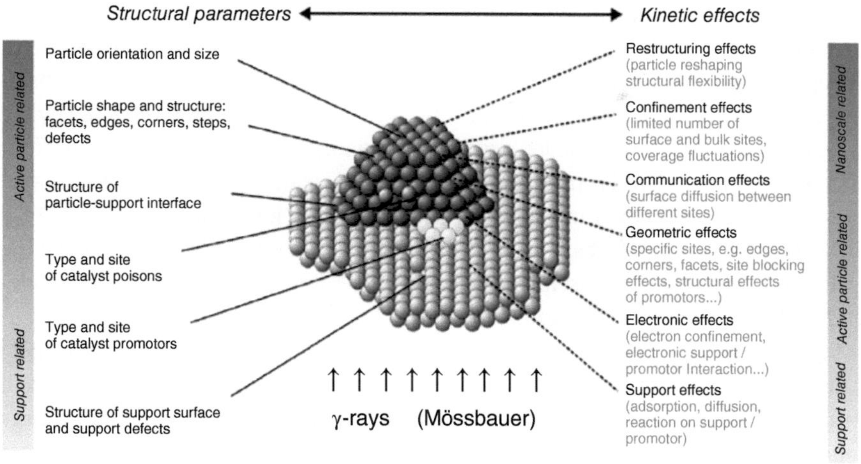

Figure 4.5 Illustration of the variety of surface sites on a supported catalyst particle irradiated with probing γ-rays for Mössbauer analysis. Source: Ref. [36]/with permission from Springer Nature.

to identify the substrate–catalyst transient state. The conditions can be illustrated in Figure 4.5.

From the point of view of the Mössbauer technique, the figure clearly demonstrates that the extracted information strongly depends on the size of particles. As was already mentioned in Section 4.1, the information (i.e. the recorded spectrum) originates from all atoms present in the particle, regardless of the extent of their participation in the catalytic process. Usually, only a smaller contribution to the spectrum originates from the important atoms located on the surface, whereas a larger part can be correlated to the inert atoms located in the bulk core of particles.

The percentage of active atoms in heterogeneous catalysts can be increased by increasing the dispersion (i.e. the fraction of atoms located on the surface of the particle). Particles with few nanometer sizes exhibit a high dispersion (>50%). The uppermost nominal limit (100%) can also be attained in particular cases. In these instances, the original properties of the bulk (e.g. metallic) character of the particular atoms are usually lost, since these single atomic centers are usually stabilized in certain covalent bonds (see single atom catalysis [37], e.g. in the case of iron in Fe-$(NC)_x$ ($x = 4$ or 5)). Atomic dispersion can be attained by other methods as well, e.g. in oxidic frameworks (zeolites) or in metal-organic frameworks (MOFs). In another method, the complex molecules used in homogeneous catalytic processes can be immobilized (~heterogenized) in zeolites (e.g. with "ship-in-the-bottle" syntheses in faujasite cages [38]).

4.3.2 Collective Effects in Particles (Magnetism)

Note that certain collective effects may emerge in small particles. Principally, these effects do not have relevance to catalytic processes; they are primarily connected to the size of the particles and may strongly influence the shape of the spectra, and thus, they also have to be considered. In practice, these effects mostly play a certain role in iron-containing substances. Distinctions can be made from two approaches. First, the separate, noninteracting single paramagnetic ions (namely, Fe^{3+}) may be considered. In their case, various hyperfine structures may appear, which is reflected by the shape obtained from line broadening to the appearance of a distinct sextet in their spectra. Paramagnetism is connected to the slow relaxation of the spins of isolated ions; thus, paramagnetism may appear, usually at very low temperatures (<77 K), while at higher temperatures, the effect diminishes. The effect also disappears if the density of Fe^{3+} ions is increased, and the single Fe^{3+} ions are not magnetically independent from each other anymore.

Magnetic effects can also be approached from the opposite direction with respect to the size, namely, moving from the large bulky particles to the dispersed small ones. In iron-containing large particles, spins can be collectively aligned in large domains; ferromagnetism may appear in bulky metallic particles, or antiferromagnetism may appear in oxides. This ordering is reflected by the split sextets in the corresponding spectra. By decreasing the size of particles, the regions in which the spins are collectively connected to each other also became smaller, the flip-flop of the principal direction of the magnetization also became more frequent, and the particles exhibit superparamagnetism. By a rough simplified approach, in the Mössbauer spectra, this effect is particularly crucial when the reciprocal frequency of the flip-flop of the direction of the average magnetization is comparable to the lifetime of the nuclear excited state (140 ns in the case of ^{57}Fe). The interpretation of the corresponding spectra is rather sophisticated; the exact description of these relaxation effects can be found in [39], for example. In a rough approach, the size range in which these effects may dominate the spectra of iron-containing particles is in the range of a few nanometers, which is the size that is very common for supported catalysts.

4.3.3 Case Studies

To illustrate the potentials of the method, some selected examples are presented from a broad choice of publications in the field of studying heterogeneous catalysts.

4.3.3.1 Metals and Alloys

Many industrial processes apply dispersed metallic particle catalysts. One of these important procedures is the conversion of coal to liquid fuels. Among these methods, a well-known procedure is the *Fischer–Tropsch process*, in which synthesis gas (a mixture of carbon monoxide and hydrogen) is converted to hydrocarbons and water by using iron-containing catalysts under pressure and an elevated temperature. Various in situ cells were already developed in the early days to study the process, in which the progress of the conversion of the metallic iron phase to different carbides was followed [26, 40]. Further studies have also been carried out due to the high industrial relevance, and the primary role can be attributed to ε-carbide, as recent operando studies attest [41]. The operando cell used for these studies is a dedicated construction, allowing for simultaneous IR spectroscopy in the FTIR mode at an uppermost ambient reactant pressure of 25 bar and a 450 °C temperature.

The metallic phase can be produced by the reduction of oxides. At the industrial scale, usually, fused oxide precursors (e.g. magnetite mixed with promoters) are activated by reduction in hydrogen. In a recent study, a mixed phase metallic iron-magnetite catalyst is formed at different reduction temperatures (350 °C, 450 °C, 550 °C, and 650 °C). In the following stage, χ Fe_5C_2 carbide develops in the Fischer–Tropsch reaction carried out at 240 °C (H_2/CO = 2 at 20 bar pressure) [42]. The compositions of the spent catalysts were also analyzed by Mössbauer spectroscopy (Figure 4.6). The catalyst activities and selectivities were followed with time on-stream measurements (Figure 4.7). This study is a good example to illustrate that, usually, a catalyst precursor is a mixed phase, and the final catalyst evolves during reaction conditions. The catalytic activity is strongly correlated with the presence of the χ Fe_5C_2 carbide phase. For demonstration, the change in the general activity and the evolution of the selectivities for various products (CH_4 and C_{5+} hydrocarbons) are shown in Figure 4.7.

FeCo alloys are also active in the FT process. To study these bimetallic catalysts, ^{57}Fe and ^{57}Co spectroscopy can be combined, and the composition changes can simultaneously be followed from the aspects of the two different metals [27].

Another example of ^{57}Co in situ emission studies can be the study of the existence of the presumed Co-Mo-S phase in the hydrotreating process used for sulfur removal; various Co_xS_y phases were successfully identified in the carbon-supported bimetallic catalysts [43]. NiW catalysts can also be used efficiently for removal of sulfur

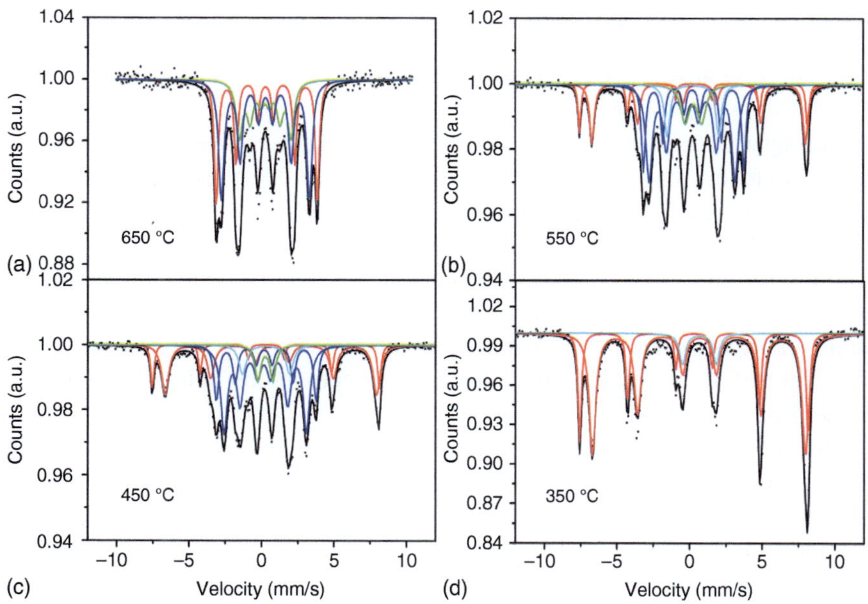

Figure 4.6 Spent fused magnetite Fischer–Tropsch catalysts following reduction in hydrogen at the temperatures indicated (650 °C (a), 550 °C (b), 450 °C (c) 350 °C (d)). Presence of magnetite is decreased by the increase in the temperature of reduction, and the conversion of metallic iron to catalytically active χ Fe_5C_2 can be deduced. Source: Reproduced from Ref. [42]/with permission from Royal Society of Chemistry.

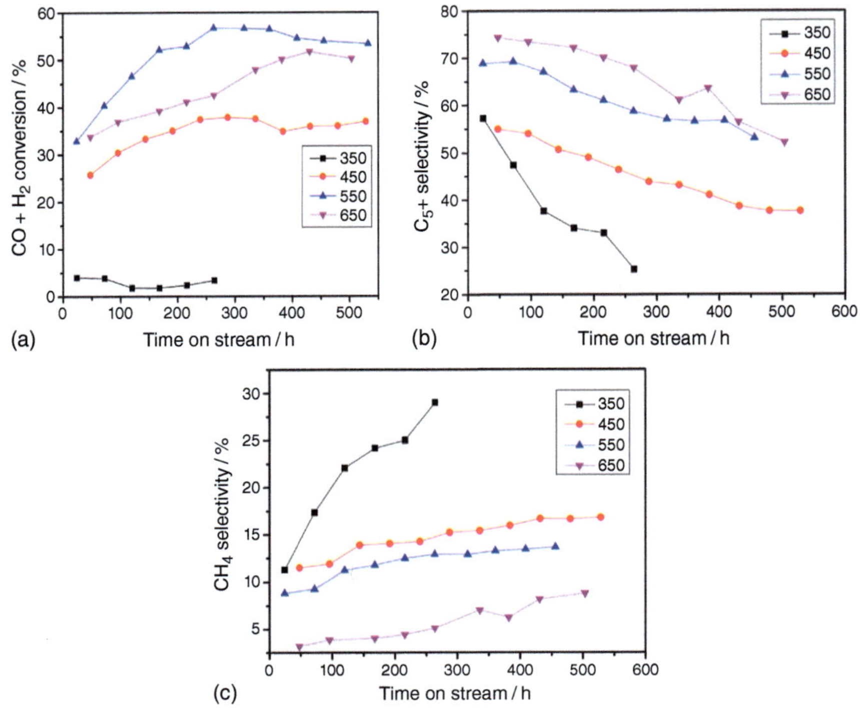

Figure 4.7 Activity (a) and selectivities for C_{5+} hydrocarbons and CH_4 (b) and (c), respectively, in the products formed on catalysts with reduction in hydrogen at the temperatures indicated prior the FT reaction. Source: Reproduced from Ref. [42]/with permission from Royal Society of Chemistry.

in analogy with the CoMo catalysts. ^{182}W Mössbauer spectroscopy was successfully applied to follow the formation of various tungsten sulfide phases under operating conditions in a dedicated high pressure in situ cell [44].

Iron can be used as a promoter added in small amounts to noble metals, e.g. Pd. For instance, almost a 100% selectivity can be attained for methanol formation from synthesis gas on silica-supported catalysts containing 2 wt% metal with a 1 : 10 Fe:Pd atomic ratio. The corresponding analysis of the Mössbauer spectra reveals that the excellent performance can be attributed to the incorporation of iron into the Pt, as in the $FePd_3$ intermetallic phase [45].

The other frequently used Mössbauer isotope tin is also an important modifier of noble metal catalysts. *Platinum-tin alloy* catalysts are used in various important processes, e.g. as reforming catalysts, to improve the octane number of gasoline, promoting dehydrogenation, isomerization, and dehydrocyclization. PtSn alloy formation can favorably be followed by Mössbauer spectroscopy, since the isomer shift of tin in these alloys characteristically depends on the composition [46]. Depending on the composition, the coverage with Pt and Sn on the different crystal faces may be significantly different, providing catalytic paths for different processes to simultaneously occur on the same particle [47]. Pt–Sn is also used to catalyze the oxidation of CO. The addition of tin to platinum in a c. 1 : 3 at/at ratio significantly increases the activity of silica-supported catalysts. Three types of alloys are identified, namely, Pt_ySn_x, PtSn, and Pt_ySn_z ($x < y < z$). A decrease in the tin content in Pt_ySn alloys and oxidation to Sn(IV) can be observed when CO is oxidized, whereas the opposite process, the reduction of Sn(IV) and partial incorporation of metallic tin into the alloy phase, is detected upon reduction at room temperature. Thus, the reversible process is Sn(IV) + Pt ⇔ Pt_ySn_x ⇔ Pt_ySn_z. It is rather impressive that reversible alloy formation can be demonstrated at as low as room temperature in the in situ cell. Further on, in a series of experiments, the bonding strengths of various Sn species can be estimated by comparing the spectral intensities of the 77 and 300 K spectra (by adopting the method proposed in [48]). The presence of a loosely bonded "Sn^{4+}-Pt" ensemble can be suggested on the surface of bimetallic particles by comparing the f_{A77K} and f_{A300K} values, and a significant role can be attributed to this ionic surface species in the activation of CO (Figure 4.8) [49].

Other noble metals, e.g. gold, have recently attracted enormous interest for catalytic processes. ^{197}Au is also a Mössbauer active nucleus, and catalyst studies can also be carried out with this isotope. The energy of the Mössbauer active transition is rather large, 77.3 keV. Thus, to perform measurements with gold, a low temperature (preferably 4.2 K with liquid helium) is necessary, and application of in situ cells has

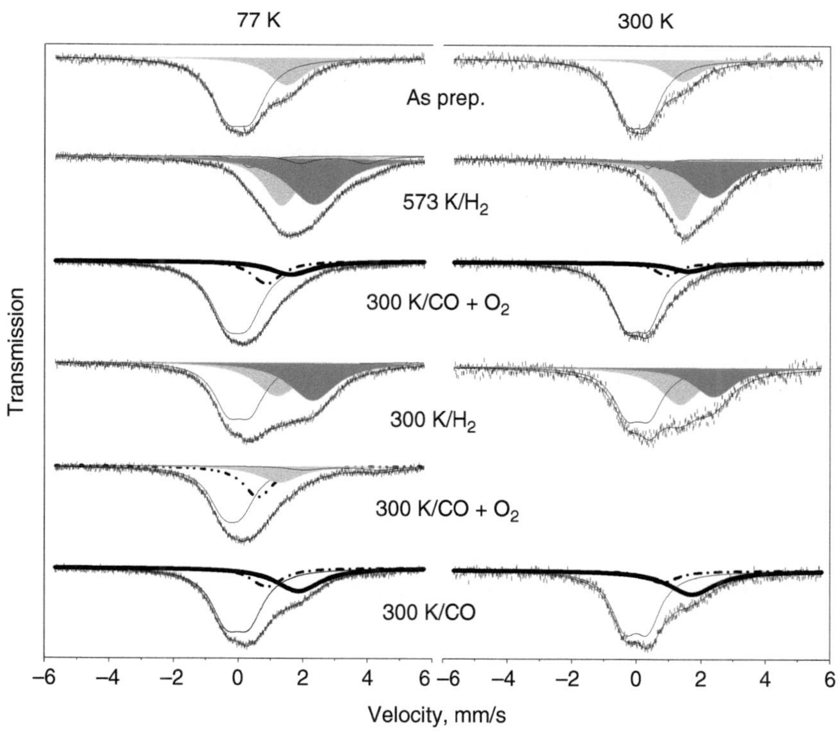

Figure 4.8 The Sn(IV) + Pt ⇔ Pt_ySn_x ⇔ Pt_ySn_z process: ($x < y < z$) reversible composition changes in silica-supported PtSn alloys at a 1 : 3 nominal Pt:Sn ratio for the treatments, as indicated. Sn(IV): thin solid line, Pt_ySn_x: light gray, PtSn: thick solid line, Pt_ySn_z: gray, Pt–Sn(IV)$_{surface}$: dot-dash. Source: Adapted from Ref. [49].

not been reported yet. An excellent overview on Mössbauer studies on gold catalysts is given in [50].

Iridium is also a frequently used catalyst material. Similar to gold, iridium also has a Mössbauer active isotope, ^{193}Ir, with a 73.0 keV transition. Correspondingly, iridium-containing Pt–Ir catalysts have also been investigated by Mössbauer spectroscopy [50].

4.3.3.2 Oxide Catalysts

In iron oxides, both Fe^{2+} and Fe^{3+} oxidation states are stable, and magnetic sextets may appear in the spectra due to antiferromagnetic interactions. Similar to metals, the addition of other modifying elements (e.g. Co and Ni) may influence the extents of the isomeric shift, magnetic field, etc. The $Fe^{3+} \Leftrightarrow Fe^{2+}$ redox process can be involved in catalytic reactions, and due to the difference in the ionic radii of Fe^{2+} and Fe^{3+}, redistribution of ions among octahedral and tetrahedral sites may take place. This redistribution can be reflected by the changes in the proportions of various magnetically split subspectra, requiring a sophisticated spectrum evaluation. Thus, to find a direct relationship between the obtained structural changes and the connected catalytic processes, certain precaution is needed during interpretation.

The $Fe^{2+} \Leftrightarrow Fe^{3+}$ conversions and redistribution among the sites of iron oxide are reported in several cases, e.g. in situ water gas shift reaction studies [51]. The applied oxide catalysts may interact with the reactants to greater extents; for instance, metallic iron and carbides may form from the reduction of a part of the iron oxide during the alkylation of phenol with methanol. The stability and activity of catalysts can be varied by adding different amounts of Co and Cu, for example, to the ferrospinel [52].

In other applications, an oxide is intentionally selected as a reaction partner, in particular, in chemical looping processes. For example, the mixture of Fe_2O_3-$MgAl_2O_4$ can preferably be used for the production of CO from CO_2, and the redistribution of iron in the solid phase was followed after 1000 redox cycles by Mössbauer spectroscopy [53].

Layered double hydroxides (LDHs) can also be used to catalyze several processes. Among these, nickel-iron oxides/hydroxides are also efficient catalysts for oxygen evolution reactions (OERs). An efficient electrocatalytic cell was designed with thin carbon paper electrodes (transparent for γ-rays) by depositing the active Ni:Fe (3 : 1) LDH onto one of the electrodes. Oxygen evolution reactions were carried out at various potentials; the presence of only Fe^{3+} was detected at the start and at the end of cycling the potential, and presence of additional Fe^{4+} was revealed in the operando spectra recorded at 1.49, 1.62, and 1.76 V potentials (Figure 4.9) [34].

Further examples of the Mössbauer study of oxide catalysts can also be found in a related review [29].

In the previous few examples, the catalytic applications of pure or bimetallic iron oxide phases were described, where the structure was constructed from extended –Fe–O–Fe–O–Fe–O– chains. The dispersion of iron can be increased by incorporating iron into the framework of another, preferably a porous oxide. An important type of these materials is zeolites with wide applications in catalytic processes,

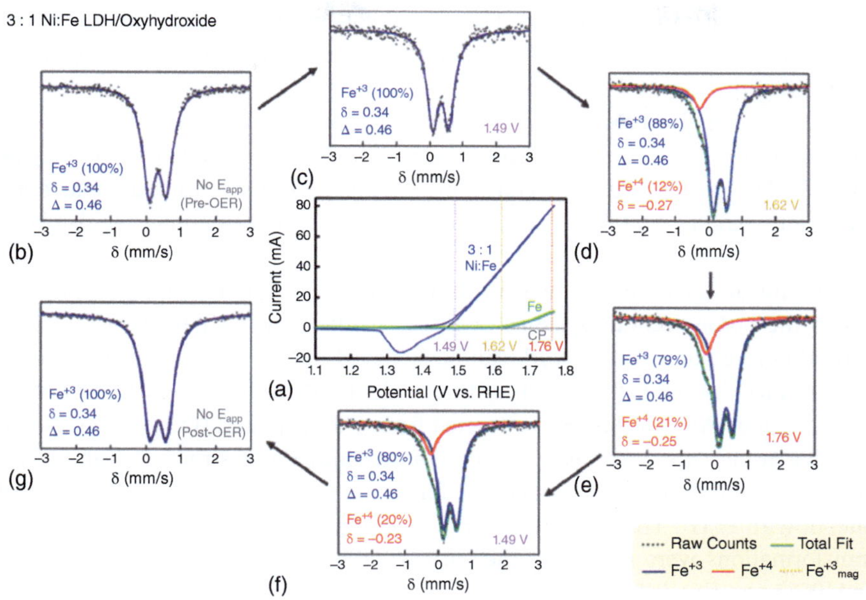

Figure 4.9 Circular voltammogram (CV) recorded on a 3 : 1 Ni:Fe layered oxyhydroxide in an operando electrocatalytic cell (a) in the center of figure. Mössbauer spectra collected in a CV cycle of oxygen evolution reactions (OERs) performed at various potentials; pre-OER: (b), 1.49 V: (c), 1.62 V: (d), 1.76 V: (e), 1.49 V: (f), and post OER: (g). Note the presence of Fe^{4+} species (in red). Source: Reproduced from Ref. [34]/with permission from American Chemical Society.

for example, in the reforming of gasoline. Their primary building units are Si- and Al-centered oxide tetrahedra. A broad variety of nanoporous crystalline structures can be constructed from this elementary unit, with pores ranging in the 0.5 to 3 nm size interval, which is same as the size range of reactant simple molecules. Iron and tin can be easily incorporated into this structure, either to tetrahedral sites, into the building units, or to other available positions in the pores, either as charge compensating extra framework ions or constituents in small clusters. The methodological burden encountered at particle studies, namely, that only a small part of the spectra is derived from the active centers located on the surface of particles, is circumvented by incorporating iron as separate ions to the framework. Theoretically, all possible centers are evenly seen. However, other conditions, such as the formation of various mixed hydrous and extra framework species, may hinder the obtention of an unambiguous interpretation. As an illustration, the case of so-called α-oxygen can be mentioned. Atomically activated oxygen can be disposed to iron ions by decomposing N_2O. This activated oxygen is able to oxidize methane to methanol, even at room temperature. There was a long debate in the literature whether these centers are dinuclear μ-oxo species or monoionic species and, further on, whether the active form of iron is Fe^{2+} or Fe^{3+}. The presence and stabilization of the Fe^{2+} component was clearly revealed under the reaction conditions of an in situ study [54]. In a related study, an operando spectrum was recorded under reaction conditions at 620 K in an N_2O atmosphere; existence of both the framework Fe^{3+} and extra framework

Fe^{2+} ions was demonstrated [55]. The partial collapse of the structure and redistribution of ions was found at 680 K, although a certain catalytic activity was still observed [56].

Sn^{4+} ions can also be successfully incorporated into zeolite structures. The substitution of tin is advantageous, since the $Sn^{4+} \Leftrightarrow Sn^{2+}$ reversible process may be connected to transferring an oxygen atom. The incorporation of tin is demonstrated for various zeolites, and their catalytic activity in oxidation reactions has also been proven [57].

Iron and tin ions are separated by a number of $[SiO_{4/2}]$ and $[AlO_{4/2}]^-$ tetrahedra in the porous zeolitic systems discussed in the previous paragraphs. Charge neutrality in a porous framework can be achieved in other arrangements as well. For instance, by the combination of three-valent Al and five-valent phosphorous ions, so-called AlPO-s can be constructed from $[AlO_{4/2}]^-$ and $[PO_{4/2}]^+$ tetrahedral building units. Iron can also be incorporated into these structures, and these FeAlPO-s exhibit an excellent activity, for example, during the hydroxylation of phenol with H_2O_2. The flexibility of the structure and the enhanced $Fe^{3+} \Leftrightarrow Fe^{2+}$ transformations were proven in the corresponding in situ Mössbauer spectra of FeAlPO-5 and FeAlPO-11 [58].

Nonporous iron phosphates can also exert a catalytic activity, since both iron and phosphorous are able to easily change their valence states. In a corresponding study, the direct oxidation of methane to methanol was performed on $FePO_4$, and the catalytic process was accompanied with the reduction of the iron and phase change and the formation of $Fe_2P_2O_7$ [59].

In all the previous examples, catalytic processes with thermal activations were considered. Another activation may also be utilized, for example, the energy of light in *photocatalytic processes*. Titania-based semiconductors are widely used with transition metal or nonmetal sensitizer dopants. The presence of vacancies and local environments of dopants significantly influence and improve the performance of the anatase form of TiO_2, as was observed with iron [60], tin [61], and europium [62] promoters.

In the previous paragraphs, the catalytic activities in different processes were observed on crystalline oxides. Curiously, *noncrystalline forms of oxides* can also exert a photocatalytic activity, as was demonstrated in the photocatalytic decomposition of methylene blue dye with soda-lime glass [63].

In the preceding paragraphs, the catalytic properties of inorganic oxides were discussed. A great variety of porous substances exists, forming an intermediate organic-inorganic intermediate type of oxide, namely, *metal-organic frameworks* (MOFs) with flexible regular structures, in which the primary $[MO_6$, M = metal$]$ octahedra are connected with chains containing carbon as well. In particular, in the case of iron, the $[FeO_6]$ octahedra can be linked with trimesic acid (benzene-1,3,5-tricarboxylate) to form the primary building $[FeO_6]$-trimesic acid trimers. Either Fe^{2+} or Fe^{3+} can be incorporated into the centers of the $[FeO_6]$ octahedra; in the former case, MIL-45(Fe) is formed, and in the latter case, MIL-100(Fe) is formed [64]. Fluorine-containing $Fe^{III}{}_3O(H_2O)_2F \bullet [C_6H_3(CO_2)_3]_2 \bullet nH_2O$

MIL-100(Fe) was also synthesized, and the corresponding Mössbauer spectra reveal the presence of three different possible environments of Fe^{III}. This porous MOF can be successfully applied as a selective catalyst in Friedel–Crafts benzylation to form diphenylmethane from benzyl chloride and benzene at 70 °C [65].

Metal-organic frameworks can also be used as catalyst precursors. For instance, $Fe_xCo_{3-x}O_4$ catalysts can be prepared by the combustion of the corresponding Prussian blue analogue, $Fe_yCo_{1-y}[Co(CN)_6]_{0.67}{\bullet}nH_2O$. This catalyst can advantageously be used in peroxymonosulfate activation to form $SO_4{}^{\bullet-}$ radicals, which are very efficient in bisphenol removal, for example. The related Mössbauer analysis demonstrates that Fe^{III} ions may occupy both the available octahedral and tetrahedral positions in the spinel structure in the most active $x = 0.8$ catalyst [66].

4.3.3.3 Catalysts with Fe–N, Fe–C, and Fe–N–C Centers

The activity of several types of catalysts is connected to the Fe–C–N or Fe–N–C centers, and a few examples are presented as follows.

Prussian blue ($Fe^{III}{}_4[Fe^{II}(CN)_6]_3$) can be deposited on TiO_2 (anatase), and the formed nanocomposite is an excellent photo-Fenton catalyst. The generation of •OH radicals is attributed to the reversible low spin $Fe^{II} \Leftrightarrow$ high-spin Fe^{III} cycling in the $Fe^{II}Fe^{III}(CN)_6$ core [67]. A Fenton-like catalytic activity can also be generated on bimetallic Fe–Co Prussian blue complex molecules, even without supporting on titania. Depending on the synthesis temperature, the shape of particles varies from spheres (0 °C) to truncated cubes (55 °C) and regular cubes (85 °C). The catalytic activity for the removal of bisphenol is strongly related to the morphology; the best activity was observed on particles with cube shapes. The influence of the Fe^{II}/Fe^{III} ratio on the particle morphology was followed by Mössbauer spectroscopy [68].

Single atom catalysts have attracted distinguished interest. Their study with Mössbauer spectroscopy is particularly advantageous over studies on bulk or supported particles, since the variety of the different positions on the surface and in the bulk of particles is *ab ovo* circumvented. One possibility is the immobilization of homogeneous catalysts by encapsulating them into zeolite cages, for example. For instance, an encaged iron phthalocyanine complex can be synthesized inside the pores of a Y zeolite. First, Fe^{2+} ions should be introduced, and then, iron phthalocyanine can be formed with the stepwise addition of four phthalonitrile molecules. The immobilized phthalocyanine can be used in the oxidation of hydroquinone to benzoquinone with hydrogen peroxide in acetic media, for example, and the formation of the μ-oxo-Fe(II)(Pc)(pyridine) complex is suggested by analyzing the Mössbauer spectra collected from the frozen reaction slurry [69].

The careful thermal decomposition of complex molecules with FeN_4 cores may also lead to the formation of Fe–N–C single atomic catalytic centers in heterogeneous catalysts. One possibility is the usage of the "template-sacrificial approach." In a particular procedure, an iron phthalocyanine complex is adsorbed to magnesia, then the complex is pyrolyzed in nitrogen, and the MgO template is removed subsequently with acid leaching by dissolution in HNO_3. The product contains dominantly single Fe–N–C centers. These catalysts exhibit an excellent

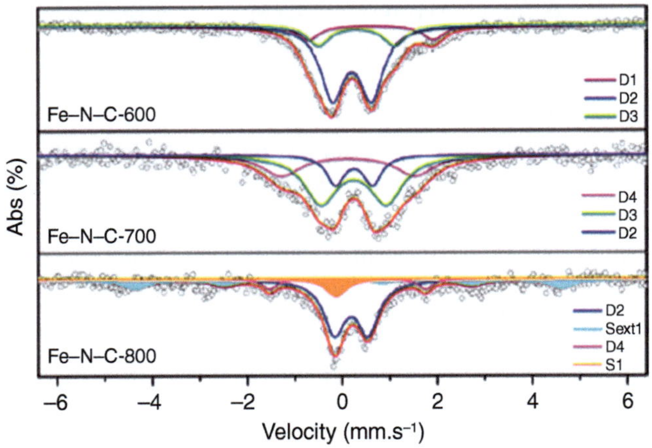

Figure 4.10 Mössbauer spectra of Fe–N–C single atom catalysts prepared by the thermal decomposition of iron phthalocyanine deposited on MgO at the temperatures indicated (°C) and the subsequent dissolution of the support. Source: Reproduced from Ref. [70]/with permission from American Chemical Society.

activity; for instance, the C—H bonds adjacent to the terminal carbon atoms can be selectively oxidized with these catalysts at room temperature. Mössbauer spectra were obtained on the Fe–N–C centers prepared by pyrolyzing the deposited complex at various temperatures (600, 700, and 800 °C). In addition to the low-spin phthalocyanine-like $Fe^{II}N_4$ (D1 in Figure 4.10) and N-($Fe^{III}N_4$)-N and high-spin X-($Fe^{III}N_4$)-Y (X,Y:O,N) with saturated coordinations (D2 and D3 in Figure 4.10), medium-spin N-($Fe^{III}N_4$) centers with unsaturated coordinations with the largest quadrupole doublet (D4) were identified. The predominant catalytic activity was attributed to the latter pentacoordinated iron center, displaying a significant contribution in the spectrum of the sample pyrolyzed at 700 °C (Figure 4.10).

Active Fe–N–C centers can be prepared with other methods as well. For instance, high energy ball milling of iron phthalocyanine and graphene nanosheets results also in a composite substance that contains coordinatively unsaturated single FeN_4 centers embedded into graphene sheets. These catalysts are active in low temperature oxidation, namely, in the oxidation of benzene to phenol at room temperature. Two characteristic components were identified in the corresponding Mössbauer spectra obtained for the pristine catalyst, which was then oxidized with H_2O_2 and exposed to benzene afterwards. A doublet was attributed to the unsaturated FeN_4 centers in graphene, with parameters similar to those of Fe^{II}-phthalocyanine, and a singlet with an isomeric shift of −0.11 mm/s was assigned to the oxidized FeN_4 center containing two attached apical oxygen atoms as well [71].

Fe–N–C catalysts are also widely used as electrocatalysts. A recent review gives an exhaustive account of their preparation and characterization with Mössbauer spectroscopy, and applications primarily include ORRs and hydrogen and OERs in CO_2 reduction reaction [72].

4.4 Biocatalysts – Enzymes

The monomolecular dispersion is assured *ab ovo* in another important class of catalysts, namely, enzymes. In enzymes, the catalytic center usually contains an active transition metal complex embedded in a thick cover made of proteins.

In living organisms, *enzymes, protein-based biocatalysts*, play an essential role in sustaining life, and they catalyze more than 5000 biochemical reactions. Many of these processes are also connected to centers containing iron. The study of enzymes has already commenced in the dawn of Mössbauer spectroscopy [73]. A report on horseradish peroxidase can be mentioned as an example, in which its activity was connected to a Fe(IV) spin-coupled oxoferryl porphyrin π-cation complex formed with a histidine coordinated to the iron, as the corresponding Mössbauer study confirmed [74]. Further details of the enzyme-catalyzed oxidation reactions, such as the activation of O_2 by heme-containing monooxygenases, were investigated later. A related study of reactive oxygen intermediates confirmed that the one-electron reduction of (Fe^{II}-O_2)HO is accompanied by the protonation of the bound O_2 to generate a low-spin (Fe^{III}-O_2H^-)HO that undergoes self-hydroxylation to form the α-*meso*-hydroxyhemin-HO product (HO = heme oxygenase). Mössbauer data reveal that ferric-hydroperoxo hemes are indeed the precursors of the reactive ferryl intermediates [75]. In the previous examples, the central iron was located in the four(five)fold nitrogen coordinated heme centers. Nonheme oxygenase enzymes, such as the aldehyde-deformylating oxygenase (ADO) that converts fatty acids to alkanes in cyanobacteria with [2Fe-2S] centers, may also play a primary role. The stages of the oxidation of dinuclear $Fe_2^{II/II}$-ADO were followed, and the presence of $Fe_2^{III/III}$-peroxyhemiacetal intermediates were confirmed in the freeze-quenched ^{57}Fe spectra [76]. The terminal oxygenase, putidamonooxin, also contains three identical clustered [2Fe-2S] subunits complete with one nonheme mononuclear (cofactor) iron complex. The cofactor can also be studied by Mössbauer spectroscopy. A corresponding early study reports that ferric iron is in the $S = 3/2$ intermediate spin state in a related ternary enzyme-substrate-NO complex, and the bonding of NO to the reduced cofactor is connected to the release of at least one other ligand [77]. An excellent recent review on Mössbauer spectroscopy performed on Fe/S proteins and various cofactors with [2Fe-2S], [3Fe-4S], and [4Fe-4S] centers is given in [78]. Synchrotron-based studies on these [4Fe4S] cofactors have also been reported recently [23].

4.5 Homogeneous Catalysts – Frozen Solutions

Polynuclear Fe-N centers can also be studied. For example, an industrially very important process, the Haber–Bosch synthesis of ammonia from N_2 and H_2 performed at a c. 200 bar pressure and between 400 and 500 °C, can be completed via an intermediate complex with a Fe_3N_2 core under ambient conditions in a tetrahydrofuran solution. A dinuclear iron complex with a Fe(μ-Cl) bond can be formed with two $MeC[C(Me)N(2,6\text{-}Me_2C_6H_3)]_2^-$, where iron ions are coordinatively linked to the four nitrogen atoms in the center. Upon adding a graphite-potassium reductant (KC_8), the original triple bond of the N_2 molecule can be replaced by Fe—N bonds in a Fe_3N_2 core formed with a third complex molecule. The reaction intermediate complex can be decomposed in the solution with hydrogen with the simultaneous formation of ammonia. In the corresponding Mössbauer spectrum of the intermediate complex, various iron sites were identified in accordance with the XRD analysis [79].

Fe-N and Fe-C catalytic centers and intermediates can be studied in *operando (quenched)* conditions *in frozen reaction mixtures*, as the following few selected examples demonstrate.

Iron complexes of tetraamido macrocyclic ligands (TAML) are excellent oxidation catalysts. The central Fe^{III} is surrounded by four nitrogen atoms in a planar structure that is open for further ligands above and below this plane. In the oxidation processes, transient high-spin ($S = 2$) and intermediate-spin ($S = 1$) Fe(IV)-TAML complexes may be formed, as described in a comprehensive review [80].

Iron centers with completely N-filled sixfold coordination spheres can also exhibit a catalytic activity. Curiously, the long-term stability of the Fe(IV) oxidation state can be observed in an iron hexahydrazide chlatrochelate under ambient conditions in aqueous media [81]. The complex can be used as a catalyst in the visible light-driven oxidation of water. The formation of intermediate Fe(V) species is proven in the process, as the 80 K Mössbauer spectra attest [82].

In another study, iron(II) catalysts were utilized for enantioselective C–H activation. Namely, iron complexes with N-heterocyclic carbene ligands (Fe/NHC) were applied in C–H alkylation. The catalytic center is formed with the appropriate Grignard reagent (N,N,N′,N′-tetramethylethane-1,2-diamine, TMEDA), and the scheme of the reaction is shown on the left of Figure 4.11. Five doublets can be distinguished in the corresponding 80 K ^{57}Fe Mössbauer spectra of the frozen reaction mixture (Figure 4.11b). The catalytic activity is attributed to a low-coordinate high-spin L_2Fe^{II}(NHC) complex with a low isomer shift and high quadrupole splitting (0.39 mm/s and 3.19 mm/s, respectively, in which the outermost doublet component is shown in green in the spectrum in Figure 4.11) [83].

Iron catalysts without nitrogen ligands are also widely used in cross-coupling reactions with Grignard reagents. In the formed intermediate, heteroleptic $R_3FeR'^-$ complexation with iron may occur in various oxidation states ranging from I to IV, depending on the size and structure of the R and R′ groups, as can be deduced from the 80 K Mössbauer spectra of frozen reaction mixtures [84].

4.5.1 Studies on Reaction Intermediates – Time-Resolved Freeze-Quenched Spectra

The extent of the obtained information on reaction intermediates can be significantly increased by recording *time-resolved freeze-quenched spectra*. Iron-catalyzed cross-coupling reactions were studied in detail by using $FeCl_2$(3,5-*t*Bu_2-SciOPP)

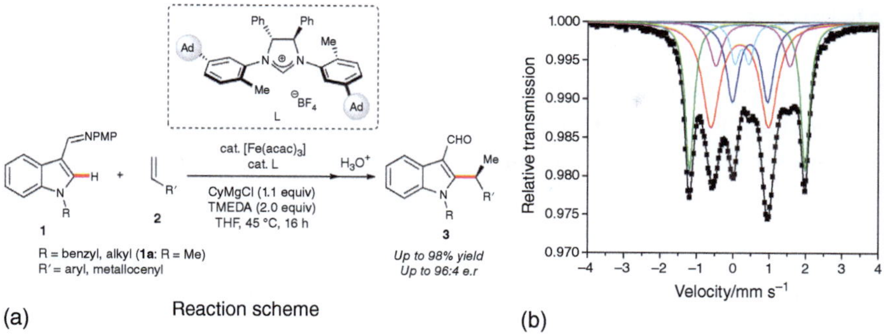

Figure 4.11 Enantioselective Grignard reaction with the Fe(II) catalyst complex corresponding to the scheme shown in (a). The catalyst complex can be identified as the outermost doublet in the 80°K spectrum of the frozen reaction mixture is shown in (b). Source: Reproduced from Ref. [83]/with permission from Royal Society of Chemistry.

(SciOPP = spin-control-intended *ortho*-phenylene bisphosphine). The combination of a $FeBr_2$(SciOPP) precatalyst and bulky alkynyl Grignard reagents, such as (triisopropylsilyl)ethylmagnesium bromide (TIPS-CC-MgBr), results in the formation of alkynylated cyclic and primary alkanes in high yields. The reaction path and stability of intermediates depend strongly on the used solvents. Various iron-containing intermediates can be identified in the spectra freeze quenched in different stages of the reaction. The exact assignment of the intermediates was also supported by single crystal X-ray studies [85].

4.6 Conclusions

Aspects of studies of catalysts performed by means of Mössbauer spectroscopy have been discussed above. Principles of the method that are important to consider for catalyst studies are discussed in certain detail. It has been demonstrated that the method can be applied to study various types of catalysts. The information obtained can even be unique if the conditions are selected upon appropriate circumspection. A variety of possibilities for the applications of the technique is illustrated in certain selected examples. For broader interpretations of the relevance of the catalytic processes, considering the results of other complementary methods is also essential.

Acknowledgment

This overview is based primarily on a series of seminars held at the Dalian Institute of Chemical Physics, China, in 2018 and 2019 during a PIFI visiting professorship awarded by the Chinese Academy of Sciences. The author is grateful to the CAS for providing this visiting professorship.

References

1 https://apps.webofknowledge.com (accessed 6 July 2020).
2 Gruverman, I.J. and Seidel, C.W. (ed.) (1976) Study of Catalysts, section in). *Mössbauer Effect Methodology*, vol. 10, 1–118. Springer Science+Business Media.
3 Dumesic, J.A. and Topsøe, H. (1977). Mössbauer specroscopy applications to heterogeneous catalysis. *Adv. Catal.* 26: 121–246.
4 Topsøe, H., Dumesic, J.A., and Mørup, S. (1980). Catalysis and surface science. In: *Applications of Mössbauer Spectroscopy* (ed. R.L. Cohen), 55–183. Academic Press.
5 Berry, F.J. (1984). Mössbauer spectroscopy in heterogeneous catalysis. In: *Mössbauer Spectroscopy Applied to Inorganic Chemistry*, vol. 1 (ed. G.J. Long), 391–442. Plenum Press.
6 Bussière, P.B. (1994). Mössbauer spectroscopy – nuclear gamma resonance. In: *Catalyst Characterization* (ed. B. Imelik and J.C. Vedrine), 215–246. Plenum Press.
7 Niemantsverdriet, J.W. (2007). Mössbauer spectroscopy. In: *Spectroscopy in Catalysis*, 121–146. Wiley-VCH.
8 Millet, J.-M.M. (2007). Mössbauer spectroscopy in heterogeneous catalysis. *Adv. Catal.* 51: 309–350.
9 Stievano, L. and Wagner, F.E. (2014). Mössbauer spectroscopy. In: *Characterization of Solid Materials and Heterogeneous Catalysts* (ed. M. Che and J.C. Vedrine), 407–451. Wiley-VCH.
10 Liu, K., Rykov, A.I., Wang, J., and Zhang, T. (2015). Recent advances in the application of Mössbauer spectroscopy in heterogeneous catalysis. *Adv. Catal.* 58: 1–142.
11 Li, X., Zhu, K., Pang, J. et al. (2018). Unique role of Mössbauer spectroscopy in assessing structural features of heterogeneous catalysts. *Appl. Catal. B Environ.* 224: 518–532.
12 Stevens, J.G., Goforth, M.A., Newman, P.C., and Fu, Y.-L. (ed.) (1986). *Catalysts Handbook*. Mössbauer Effect Data Center.
13 https://en.wikipedia.org/wiki/M%C3%B6ssbauer_spectroscopy.
14 Greenwood, N.N. and Gibb, T.C. (1971). *Mössbauer Spectroscopy*. London: Chapman & Hall Ltd.
15 Murad, E. and Cashion, J. (2004). *Mössbauer Spectroscopy of Environmental Materials and Their Industrial Utilization*. Springer Science+Business Media.
16 Gütlich, P., Bill, E., and Trautwein, A.X. (2011). *Mössbauer Spectroscopy and Transition Metal Chemistry*. Springer.
17 Long, G.J. (1984). Basic concepts of Mössbauer spectroscopy. In: *Mössbauer Spectroscopy Applied to Inorganic Chemistry*, vol. 1 (ed. G.J. Long), 7–26. Springer.
18 Fultz, B. (2011). Mössbauer spectrometry. In: *Characterization of Materials* (ed. E. Kaufmann). Wiley.
19 mossbauer.info (accessed 18 July 2020).
20 Wang, J., Jin, C.J., Liu, X. et al. (2012). Mössbauer spectroscopy database: past, present, future. *Hyperfine Interact.* 204: 111–117.

21 Stievano, L., Santucci, S., Lozzi, L. et al. (1998). ^{197}Au Mössbauer study of gold particles obtained by evaporation of metallic gold on Mylar. *J. Non-Cryst. Solids* 232–234: 644–649.

22 Varnell, J.A., Tse, E.C.M., Schulz, C.E. et al. (2016). Identification of carbon encapsulated iron nanoparticles as active species in non-precious metal oxygen reduction catalysts. *Nat. Commun.* 7: 12582.

23 Mebs, S., Haumann, M., Stripp, S.T. et al. (2019). Differential protonation at the catalytic six-iron cofactor of [FeFe]-hydrogenases revealed by ^{57}Fe nuclear resonance X-ray scattering and quantum mechanics/molecular mechanics analyses. *Inorg. Chem.* 58: 4000–4013.

24 Gee, L., Lin, C.-Y., Lerche, M. et al. (2016). Synchrotron-based nickel Mössbauer spectroscopy. *Inorg. Chem.* 55: 6866–6872.

25 Gurlo, A. and Riedel, R. (2007). *In situ* and *operando* spectroscopy for assessing mechanisms of gas sensing. *Angew. Chem. Int. Ed.* 46: 3826–3848.

26 Shen, W.M., Dumesic, J.A., and Hill, G.C. (1981). Mössbauer spectroscopy cell for *in situ* catalyst characterization and reaction kinetic studies at high pressures. *Rev. Sci. Instrum.* 52: 858–862.

27 Topsøe, H. (2003). Developments in operando studies and *in situ* characterization of heterogeneous catalysts. *J. Catal.* 216: 155–164.

28 Crajé, M.W.J., van der Kraan, A.M., van de Loosdrecht, J. et al. (2002). The application of Mössbauer emission spectroscopy to industrial cobalt based Fischer-Tropsch catalysts. *Cat. Today* 71: 369–379.

29 Bibicu, I. and Frunza, L. (2006). Characterization of oxide catalysts by *in situ* Mössbauer spectroscopy. *Romanian Rep. Phys.* 58: 427–453.

30 Wezendonk, T.A., Nasalevich, M.A., Warringa, Q.S.E. et al. (2016). Elucidating the nature of Fe species during pyrolysis of the Fe-BTC MOF into highly active and stable Fischer-Tropsch catalysts. *ACS Catal.* 6: 3236–3247.

31 Tsakoumis, N.E., York, A.P.E., Chen, D., and Rønning, M. (2015). Catalyst characterisation techniques and reaction cells operating at realistic conditions; towards acquisition of kinetically relevant information. *Catal. Sci. Technol* 5: 4859–4883.

32 Bouwkamp-Wijnoltz, A.L., Visscher, W., van Veen, J.A.R. et al. (2002). On active-site heterogeneity in pyrolyzed carbon-supported iron porphyrin catalysts for the electrochemical reduction of oxygen: an *in situ* Mössbauer study. *J. Phys. Chem. B* 106: 12993–13001.

33 Greaves, T.L., Cashion, J.D., Benci, A.L. et al. (2002). Multiple Mössbauer spectra data acquisition with an *in situ* cyclic voltammetry system. *Hyperfine Interact. C* 5: 21–24.

34 Chen, J.Y.C., Dang, L., Liang, H. et al. (2015). Operando analysis of NiFe and Fe oxyhydroxide electrocatalysts for water oxidation: detection of Fe^{4+} by Mössbauer spectroscopy. *J. Am. Chem. Soc.* 137: 15090–15093.

35 Li, X., Wang, H.-Y., Yang, H. et al. (2018). *In situ/operando* characterization techniques to probe the electrochemical reactions for energy conversion. *Small Methods* 2: 1700395.

36 Libuda, J. and Freund, H.-J. (2005). Molecular beam experiments on model catalysts. *Surf. Sci. Rep.* 57: 157–298.

37 Wang, A., Li, J., and Zhang, T. (2018). Heterogeneous single-atom catalysis. *Nat. Rev. Chem.* 2: 65–81.

38 Parton, R.F., Utterhoeven, L., and Jacobs, P.A. (1991). Iron phthalocyanine encaged in zeolite Y and VPI-5 molecular sieves as catalysts for the oxyfunctionalization of n-alkanes. *Stud. Surf. Sci. Catal.* 59: 395–403.

39 Blume, M. and Tjon, J.A. (1968). Mössbauer spectra in a fluctuating environment. *Phys. Rev.* 165: 446–456.

40 Niemantsverdriet, J.W., van der Kraan, A.M., van Dijk, W.L., and van der Baan, H.S. (1980). Behavior of metallic iron catalysts during Fischer-Tropsch synthesis studied with Mössbauer spectroscopy, X-ray diffraction, carbon content determination, and reaction kinetic measurements. *J. Phys. Chem.* 84: 3363–3370.

41 Wang, P., Chen, W., Chiang, F.-K. et al. (2018). Synthesis of stable and low-CO_2 selective ε-iron carbide Fischer-Tropsch catalysts. *Sci. Adv.* 4: eaau2947.

42 Zhang, J., Abbas, M., and Chen, J. (2017). The evolution of Fe phases of a fused iron catalyst during reduction and Fischer-Tropsch synthesis. *Catal. Sci. Technol.* 7: 3626–3636.

43 Crajé, M.W.J., Gerkema, E., de Beer, V.H.J., and van der Kraan, A.M. (1990). Stability of the so-called "Co-Mo-S" phase in a carbon-supported Co-Mo sulfide catalyst at very low Co/Mo ratio. *Hyperfine Interact.* 57: 1795–1799.

44 Dugulan, A.I., Hensen, E.J.M., and van Veen, J.A.R. (2010). Effect of pressure on the sulfidation behavior of NiW catalysts: a ^{182}W Mössbauer spectroscopy study. *Catal. Today* 150: 224–230.

45 Lázár, K. (2012). Study of catalysts by Mössbauer spectroscopy. *Mössbauer Eff. Ref. Data J.* 35: 135–143.

46 Olivier-Fourcade, J. (2004). Investigation of redox properties of different $PtSn/Al_2O_3$ catalysts. *ChemPhysChem.* 5: 1734.

47 Kappenstein, C., Gue'rin, M., Lazar, K. et al. (1998). Characterization and activity in n-hexane rearrangement reactions of metallic phases on $Pt\text{-}Sn/Al_2O_3$ catalysts of different preparation. *J. Chem. Soc. Faraday Trans.* 94: 2463–2473.

48 Parish, R.V. (1984). Structure and bonding in tin compounds. In: *Mössbauer Spectroscopy Applied to inorganic Chemistry* (ed. G.J. Long), 527–575. Plenum Press.

49 Margitfalvi, J.L., Borbáth, I., Hegedűs, M. et al. (2002). Low temperature oxidation of CO over tin-modified Pt/SiO_2 catalysts. *Catal. Today* 73: 343–353.

50 Stievano, L. and Wagner, F.E. (2005). Investigation of catalytic materials using ^{197}Au Mössbauer spectroscopy. In: *Industrial Applications of the Mössbauer Effect*, AIP Conf. Proc., vol. 765 (ed. M. Gracia, J.F. Marco, and F. Plazaola), 3–12.

51 Cherkezova-Zheleva, Z. and Mitov, I. (2010). *In situ* Mössbauer investigation of iron oxide catalyst in water gas shift reaction – impact of oxyreduction potential and temperature. *J. Phys. Conf. Ser.* 217: 012044.

52 Lázár, K., Mathew, T., Koppány, Z. et al. (2002). $Cu_{1-x}Co_xFe_2O_4$ ferrospinels in alkylation: structural changes upon reaction. *Phys. Chem. Chem. Phys.* 4: 3530–3536.

53 Buelens, L.C., Van Alboom, A., Poelman, H. et al. (2019). Fe_2O_3-$MgAl_2O_4$ for CO production from CO_2: Mössbauer spectroscopy and *in situ* X-ray diffraction. *ACS Sustain. Chem. Eng.* 7: 9553–9565.

54 Dubkov, K.A., Ovanesyan, N.S., Shteinman, A.A. et al. (2002). Evolution of iron states and formation of α-sites upon activation of FeZSM-5 zeolites. *J. Catal.* 207: 341–352.

55 Lázár, K., Pozdnyakova, O., Wootsch, A. et al. (2006). Iron ion sin ZSM-5 zeolite: Fe^{3+} in framework, Fe^{2+} in extra-framework positions in catalytic N_2O decomposition. *Hyperfine Interact.* 167: 779–784.

56 Guczi, L. and Lázár, K. (2009). Mössbauer spectroscopy in heterogeneous catalysis: a brief overview and recent study on paramagnetic features detected in catalytic centres of Fe-ZSM-5. *React. Kinet. Catal. Lett.* 96: 335–343.

57 Lázár, K., Szeleczky, A.M., Mal, N.K., and Ramaswamy, A.V. (1997). *In situ* ^{119}Sn-Mössbauer spectroscopic study on MFI, MEL and MTW tin silicalites. *Zeolites* 19: 123–127.

58 Lázár, K. and Cejka, J. (1999). Valency and coordination of iron in FeAlPO molecular sieves: an *in situ* Mössbauer study. *Stud. Surf. Sci. Catal.* 125: 213–220.

59 Dasireddy, V.D.B., Khan, F.B., Hanzel, D. et al. (2017). Application of Mössbauer spectroscopy in industrial heterogeneous catalysis: effect of oxidant on $FePO_4$ material phase transformations in direct methanol synthesis. *Hyperfine Interact.* 238: 29.

60 Bajnóczi, É.G., Balazs, N., Mogyorosi, K. et al. (2011). The influence of the local structure of Fe(III) on the photocatalytic activity of doped TiO_2 photocatalysts – an EXAFS, XPS and Mössbauer spectroscopic study. *Appl. Catal. B Environ.* 103: 232–239.

61 Ganeshraja, A.S., Nomura, K., and Wang, J. (2016). ^{119}Sn Mössbauer studies on ferromagnetic and photocatalytic Sn-TiO_2 nanocrystals. *Hyperfine Interact.* 237: 139.

62 Diamandescu, L., Vasiliu, F., Tarabasanu-Mihaila, D. et al. (2008). Structural and photocatalytic properties of iron- and europium-doped TiO_2 nanoparticles obtained under hydrothermal conditions. *Mater. Chem. Phys.* 112: 146–153.

63 Ali, A.S., Nomura, K., Homonnay, Z. et al. (2019). The relationship between local structure and photo-Fenton catalytic ability of glasses and glass-ceramics prepared from Japanese slag. *J. Radioanal. Nucl. Chem.* 322: 751–761.

64 Birsa Čelič, T., Rangus, M., Lázár, K. et al. (2012). Spectroscopic evidence for the structure directing role of the solvent in the synthesis of two iron carboxylates. *Angew. Chem. Int. Ed.* 51: 12490–12494.

65 Horcajada, P., Surblé, S., Serre, C. et al. (2007). Synthesis and catalytic properties of MIL-100(Fe), an iron(III) carboxylate with large pores. *Chem. Commun.* 2007: 2820–2822.

66 Li, X., Wang, Z., Zhang, B. et al. (2016). $Fe_xCo_{3-x}O_4$ nanocages derived from nanoscale metal-organic frameworks for removal of bisphenol A by activation of peroxymonosulfate. *Appl. Catal. B Environ.* 181: 788–799.

67 Li, X., Wang, J., Rykov, A. et al. (2015). Prussian blue/TiO_2 nanocomposites as a heterogeneous photo-Fenton catalyst for degradation of organic pollutants in water. *Catal. Sci. Technol.* 5: 504–514.

68 Liu, J., Li, X., Zhao, C. et al. (2018). Shape-controlled synthesis of metal-organic frameworks with adjustable Fenton-like catalytic activity. *ACS Appl. Mater. Interfaces* 44: 38051–38056.

69 Lázár, K., Szeleczky, A.M., Notheisz, F., and Zsigmond, Á. (1995). Encaged iron phthalocyanine for oxygen transfer: catalytic and Mössbauer spectroscopic study. *Stud. Surf. Sci. Catal.* 94: 720–727.

70 Liu, W., Zhang, L., Liu, X. et al. (2017). Discriminating catalytically active FeN_x species of atomically dispersed Fe-N-C catalyst for selective oxidation of the C-H bond. *J. Am. Chem. Soc.* 139: 10790–10798.

71 Deng, D., Chen, X., Yu, L. et al. (2015). A single iron site confined in a graphene matrix for the catalytic oxidation of benzene at room temperature. *Sci. Adv.* 1: e1500462.

72 Kramm, I.E., Ni, L., and Wagner, S. (2019). ^{57}Fe Mössbauer spectroscopy characterization of electrocatalysts. *Adv. Mater.* 2019: 1805623.

73 Dickson, D.P.E. (1984). Applications to biological systems. In: *Mössbauer Spectroscopy Applied to Inorganic Chemistry* (ed. G.J. Long), 339–390. Plenum Press.

74 Schulz, C.E., Rutter, R., Sage, J.T. et al. (1984). Mössbauer and electron paramagnetic resonance studies of horseradish peroxidase and its catalytic intermediates. *Biochemistry* 23: 4743–4754.

75 Garcia-Serres, R., Davydov, R.M., Matsui, T. et al. (2007). Direct reaction pathways followed upon reduction of oxy-heme oxygenase and oxy-myoglobin as characterized by Mössbauer spectroscopy. *J. Am. Chem. Soc.* 129: 1402–1412.

76 Rajakovich, L.J., Nørgaard, H., Warui, D.M. et al. (2015). Rapid reduction of the diferric-peroxihemiacetal intermediate in aldehyde-deformylating oxagenase by a Cyanobacterial Ferredoxin: evidence for a free-radical mechanism. *J. Am. Chem. Soc.* 137: 11695–11709.

77 Bill, E., Bernhardt, F.H., Trautwein, A.X., and Winkler, H. (1985). Mössbauer investigation of the cofactor iron of putidamonooxin. *Eur. J. Biochem.* 147: 177–182.

78 Pandelia, M.-E., Lanz, N.D., Booker, S.J., and Krebs, C. (2015). Mössbauer spectroscopy of Fe/S proteins. *Biochim. Biophys. Acta* 1853: 1395–1405.

79 Rodriguez, M., Bill, E., Brennessel, W.W., and Holland, P.L. (2011). N_2 reduction and hydrogenation to ammonia by a molecular iron-potassium complex. *Science* 334: 780–783.

80 Chanda, A. High-valent iron complexes with tetraamido macrocyclic ligands: structures, Mössbauer spectroscopy and DFT calculations. *J. Inorg. Biochem.* 100: 606–619.

81 Tomyn, S., Shylin, S.I., Bykov, D. et al. (2016). Indefinitely stable iron(IV) cage complexes, formed in water by air oxidation. *Nat. Commun.* 8: 14099.

82 Shylin, S.I., Pavliuk, M.V., D'Amario, L. et al. (2019). Efficient visible light-driven water oxidation catalysed by an iron(IV) clathrochelate complex. *Chem. Commun.* 55: 3335–3338.

83 Loup, J., Parchomyk, T., Lülf, S. et al. (2019). Mössbauer and mass spectrometry support for iron(II) catalysts in enantioselective C-H activation. *Dalton Trans.* 48: 5135.

84 Parchomyk, T., Demeshko, S., Meyer, F., and Koszinowski, K. (2018). Oxidation states, stability and reactivity of organoferrate complexes. *J. Am. Chem. Soc.* 140: 9709–9720.

85 Kneebone, J.L., Brennessel, W.W., and Neidig, M.L. (2017). Intermediates and reactivity in iron-catalyzed cross-couplings of alkynyl Grignards with alkyl halides. *J. Am. Chem. Soc.* 139: 6988–7003.

5

Application of Mössbauer Spectroscopy in Studying Catalysts for CO Oxidation and Preferential Oxidation of CO in H_2

Kuo Liu[1,4], Junhu Wang[2,3], and Tao Zhang[2,3]

[1] *Research Center for Eco-Environmental Sciences, Chinese Academy of Sciences, 18 Shuangqing Road, Beijing 100085, China*
[2] *State Key Laboratory of Catalysis, Dalian Institute of Chemical Physics, Chinese Academy of Sciences, No. 457 Zhongshan Road, Dalian 116023, China*
[3] *Mössbauer Effect Data Center, Dalian Institute of Chemical Physics, Chinese Academy of Sciences, No. 457 Zhongshan Road, Dalian 116023, China*
[4] *Beijing Key Laboratory of Ionic Liquids Clean Process, Institute of Process Engineering, Chinese Academy of Sciences, 1 North 2nd Street, Zhongguancun, Beijing 100190, China*

5.1 Introduction

Polymeric electrolyte membrane fuel cells (PEMFCs) convert chemical energy to electricity and are suitable for application in portable and stationary power. PEMFCs apply platinum-based electrodes and polymer membranes as electrolyte and often operate at relatively low temperatures of below 100 °C. H_2 was used typically as the fuel for PEMFCs. However, H_2 is often produced from the reforming of hydrocarbons and is not pure but includes H_2O and CO. One of the drawbacks of PEMFCs is the platinum-based electrodes are sensitive to CO, and even a small amount of CO in the gas feed can poison the platinum-based electrodes and the efficiency of PEMFCs will be decreased [1]. In order to prevent poisoning of the electrode, the concentration of CO in H_2 must be reduced to below 10 ppm. Water gas shift reaction ($H_2O + CO \leftrightarrow CO_2 + H_2$) can reduce the CO content to ~1 vol.% [2, 3]. To further lower the CO content in H_2 to <10 ppm, different technologies for CO removal from H_2 have been developed. The preferential oxidation of CO in H_2 (PROX) is an effective way of eliminating CO from H_2, involving following competitive reactions:

$$CO + 0.5O_2 \rightarrow CO_2 \quad \Delta H_{298^\circ} = -283\ \text{kJ mol}^{-1}$$

$$H_2 + 0.5O_2 \rightarrow H_2O \quad \Delta H_{298^\circ} = -242\ \text{kJ mol}^{-1}$$

The catalysts for the PROX reaction should have high activity and selectivity for CO oxidation at the operation temperatures (~80 °C). As shown in Figure 5.1, the Au-based catalysts exhibited excellent activity for CO oxidation at low temperatures below 80 °C, whereas the Cu-based catalysts, e.g. $CuO–CeO_2$ and $CuO–Fe_2O_3$, exhibit high CO conversion above 100 °C. Also, the platinum group metal (PGM)

Mössbauer Spectroscopy: Applications in Chemistry and Materials Science, First Edition. Edited by Yann Garcia, Junhu Wang, and Tao Zhang.

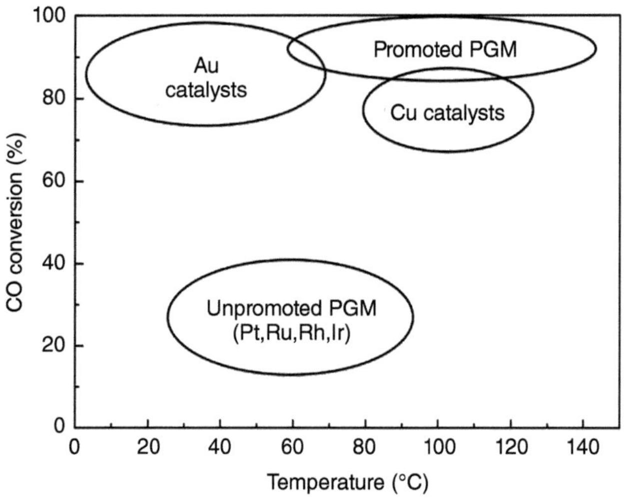

Figure 5.1 CO conversion vs. reaction temperature over different catalysts. Source: Reproduced from Ref. [4]/with permission from American Chemical Society.

supported on inert support materials, e.g. Pt/Al_2O_3 and Ir/SiO_2, exhibited low activity during PROX reaction [4]. Therefore, the Au-based, Cu-based, and PGM on inert support catalysts are less attractive in the industrial application due to either their low activity in the operation temperature range of 80–100 °C or the poor resistance to CO_2 and H_2O. The reducible metal oxide promoted PGM catalysts are promising candidates for PROX due to their excellent activities at 80–120 °C, as shown in Figure 5.1. Fe_2O_3 is one of the excellent reducible metal oxide promoters. The CO conversion and CO_2 selectivity on Fe_2O_3-promoted Pt-based and Ir-based catalysts are both >90% at the operation temperature of PEMFCs (around 80 °C) [5–8], indicating that Fe_2O_3-promoted PGM catalysts are promising PROX catalysts for industrial application.

It is widely accepted that a competitive adsorption between O_2, CO, and H_2 is present on the PGM supported on inert support materials, and at low temperatures, CO covers most of the surface active sites and hinders the adsorption of O_2 and H_2, leading to a low CO conversion at low temperatures [9, 10]. Only when CO desorbs from the surface at high temperatures, can O_2 and H_2 adsorb on the active sites. The adsorption of O_2 is always an important reaction step for PROX reaction on the PGM supported on inert support materials. On the contrary, on the reducible metal oxide-promoted PGM catalysts, CO adsorbs on the PGM site while O_2 is activated on the reducible metal oxide [11–13], and in this way, the competitive adsorption is absent and the activation of O_2 is not the rate-limiting step. In this way, the CO oxidation is facilitated by the reducible metal oxides. Therefore, it is essential to study the effects and properties of the reducible metal oxides.

With the development of theoretical calculation and characterization techniques in recent years, it is possible to identify the active site components on the surface of heterogeneous catalysts. Mössbauer spectroscopy is an effective tool for quantitatively and qualitatively determining different Fe species, which is suitable for studying the Fe-containing materials [14–18]. Isomer shift (δ, originating from the interaction between the s-electrons and the nucleus), quadrupole splitting (ΔE_Q, originating from the interaction between an electric field gradient and the electric quadrupole moment of the nucleus), and magnetic hyperfine field (originating from the interaction between the magnetic field (MF) at the nucleus and the nuclear magnetic dipole moment) are three important parameters obtained from the Mössbauer spectra, providing information on the oxidation state, electron spin configuration, and the molecular symmetry [14–16]. Under some circumstances, it is necessary to identify low-valent Fe species, which are sometimes sensitive to the oxygen in air. To protect the low-valent Fe species from being oxidized by air, quasi-in situ and in situ Mössbauer spectroscopies are developed and are proved to be effective methods for identifying the low-valent Fe species after different treatments [17, 19]. In this chapter, the application of Mössbauer spectroscopy in investigating the active sites of the PROX catalysts in the past two decades is reviewed comprehensively. In the absence of H_2, the PROX reaction would be CO oxidation, and thus, Mössbauer spectroscopy studies on CO oxidation was also summarized in this chapter.

5.2 Application of Mössbauer Spectroscopy in CO Oxidation

5.2.1 ^{57}Fe Mössbauer Spectroscopy

^{57}Fe Mössbauer spectroscopy is effective in clarifying the chemical state, the valence, and the structure of the Fe species in CO oxidation catalysts quantitatively. Iron oxide supported Au catalysts are well known for their excellent activities of CO oxidation, and it is worthwhile to study the effects of Fe species. Wagner et al. [20] prepared a series of iron oxides supported Au catalysts (Au loading $0.17 \sim 1.17$ wt.%), but no correlation between the CO oxidation activity and the metallic Au contents was found. ^{57}Fe Mössbauer results at 293 K show that the catalysts consist of hematite ($\delta = 0.26 \sim 0.27$ mm s^{-1}, $\Delta E_Q = -0.21 \sim -0.23$ mm s^{-1}, $B = 49.1 \sim 51.9$ T), ferrihydrite ($\delta = 0.23 \sim 0.24$ mm s^{-1}, $\Delta E_Q = 0.63 \sim 0.69$ mm s^{-1}), and goethite, and the content of these species varied with the preparation methods and different Au concentrations. A positive correlation between the activity and the ferrihydrite content was found, indicating that ferrihydrite was crucial for CO oxidation. OH groups were found to be important and could participate in CO oxidation via reacting with CO to form reactive formates that could be oxidized by oxygen species to produce CO_2 [21, 22]. The support Fe_2O_3 after calcination at 200 and 400 °C exhibited higher CO oxidation activities than that after calcination at 600 °C. It should be noted that the composition of Fe_2O_3 after calcination was similar to that of Au/Fe_2O_3 after calcination at the same temperature, and also, the chemical composition could not affect CO oxidation on these catalysts [22]. The stability of Au/Fe_2O_3 (Au/Fe atomic ratio of 1/50) prepared by coprecipitation method was found to decrease with the increasing calcination temperature from 100 to 400 °C [23] due to the transformation of ferrihydrite to hematite. Šmit et al.

[24] also found hematite was the main component in Au/Fe_2O_3 with a 2 wt.% Au loading after heating at 433 K for 4 h. Also, the sintering of metallic Au, the decrease of surface area, the conversion of gold oxide to metallic Au, the weakened interaction between Au and Fe species and the generation of carbonate-like species might contribute to the deactivation of Au/Fe_2O_3 [23]. The calcination atmosphere also affects CO oxidation of Au/FeO_x [25]. Au/FeO_x catalysts synthesized in Ar consisted of mainly magnetite and maghemite, while those prepared in air contained only maghemite. Catalysts having more maghemite possessed greater amounts of weakly bound oxygen species, contributing to higher CO conversions. Pretreatment in H_2 could increase the OH groups, promoting CO oxidation [25]. Hydroxyapatite (HAP) was found to promote the sinter-resistant ability of FeO_x in Au/FeO_x, stabilize the γ-Fe_2O_3 crystalline phase, and prohibit the transformation from γ-Fe_2O_3 to α-Fe_2O_3 when the catalysts were calcined, leading to a high activity [26].

Mesoporous silica SBA-15 with 14 wt.% Fe_2O_3 was found to consist of superparamagnetic nanosized particles, whereas the SBA-15 with a higher Fe_2O_3 content (54 wt.%) was composed of superparamagnetic particles and larger Fe_2O_3. When Au was supported on these supports, catalysts with the most OH groups were the most reactive, and activity increased with the decreased Fe_2O_3 contents [27]. FeSiBEA zeolites with 1 wt.% Fe synthesized by a two-step post synthesis method performed much better than 1 wt.% Fe/SiO_2 in CO oxidation, and FeSiBEA prepared in acidic (pH = 2.6) condition exhibited a better activity than that prepared in alkaline (pH = 10) condition [28]. Mössbauer results show that Fe species in FeSiBEA prepared in acidic condition existed as a less distorted and a strongly distorted pseudotetrahedral Fe(III) species, and the later Fe(III) species were less stable and had a higher capability to be reduced by CO to form active sites than the other Fe species in FeSiBEA or Fe/SiO_2.

γ-Al_2O_3 supported iron catalysts were prepared using the precursor iron tris-acetylacetonate by an impregnation method [29]. The catalytic activity decreased with the increasing Fe content from 1 to 10 wt.%. By combining Mössbauer results, diffuse reflectance optical absorption (DRUVS), X-ray diffraction (XRD), X-ray absorption near edge structure (XANES), and extended X-ray absorption fine structure (EXAFS), it was found that the amount of tetrahedrally coordinated Fe^{3+} centers decreased while Fe_xO_y particles increased with the increasing Fe contents. $AlFeO_3$ or the tetrahedrally coordinated Fe^{3+} centers on the surface of γ-Al_2O_3 or in its structure were suggested to be important for CO oxidation. Carriazo et al. [30] modified a Colombian bentonite using intercalation with poly-hydroxocationic solutions containing $Al^{3+} + Fe^{3+}$ or $Al^{3+} + Ce^{3+} + Fe^{3+}$. Mössbauer spectra of all the catalysts consisted of hematite structure ($\delta = 0.21\ mm\ s^{-1}$, $\Delta E_Q = -0.28 \sim -0.20\ mm\ s^{-1}$, hyperfine MF = 500 ~ 515 kOe) and Fe^{3+} ions in octahedra coordination in the phyllosilicate structure ($\delta = 0.30\ mm\ s^{-1}$, $\Delta E_Q = 0.9 \sim 0.98\ mm\ s^{-1}$). Iron atoms were suggested to be active centers for CO oxidation. However, the activities of CO oxidation on these catalysts were low.

Spinel materials with the addition of Fe species were also used in CO oxidation. In $Fe_{3-x}Mn_xO_4$ ($x = 0 \sim 0.53$) spinels, by combining XRD and Mössbauer spectroscopy results, pure spinel phases were found in the presence of Mn [31], while Fe_3O_4 in octahedral ($\delta = 0.17\,mm\,s^{-1}$, hyperfine field 49.0 T) and tetrahedral ($\delta = 0.54\,mm\,s^{-1}$, hyperfine field 46.0 T) sites were present in the absence of Mn. Mn incorporated in the octahedral site replacing Fe^{2+} and then the tetrahedral site with increasing Mn content. All the $Fe_{3-x}Mn_xO_4$ catalysts exhibited similar CO oxidation activity below 200 °C, but the activity increased with the contents of Mn above 200 °C. The catalyst with the highest Mn content was oxidized to maghemite partially at 150 °C, but was completely oxidized to hematite at 300 °C. Catalyst with a lower Mn content ($x = 0$ or 0.21) contained maghemite and a small amount of magnetite at both 150 and 300 °C. Thus, hematite was important for CO oxidation at higher temperatures. On Pd/$FeMnO_3$, spinel structures of γ-$FeMnO_3$ and α-$FeMnO_3$ were identified based on the XRD and Mössbauer results [32], and no synergistic effect between Mn^{3+} and Fe^{3+} cations was present during CO oxidation. Mn-substituted spinel Mn^{3+}–O–Fe/Al_2O_3 exhibited a higher CO oxidation activity than γ-Fe_2O_3/Al_2O_3 [33]. Mn^{3+}–O–Fe/Al_2O_3 contained $Mn_xFe_{2-x}O_4$ on amorphous $Fe_xAl_{2-x}O_3$, while γ-Fe_2O_3/Al_2O_3 consisted of γ-Fe_2O_3 and γ-$Fe_xAl_{2-x}O_3$. The mixed spinel phases provided more labile oxygen species and more active sites than the clusters of the binary Mn oxide, leading to a higher activity of CO oxidation. Similarly, Ni or Co substituted magnetite having spinel structure exhibited higher CO oxidation activities than pure magnetite [34].

Biogenic iron oxides were produced from the metabolic activities of the iron-oxidizing bacteria of the group *Sphaerotilus-Leptothrix*, and the natural biomass was obtained from the freshwater flow at Vitosha Mountain [35]. Based on the XRD and Mössbauer results, the natural biomass contained ultra-dispersed ($\delta = 0.36\,mm\,s^{-1}$, $\Delta E_Q = 0.55\,mm\,s^{-1}$) and highly dispersed ($\delta = 0.32\,mm\,s^{-1}$, $\Delta E_Q = -0.07\,mm\,s^{-1}$, $H = 27.0\,T$) α-FeOOH species. After CO oxidation, the natural biomass consisted of mainly highly dispersed α-Fe_2O_3 (hematite), which was suggested as the active phase for CO oxidation at 200–250 °C. Cultivating the bacteria in different nutrient media would lead to the formation of new Fe species. Samples cultivated in the media of Hay-isolation, isolation medium (ISSL, containing glucose, $(NH_4)_2SO_4$, $Ca(NO_3)_2$, K_2HPO_4, $MgSO_4$, KCl, $CaCO_3$, vitamin B12, vitamin B1, distilled H_2O, Fe grit, $FeCO_3$, and $FeCl_2$), and Winogradsky were found to contain γ-FeOOH and superparamagnetic (SPM) Fe^{3+}, while the sample cultivated in Fedorov medium was the most complex, containing Fe_3O_4, α-FeOOH, γ-FeOOH, and SPM Fe^{3+}. The biomass obtained by cultivating *Leptothrix* genus bacteria in Adler and Lieske media consisted of γ-Fe_2O_3, α-FeOOH, γ-FeOOH, and SPM Fe^{3+} with different ratios [36], while that in Fedorov contained γ-Fe_2O_3 with the Fe^{3+} located at the octahedral and tetrahedral positions, α-FeOOH in two different particle sizes, γ-FeOOH, and SPM Fe^{3+}. By cultivating *Leptothrix* sp. in Lieske's medium with the Al foil covered by anodic Al_2O_3, γ-FeOOH was found to be deposited on the lamellae [37, 38]. When the content of Fe on the surface was reduced to 0.6 wt.% or 0.9 wt.%, the catalytic activity decreased [38]. The material containing a lower Fe content (0.9 wt.% or 0.6 wt.%) consisted of γ-FeOOH ($\delta = 0.21 \sim 0.22\,mm\,s^{-1}$, $\Delta E_Q = 0.33 \sim 0.34\,mm\,s^{-1}$), while that with 9.2 wt.% Fe consisted of not only γ-FeOOH but also superparamagnetic oxide/hydroxide phases ($\delta = 0.36\,mm\,s^{-1}$, $\Delta E_Q = 0.63\,mm\,s^{-1}$). Although the materials prepared exhibit lower CO oxidation activities, the materials show potential in the application as catalyst supports.

5.2.2 ^{119}Sn Mössbauer Spectroscopy

A synergetic effect between Sn and Pt has been found in Sn–Pt/SiO_2 to enhance the CO oxidation activity. Margitfalvi et al. [39–41] prepared Sn–Pt/SiO_2 catalysts using a controlled surface reaction (CSR) method. Multilayer organometallic complex decomposed in the reductive (denoted as (H)) and oxidative (denoted as (O)) atmospheres by temperature-programmed reaction (TPRe) treatment from 25 to 300 °C. After reduction in H_2, Sn–Pt/SiO_2 (0.15 g) could completely convert CO to CO_2 at >50 °C in an atmosphere with 40 Torr O_2 and 16 Torr CO in He at a space velocity 30,000 mL g_{cat}^{-1} h^{-1}. The difference in T_{50} values on Pt–Sn/SiO_2 (type H) and Pt–Sn/SiO_2 (type O) was insignificant, and Pt–Sn/SiO_2 (type O) was sensitive to the reduction temperature in H_2. According to the in situ Mössbauer results, the as received (type H) catalyst prepared using the precursor tin tetraethyl consisted of Sn^{4+} ($\delta = 0.09$ mm s^{-1}, $\Delta E_Q = 0.64$ mm s^{-1}, FWHM = 0.86 mm s^{-1}), Sn dissolved in Pt (Sn–Pt alloy (a), $\delta = 1.20 \sim 1.56$ mm s^{-1}), and Sn-rich Pt_2Sn_3 and $PtSn_4$ phases (Sn–Pt alloy (b), $\delta = 1.98 \sim 2.39$ mm s^{-1}), while the as received (type O) catalyst consisted of only Sn^{4+} ($\delta = 0.08$ mm s^{-1}, $\Delta E_Q = 0.72$ mm s^{-1}, FWHM = 0.99 mm s^{-1}). After H_2 reduction at 300 °C, PtSn(a) and PtSn(b) were observed in Sn–Pt/SiO_2. After CO oxidation, the amount of PtSn(a) and PtSn(b) alloy species decreased, forming Sn^{4+} oxide (Sn^{4+}(ox), $\delta = 0.00 \sim 0.01$ mm s^{-1}, $\Delta E_Q = 0.70 \sim 0.72$ mm s^{-1}), well-dispersed surface Sn^{4+} species (Sn^{4+}(sf), $\delta = 0.80 \sim 0.93$ mm s^{-1}), and PtSn (1 : 1) alloy ($\delta = 1.58 \sim 1.60$ mm s^{-1}). When the catalyst was treated in H_2 at room temperature, Sn^{4+}(sf) and Pt–Sn (1 : 1) alloy disappeared, whereas PtSn(a) and PtSn(b) alloy species appeared. Exposure of the reduced catalyst to CO and O_2 gas mixture at room temperature led to the reoxidation of PtSn(a) and PtSn(b) alloy species, forming Pt species and Sn^{4+}(sf) again, indicating that the Pt and Sn species in close contact were involved in CO oxidation and could be reconstructed [40, 41]. The interconversion PtSn(b) $\leftrightarrow$ Sn^{4+} + Pt took place at room temperature during CO oxidation and the subsequent treatment in H_2 [42]. The "Sn^{4+}-Pt" ensemble sites formed during CO oxidation were responsible for CO activation via the electron transfer from CO to Sn^{4+} species.

The Au/SnO_x–Al_2O_3 catalysts prepared by direct anionic exchange and deposition-precipitation with urea (DPU) methods could achieve ~50% and >90% CO conversion at room temperature and above 75 °C, respectively, in an atmosphere containing 44 Torr O_2 and 16 Torr CO in He with the flow rate 70 mL min^{-1} and 0.075 g catalyst [43, 44]. Mössbauer spectra of the as received Au/SnO_x–Al_2O_3 (4.2 wt.% Sn, prepared by DPU) catalyst consisted of two kinds of Sn^{4+}, i.e. Sn^{4+}(1) assigned by more "bulky" species ($\delta = 0.02$ mm s^{-1}, ΔE_Q = 0.47 mm s^{-1}) and Sn^{4+}(2) at the perimeter interface of the SnO_2 nanoparticles ($\delta = 0.06$ mm s^{-1}, $\Delta E_Q = 1.07$ mm s^{-1}). After reduction by H_2 at 350 °C, some Sn^{4+}(1) remained while Sn^{4+}(2) was absent, and "bulky" SnO species (Sn^{2+}(1), $\delta = 2.75$ mm s^{-1}, $\Delta E_Q = 2.26$ mm s^{-1}), Sn–O–Al-like species (Sn^{2+}(2), $\delta = 3.12$ mm s^{-1}, $\Delta E_Q = 2.18$ mm s^{-1}), and a small amount of metallic Sn^0 (δ = 2.56 mm s^{-1}) appeared. When the Sn content was decreased to 3 wt.% in Au/SnO_x–Al_2O_3, AuSn alloy ($\delta = 2.41$ mm s^{-1}) was observed [44]. The presence of AuSn alloy indicated that Au and Sn species were in close contact. Sn was inclined to segregate on the surface of bimetallic AuSn particles when the Sn content was increased to 4.2 wt.%, leading to the disappearance of the AuSn alloy. In the case of SnO_x–Al_2O_3, no metallic Sn^0 formed after the same reduction treatment, indicating that Au promoted the reduction of the Sn^{2+} species. O_2 was found to adsorb and activate on both the reduced Sn species and Au particles. Also, the reduction of the Sn species led to partial negative charge of Au, accelerating the oxygen adsorption in the form of superoxide (O_2^-). The oxidation treatment of Au/SnO_x–Al_2O_3 resulted in a decrease of the oxygen vacancies, and further, a lower CO oxidation activity.

5.2.3 ^{197}Au Mössbauer Spectroscopy

^{197}Au Mössbauer spectroscopy can determine the chemical state and the valence of Au in the catalysts [45, 46]. Wagner et al. [20] prepared iron oxides supported Au catalysts by wetness impregnation, coprecipitation, and inverse coprecipitation methods. According to the ^{197}Au Mossbauer spectra, these catalysts were found to consist of metallic Au (a single line with a δ of $-1.23\ mm\ s^{-1}$) and gold oxide ($\delta = 1.15 \sim 1.37\ mm\ s^{-1}$, $\Delta E_Q = 1.4 \sim 1.75\ mm\ s^{-1}$). The catalyst prepared by wetness impregnation contained mainly Au oxide (98%). Also, the catalyst prepared by coprecipitation with a higher Au content (1.17 wt.%) had 90% Au oxide and 10% metallic Au. No correlation between the activity of CO oxidation and the abundance of metallic Au was found in their study. On the other hand, Hutchings et al. [47] prepared a series of Au/Fe_2O_3 using a coprecipitation method treated in different ways. The ^{197}Au Mössbauer spectra of Au/Fe_2O_3 calcined at 400 °C exhibited only a singlet ascribed to metallic Au ($\delta = -1.14\ mm\ s^{-1}$), while the spectra of the catalyst after drying consisted of two doublets with $\delta = 0.06$ and $1.37\ mm\ s^{-1}$, $\Delta E_Q = 2.08$ and $1.82\ mm\ s^{-1}$, respectively, indicative of the Au(III)-oxyhydroxide species. The activity of CO oxidation on Au/Fe_2O_3 calcined at 400 °C was lower than that without calcination at high temperatures. By combining X-ray photoelectron spectroscopy (XPS), EXAFS, in situ XANES, and ^{197}Au Mössbauer results, the authors suggest that the cationic Au is important for CO oxidation at 298 K. Similar conclusions have been drawn by Daniells et al. [46] and Finch et al. [48]. Calcination or treatment of Au/Fe_2O_3 at 200, 400, and 600 °C resulted in the formation of only metallic Au [21, 24], and the particle size of Au increased from 1.5 to 10 nm with the increasing calcination temperature from 400 to 600 °C. It is impossible for CO to adsorb on Au particles with 1.5 nm, but CO could adsorb on Au with the particle size of 4 nm. CO adsorption took place on Au particles and the OH groups on the Au particles would react with CO to produce hydroxycarbonyl species, which would be oxidized to bicarbonate by O_2 from the Fe_2O_3 support, and further, produced CO_2 [46].

Au supported on other supports, e.g. MnO_x, $Mg(OH)_2$, and TiO_2, were also studied by Mössbauer spectroscopy. Au/MnO_x prepared by a coprecipitation method exhibited 76% CO conversion during CO oxidation at 25 °C at a space velocity $20\,000\ h^{-1}$ in an atmosphere containing 1% CO and 0.5% O_2 in N_2 [49]. According to the ^{197}Au Mössbauer spectroscopy of the Au-Mn coprecipitates, a single line ascribed to metallic gold (Au^0, $\delta = -1.24 \pm 0.01\ mm\ s^{-1}$) was found. Metallic (Au(0)) was formed during the coprecipitation process instead of during drying. Kobayashi et al. [50]

prepared Au/$Mg(OH)_2$ and Au/TiO_2 by deposition–precipitation. Au/$Mg(OH)_2$ exhibited a higher CO oxidation activity than Au/TiO_2. ^{197}Au Mössbauer spectra of Au/$Mg(OH)_2$ after calcination consisted of a singlet with the δ nearly 0 and three doublets with positive δ. The authors attributed the singlet to the cores of the Au particles, and one of the doublets with a small positive δ and large ΔE_Q was attributed to Au^I atoms. The other two doublets with larger δ were assigned to Au^{III} atoms. However, no Au^I was present in Au/TiO_2. The Au^I atoms at the interface between Au and $Mg(OH)_2$ were important for CO oxidation. Similar conclusions have been drawn on Au–titania–zirconia [51]. AuHY zeolites prepared by adsorption/impregnation of anionic Au or by ion exchange of cationic Au were found to contain mainly Au(I) or Au(III), and Au(III)/HY was inactive for CO oxidation initially. However, the activity appeared with increasing reaction time, and some degree of Au(III) reduction took place [52].

5.2.4 ^{193}Ir Mössbauer spectroscopy

^{193}Ir Mössbauer spectroscopy is useful in determining the chemical state and structural properties of Ir species. Sawicki et al. [53] prepared PtIr catalysts using incipient wetness impregnation method. The polystyrene–divinylbenzene (SDB) granules supported 9 wt.% Pt and 1 wt.% Ir catalyst (400 mg) exhibited a high CO conversion (almost 100%) at 25 °C in an atmosphere of 750 ppm CO in humid air (relative humidity 50%) at a flow rate of $1\,L\,min^{-1}$. Mössbauer spectroscopy (73.0 keV γ-rays of ^{193}Ir) of the samples was measured at liquid He temperature. The source isotope ^{193}Os was produced by the neutron activation of ^{192}Os metal in a thermal flux. The Mössbauer spectra of the compound H_2IrCl_6 consisted of a spectrum with $\delta = -1.00\,mm\,s^{-1}$ and $\Delta E_Q = 0.30\,mm\,s^{-1}$ attributed to tetravalent Ir(IV) species. The trivalent Ir(III) in $IrCl_3$ led to a spectrum with $\delta = -1.94\,mm\,s^{-1}$ and $\Delta E_Q = 0.54\,mm\,s^{-1}$, while tetravalent Ir(IV) in IrO_2 exhibited a spectrum with $\delta = -0.84\,mm\,s^{-1}$ and $\Delta E_Q = 2.70\,mm\,s^{-1}$. The Mossbauer spectra of Pt250H/9Ir250H polymer-supported catalyst (a sample impregnated with Pt followed by reduction in H_2 at 250 °C and then impregnated with Ir followed by reduction in H_2 at 250 °C) consisted of an asymmetric quadrupole-split doublet ($\delta = -0.86\,mm\,s^{-1}$ and $\Delta E_Q = 2.47\,mm\,s^{-1}$) and a broadened single line ($\delta = -0.04\,mm\,s^{-1}$ and $\Delta E_Q = 1.06\,mm\,s^{-1}$), assigned to highly non-stoichiometric IrO_{2-x} and Ir-rich Pt–Ir alloy clusters $Pt_{0.05}Ir_{0.95}$, respectively. In the as-deposited Ir and Ir–Pt codeposition samples, Ir existed mostly as Ir(III) in $[IrCl_6]^{3-}$ ($\delta = -2.1\,mm\,s^{-1}$, $\Delta E_Q = 0.8\,mm\,s^{-1}$) and Ir(IV) in $[IrCl_6]^{2-}$ ($\delta = -0.5\,mm\,s^{-1}$). In the H_2-treated Ir-containing samples, Ir existed as IrO_2 ($\delta = -0.8\,mm\,s^{-1}$, $\Delta E_Q = 2.6\,mm\,s^{-1}$), while $Ir(CO)_n$ ($\delta = 0.7\,mm\,s^{-1}$, $\Delta E_Q = 2.7\,mm\,s^{-1}$) and IrO_{2-x} ($\delta = -0.5\,mm\,s^{-1}$, $\Delta E_Q = 2.5\,mm\,s^{-1}$) were present in the CO treated 1 wt.% Ir sample. After reduction in H_2 or CO atmosphere at 200 or 250 °C, Ir existed as IrO_{2-x} ($\delta = -1.0 \sim -0.8\,mm\,s^{-1}$, $\Delta E_Q = 1.9 \sim 2.7\,mm\,s^{-1}$) and metallic Pt–Ir phase ($\delta = -0.7 \sim -0.2\,mm\,s^{-1}$) in the Ir–Pt containing samples. Ir species was suggested to provide atomic oxygen that would react with CO molecules adsorbed on the adjacent metallic Pt site.

5.3 Application of Mössbauer Spectroscopy in PROX

5.3.1 PtFe-Containing Catalysts

PtFe nanoparticles and Al_2O_3 supported $PtFe_x$ (ratio of Fe/Pt (x) = 1, 2, and 3) alloy nanoparticle catalysts were prepared using the precursors $Pt(acac)_2$ (acac = acetylacetonato) and $Fe(acac)_3$ in ethylene glycol [54]. As shown in Figure 5.2, the $PtFe/Al_2O_3$, $PtFe_2/Al_2O_3$, and $PtFe_3/Al_2O_3$ exhibited higher CO conversion than Pt/Al_2O_3, and >40% CO conversion was obtained at temperatures 80 and 100 °C on $PtFe_2/Al_2O_3$ (40% H_2, 2% CO, 1% O_2, He balance, and space velocity 40 000 $cm^3\ g^{-1}\ h^{-1}$), indicating that Fe plays the part of promoting the CO oxidation reaction under H_2-rich conditions. The presence of Fe facilitated the CO_2 selectivity as well. ^{57}Fe Mössbauer spectroscopy at room temperature was conducted to study the as-prepared catalysts. As shown in Figure 5.2c, these $PtFe_x/Al_2O_3$ catalysts consisted of one alloyed iron species, one ferrous, and one ferric component. The appearance of the ferrous component (Fe^{2+}) was due to the presence of Al_2O_3 support, since Fe^{2+} was absent in the Mössbauer spectrum of PtFe nanoparticles. Low-valent iron (Fe^{2+}) existed in the as-prepared catalysts possibly due to soft reducibility of the ethylene glycol and oleic acid or oleylamine. The amount of Fe^{2+} increased with higher amount of Fe species, i.e. $PtFe/Al_2O_3 < PtFe_2/Al_2O_3 < PtFe_3/Al_2O_3$ (Table 5.1). The δ and ΔE_Q values of the Fe species in the alloyed state decreased from 0.35 to 0.18 mm^{-1} and from 0.35 to 0 $mm\ s^{-1}$, respectively, with increasing Fe

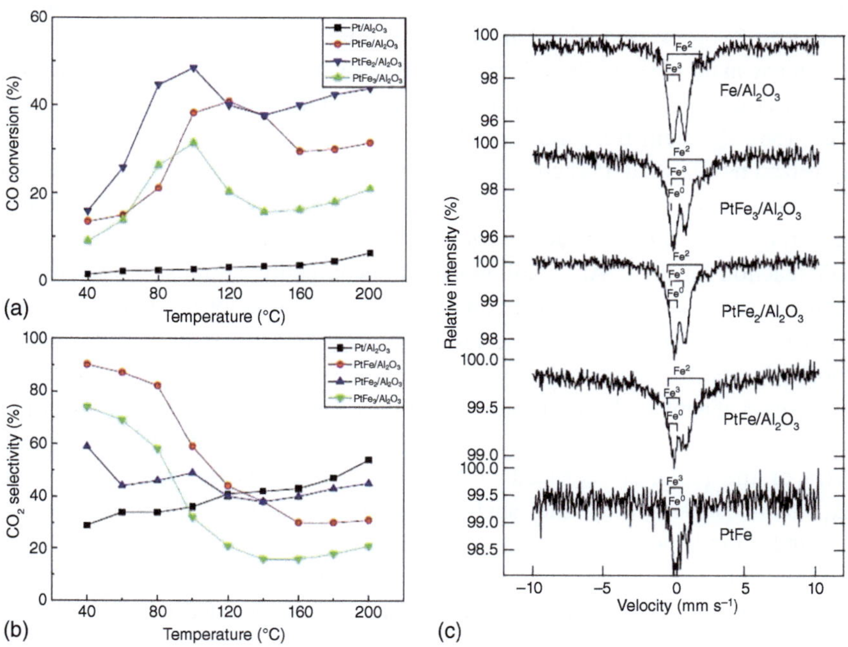

Figure 5.2 (a) CO conversion, (b) CO_2 selectivity during PROX, and (c) ^{57}Fe Mössbauer spectroscopy study over $PtFe/Al_2O_3$, $PtFe_2/Al_2O_3$, and $PtFe_3/Al_2O_3$ catalysts. Source: Reproduced from Ref. [54]/with permission from Springer Nature.

Table 5.1 ^{57}Fe Mössbauer parameters of as-prepared $PtFe/Al_2O_3$, $PtFe_2/Al_2O_3$, and $PtFe_3/Al_2O_3$ catalysts.

Composition	Oxidation state of Fe species	δ ($mm\,s^{-1}$)	ΔE_Q ($mm\,s^{-1}$)	Spectral area (%)
$PtFe_3/Al_2O_3$	Fe^{3+}	0.32	1.04	60
	Fe^{2+}	1.05	1.95	15
	Alloy	0.18	0	25
$PtFe_2/Al_2O_3$	Fe^{3+}	0.33	0.87	75
	Fe^{2+}	1.16	2.02	10
	Alloy	0.33	0.32	15
$PtFe/Al_2O_3$	Fe^{3+}	0.30	0.91	50
	Fe^{2+}	1.15	1.99	3
	Alloy	0.35	0.35	47

δ: isomer shift; ΔE_Q: electric quadrupole splitting. The uncertainty of spectral area is ±5% of the reported value.
Source: Ref. [54]/with permission from Springer Nature.

contents, indicating that both the s-density and the symmetry of charge distribution at the Fe nuclear position in $PtFe_3/Al_2O_3$ are higher than those in $PtFe/Al_2O_3$ or $PtFe_2/Al_2O_3$. Also, $PtFe_x$ ($x = 1$, 2, and 3) nanoparticles on Al_2O_3 supports remained in the alloyed state. The Fe species in the Al_2O_3 supported PtFe catalysts synthesized by conventional methods in the literature is mostly reported to be in the ferric state [55], while the metallic (Fe^0) or ferrous (Fe^{2+}) components are absent after calcination. Reduction in H_2 or CO can lead to the appearance of Fe^0 or Fe^{2+} [56]. Different from those catalysts, the as-prepared $PtFe_x/Al_2O_3$ catalysts synthesized by Yin et al. [54] contained ferric, ferrous, and alloyed Fe species, and the states of the Fe species were stable even after being exposed to air for 1 month. Thus, $PtFe_x/Al_2O_3$ were also active without the pretreatment of H_2 reduction. The PROX reaction on $PtFe_x/Al_2O_3$ follows the noncompetitive dual site mechanism, and the Pt site was for CO adsorption and the Fe site was for O_2 activation. The presence of alloyed Pt and Fe species might benefit the catalytic performance of PROX.

In the case of activated carbon supported PtFe catalyst prepared by impregnation simultaneously with Pt and Fe salts (10 wt.% Fe and 0.5 wt.% Pt), the Mössbauer spectra of the material consisted of two sets of doublets ascribed to the core ($\Delta E_Q = 0.66 \sim 0.69\,mm\,s^{-1}$) and shell ($\Delta E_Q = 1.13 \sim 1.25\,mm\,s^{-1}$) layers of the supported ultradispersed Fe oxide particles (<10 nm), respectively [57]. After PROX reaction, the ratio of core to surface Fe species did not change significantly, but Fe^{2+} appeared. When the activated carbon was impregnated with Fe salt and subsequently Pt salt, the catalyst was found to consist not only the core and shell layers of ultradispersed Fe oxide particles but also Fe^{2+}-ions ($\delta = 0.97\,mm\,s^{-1}$, $\Delta E_Q = 2.33\,mm\,s^{-1}$). After the PROX reaction, the amount of core Fe species and the Fe^{2+} ions increased while the surface Fe species decreased, indicating the partial agglomeration of the Fe oxide particles. However, the PROX activity was lower on these PtFe-containing catalysts than on the activated carbon supported Pt catalyst, which was ascribed to the increased surface oxygen species due to the presence of iron oxide and the acceleration of H_2 oxidation during PROX.

Schmal et al. [58] prepared $Fe_xZr_{(1-x)}O_2$ using the coprecipitation method, and then, $Pt/Fe_xZr_{(1-x)}O_2$ ($x=0.25$, 0.5, 0.75) were prepared by incipient wetness impregnation. ^{57}Fe Mössbauer spectra of Pt/Fe_2O_3 measured at room temperature show that α-Fe_2O_3 was the only iron species, and XRD shows good crystalline phase of α-Fe_2O_3. The spectra of $Pt/Fe_{0.25}Zr_{0.75}O_2$ and $Pt/Fe_{0.5}Zr_{0.5}O_2$ were well fitted by one sextet assigned to α-Fe_2O_3 and one doublet given rise by superparamagnetic α-Fe_2O_3 (small crystals) or paramagnetic Fe^{3+} ions in the lattice of segregated ZrO_2 (Figure 5.3a). For the sextet, it should be noted that the hyperfine MF values were smaller than that of the hematite "bulk" α-Fe_2O_3. Both the presence of small particles of α-Fe_2O_3 crystallites and ZrO_2 and the substitution of Fe^{3+} ions in the α-Fe_2O_3 lattice by Zr^{4+}could diminish the hyperfine MF. The doublet with a higher ΔE_Q value than superparamagnetic α-Fe_2O_3 was assigned to the paramagnetic Fe^{3+} ions in the lattice of segregated ZrO_2. The concentration of the doublet increased with increasing ZrO_2 content, suggesting that a greater amount of ZrO_2 was segregated and more Fe^{3+} ions entered inside the ZrO_2 lattice. On the other hand, the ^{57}Fe Mössbauer spectrum of $Pt/Fe_{0.75}Zr_{0.25}O_2$ at room temperature was well fitted with a doublet assigned to paramagnetic Fe^{3+} ions inside ZrO_2 lattice and two sextets ascribed to the same phase as that in $Pt/Fe_{0.25}Zr_{0.75}O_2$ and the higher content of Fe^{3+} ions substituted by Zr^{4+}, respectively. After reduction at 500 °C, ^{57}Fe Mössbauer spectra of $Pt/Fe_{0.5}Zr_{0.5}O_2$, $Pt/Fe_{0.75}Zr_{0.25}O_2$, and Pt/Fe_2O_3 exhibit two sextets (Figure 5.3b) ascribed to $Fe^{\text{“}2.5+\text{”}}$ ions in octahedral sites and Fe^{3+} in tetrahedral sites in Fe_3O_4 [59]. Pt/Fe_2O_3 exhibited a better performance than $Pt/Fe_xZr_{(1-x)}O_2$ and Pt/ZrO_2 (Figure 5.3c). The turnover frequencies (TOF) of the catalysts at 70 °C follows the order of $Pt/ZrO_2 < Pt/Fe_{0.25}Zr_{0.75}O_2 < Pt/Fe_{0.75}Zr_{0.25}O_2 < Pt/Fe_2O_3$. Mössbauer spectra showed that Fe_3O_4 formed after reduction at 500 °C, and the content of Fe_3O_4 decreased with the increasing ZrO_2 content. Also, Fe_3O_4 in the structure became less stoichiometric in the presence of more ZrO_2. Consequently, the authors proposed that Fe^{3+} ions might interact with Pt^0 in close contact with them, generating new sites, accelerating CO and H_2 adsorption, forming superoxides (O^{2-}) on the neighboring particles, and further, facilitating the vacancy formation and the oxygen mobility. These changes promoted the oxidation reactions of either CO or H_2.

5.3.2 Au-Based Catalysts

Au-based catalysts performed excellently in CO oxidation and PROX and attracted widespread interest. Zepeda et al. [60] modified HMS (hexagonal mesoporous silica) using Ce^{4+}, Ti^{4+}, and Fe^{3+} ions, and prepared mesoporous Au/HMS-Ce, Au/HMS-Ti, and Au/HMS-Fe catalysts using a deposition-precipitation method. The PROX activity follows the order of Au/HMS–Fe > Au/HMS–Ce > Au/HMS–Ti > Au/HMS (Figure 5.4a). The crystal size of Au on different catalysts was calculated by Debye–Scherrer equation based on the XRD line at 38°, following the order of

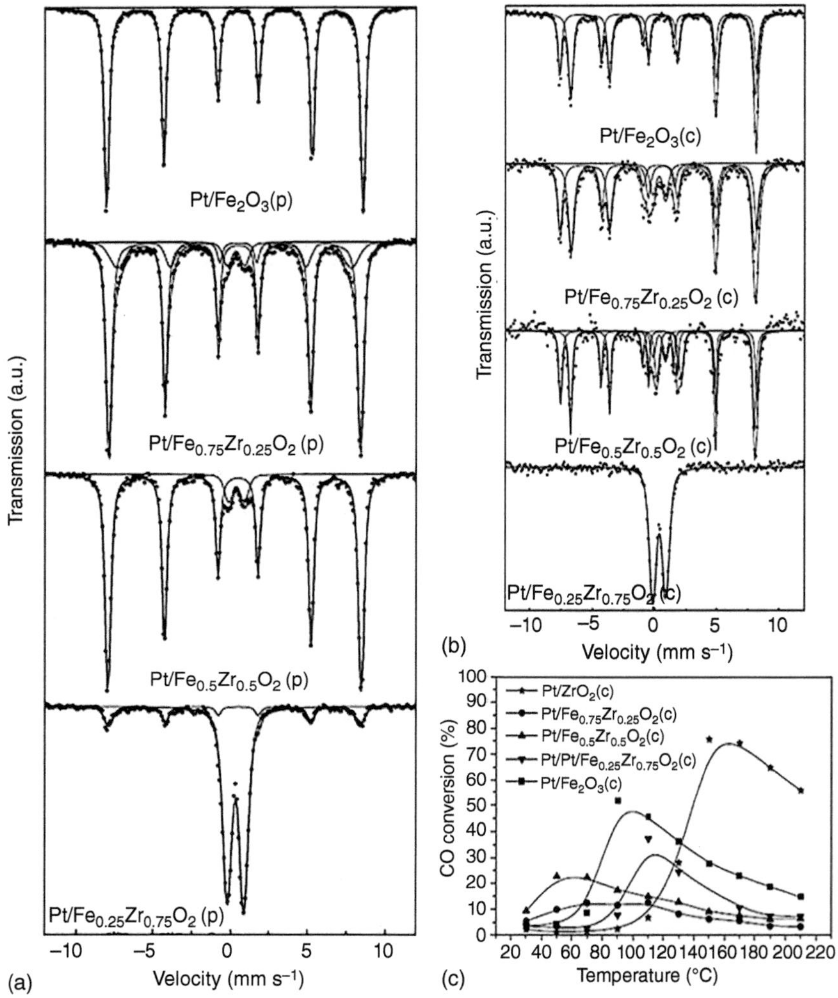

Figure 5.3 ^{57}Fe Mössbauer spectroscopy conducted at 25 °C on (a) fresh and (b) reduced Pt/Fe_2O_3 and $Pt/Fe_xZr_{(1-x)}O_2$ ($x = 0.25, 0.5, 0.75$) catalysts and (c) CO conversion on Pt/Fe_2O_3, $Pt/Fe_xZr_{(1-x)}O_2$, and Pt/ZrO_2 [58]. Reaction conditions: 1% CO, 1% O_2, 60% H_2, He balance, 100 mg catalyst and 100 mL/min flow rate. Source: Reproduced from Ref. [58]/with permission from Elsevier.

Au/HMS-Ce (7.4 nm) > Au/HMS (6.3 nm) > Au/HMS-Ti (5.2 nm) > Au/HMS-Fe (4.6 nm). High-resolution transmission electron microscopy (HRTEM) results show that the population of Au with sizes in the range of 0.5–3 nm is the largest on Au/HMS-Fe and the lowest on Au/HMS-Ce. Therefore, the presence of Fe enhanced the dispersion of Au. Au/Si ratio of Au/HMS-Fe after the PROX reaction was the highest among all catalysts, and the surface exposure of Au follows the order of Au/HMS-Fe (0.0058) > Au/HMS-Ti (0.0052) > Au/HMS-Ce (0.0048) > Au/HMS (0.0044). To make clear if the support could supply oxygen during the PROX reaction, the oxygen storage capacities (OSC) of the spent catalysts

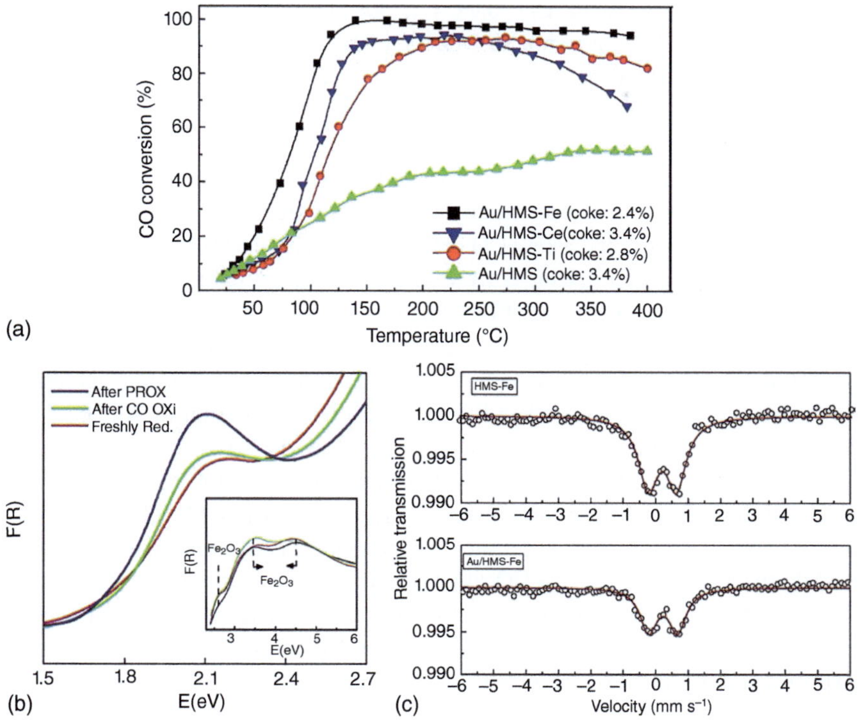

Figure 5.4 (a) CO conversion on Au/HMS-Fe, Au/HMS-Ce, Au/HMS-Ti, and Au/HMS during PROX; (b) UV-vis results of the reduced and spent Au/HMS-Fe; and (c) ^{57}Fe Mössbauer spectroscopy of the reduced Au/HMS-Fe. Source: Reproduced from Ref. [60]/with permission from Elsevier.

were studied, and it was found that the OSC of the spent catalysts followed the order of Au/HMS-Fe > Au/HMS-Ce > Au/HMS-Ti > Au/HMS. On Au/HMS-Ce, the Ce^{3+}/Ce^{4+} ratio decreased after the PROX reaction, indicating some surface Ce^{3+} ions were oxidized during the reaction. The reduced Au/HMS-Fe catalyst exhibited three signals from the UV-vis spectra at 2.6, 3.5, and 4.5 eV (Figure 5.4b) usually ascribed to the charge transfer from oxygen to Fe^{2+} (2.6 eV) and to Fe^{3+} (3.5 and 4.5 eV). However, according to Mössbauer spectrum of the reduced Au/HMS-Fe catalyst (Figure 5.4c), the iron species with the δ and ΔE_Q values of 0.33 ± 0.05 and 0.89 ± 0.05 mm s^{-1} were assigned to Fe^{3+}. Consequently, the UV-vis band at 2.6 eV in Zepeda et al.'s study was ascribed to the charge transfer from oxygen to Fe^{3+}. In comparison with the reduced catalyst, the spent Au/HMS-Fe showed non-Fe leaching from the HMS supports during the reaction, indicative of the strongest metal–support interaction among the three catalysts. Au/HMS-Fe has the highest OSC and the largest surface exposure of the Au species during the PROX reaction, and the largest amount of small metallic Au nanoparticles after reduction. All the unique properties above led to a higher PROX activity on Au/HMS-Fe than on Au/HMS-Ce, Au/HMS-Ti, or Au/HMS.

Au/Fe_2O_3 catalysts prepared using coprecipitation followed by a two-stage calcination procedure at 400 °C and then 500 °C were effective in PROX in an atmosphere containing 0.9% CO, 0.9% O_2, 50% H_2, 4.7% H_2O, and 22% CO_2, achieving >99.5% CO conversion as well as >50% CO_2 selectivity in the temperature range of 80–100 °C at a gas hourly space velocity (GHSV) 12 000 h^{-1} [61]. TEM results show that the Au particle size grows with increasing temperature of calcination, and the catalyst calcined at 600 °C has a large average Au particle size of 16.1 nm. Also, the Brunauer-Emmett-Teller (BET) surface area of the catalyst calcined at 600 °C was small. The reasons above were why Au/Fe_2O_3 calcined at 600 °C exhibited a low CO conversion during PROX. ^{197}Au Mössbauer spectra of Au/Fe_2O_3 after drying at 25 °C and 120 °C show the presence of Au(III) oxyhydroxide (two doublets with IS = 0.06 mm s^{-1}, ΔE_Q = 2.08 ~ 2.29 mm s^{-1} and δ = 1.37 ~ 1.48 mm s^{-1}, ΔE_Q = 1.82 ~ 2.01 mm s^{-1}). After the two-stage calcination (400 °C followed by 550 °C) and calcination at 600 °C, Au was in the state of metallic Au (δ = −1.27 ~ −1.29 mm s^{-1}). For the catalyst after calcination at 400 °C, ^{197}Au Mössbauer spectra were composed of a singlet with an δ = −1.14 mm s^{-1}, the value of which was different from those after the higher temperature calcination. This phenomenon was ascribed to the presence of Au(III) or Au(I). Metallic Au species is active for CO oxidation, while cationic Au species are active for the reverse water gas shift reaction ($CO_2 + H_2 \rightarrow CO + H_2O$), which will result in the production of CO and decrease the CO conversion during PROX. The two-stage calcination procedure applied in the preparation of Au/Fe_2O_3 led to the absence of the cationic Au species and inhibited the reverse water gas shift reaction [61].

5.3.3 IrFe-Containing Catalysts

5.3.3.1 Porous Carbon Supported IrFe Catalysts

Zhao et al. [62] supported IrFe on different carbon materials using coimpregnation method, with an Ir loading of 3 wt.% and an Ir/Fe atomic ratio of 1/5. The mesoporous carbon (MC), NORIT active carbon (AC), carbon black (XC-72), and CMK-3 were impregnated into the aqueous mixture of $Fe(NO_3)_3$ and H_2IrCl_6. The MC supported IrFe catalyst (IrFe/MC) exhibited a better performance than the IrFe catalysts supported on AC (IrFe/AC), XC-7 (IrFe/XC-72), and CMK-3 (IrFe/CMK-3). The CO conversion follows the order of IrFe/MC > IrFe/AC > IrFe/XC-72 > IrFe/CMK-3. An optimum CO conversion of 83.5% was achieved at 80 °C on IrFe/MC, with the CO_2 selectivity 89% at a space velocity of 40 000 mL $g_{cat}{}^{-1}$ h^{-1} in a gas mixture of 40 vol.% H_2, 2 vol.% CO, 1 vol.% O_2, and He balance. The results of XRD, Mössbauer spectroscopy, temperature programmed reduction by H_2 (H_2-TPR), and XPS demonstrate that the iron species are presented as Fe^{3+}. According to XRD results, the signals of the iron species in IrFe/MC were strengthened, indicating that the iron species were well crystallized. Furthermore, the Mössbauer spectra show that IrFe/MC exhibits the smallest ΔE_Q (0.79 mm s^{-1}) and line width (0.46 mm s^{-1}) among all the catalysts, indicative of better crystal order form and lattice symmetry of the Fe species in IrFe/MC. Large amounts of quinone, carbonyl, and carboxyl groups were detected on MC since it was prepared from resin, benefiting the anchoring of the Ir and Fe metal ions. In turn, the interaction between the metal species and C=O groups benefited the crystallization of iron oxides, leading to a better crystallization of Fe species in IrFe/MC. Because of the well-crystallized Fe^{3+} species, more lattice oxygen was observed in IrFe/MC. Also, less positive Ir species were proved to be present in IrFe/MC and IrFe/AC. All the factors above led to a more excellent PROX activity on carbon materials supported IrFe catalysts.

5.3.3.2 SiO_2 and Al_2O_3 Supported IrFe Catalysts

Some low-valent Fe species are sensitive to oxygen and can be oxidized to higher valent Fe species by air. Therefore, Mössbauer studies on the PROX reaction that are not in situ cannot exclude the influence of air on the iron state of the PROX catalysts after H_2 reduction or the PROX reaction. Quasi-in situ ^{57}Fe Mössbauer spectroscopy applies a quartz reactor that can protect the samples from exposure to air after different kinds of treatments. The detailed description of the reactor is shown in the reference [17]. In this section, the application of quasi-in situ ^{57}Fe Mössbauer spectroscopy in studying the PROX catalysts is summarized.

Preparation and Activity SiO_2 supported IrFe catalyst with an atomic ratio of Fe to Ir corresponding to 5/1 was prepared by a coimpregnation method. The concentration of H_2 and H_2O in the gas feed could affect the activity of PROX [63], and the conversion of CO oxidation in the absence of H_2 (H_2 concentration 0%) was below 10% in the temperature range of 80–120 °C (Figure 5.5a). Nevertheless, adding 2% H_2 in the gas feed increased the conversion of CO oxidation dramatically, reaching ~80% above 120 °C. Further increasing the H_2 concentration to 10% and 40% increased the conversion of CO oxidation to ~80% at 100 °C, indicating that the PROX reaction has been enhanced significantly by the excessive H_2. H_2O was also reported to promote the CO oxidation at low temperatures [63]. The promotional role of H_2 was not so dramatic on Ir/SiO_2 (Figure 5.5b) as that on Ir–Fe/SiO_2, indicating that the positive effect of H_2 correlated closely with the Fe species.

The impregnation sequence of Ir and Fe during preparation can also influence the PROX activity. The Ir–Fe/Al_2O_3 catalysts prepared by coimpregnation method (denoted as Ir–Fe/Al_2O_3), by impregnation in aqueous $Fe(NO_3)_3$ solution and then aqueous chloroiridic acid solution (denoted as Ir/Fe/Al_2O_3), and by impregnation in aqueous chloroiridic acid solution and then aqueous $Fe(NO_3)_3$ solution (denoted as Fe/Ir/Al_2O_3) were prepared. The atomic Fe/Ir ratios of the Al_2O_3 supported Ir and Fe catalysts were 5/1. Ir–Fe/Al_2O_3 exhibited the best PROX activity with a CO conversion of around 80% at 80–120 °C, while Ir/Fe/Al_2O_3 was the worst catalyst for PROX (<25% at 80–120 °C) (Figure 5.5c) [64].

Quasi-In Situ Mössbauer Studies It is widely accepted that CO oxidation follows Langmuir–Hinshelwood mechanism and a competitive adsorption between O_2, H_2, and CO is present since all these gas molecules adsorb on the noble metal sites on single metal catalysts [65]. The activation of O_2 on these catalysts is important during PROX reaction. However, the competitive adsorption is absent on the bimetallic catalysts (e.g. IrFe/SiO_2), since they possess two kinds of active sites for activating O_2 and CO, respectively [11, 12, 66–68]. On IrFe/SiO_2, Fe weakened the adsorption of CO on Ir but promoted O_2 adsorption greatly, and quasi-in situ Mössbauer

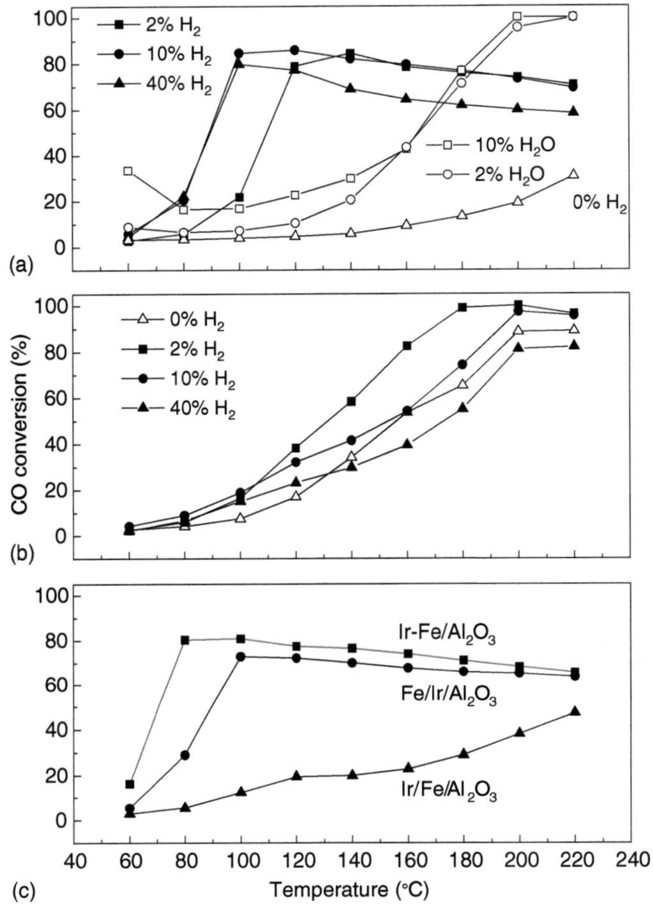

Figure 5.5 CO conversions over (a) Ir–Fe/SiO_2, (b) Ir/SiO_2, and (c) Ir–Fe/Al_2O_3, Fe/Ir/Al_2O_3, and Ir/Fe/Al_2O_3 catalysts [63, 64]. GHSV = 40 000 mL h^{-1} $g_{Cat}{}^{-1}$. Source: (a) Reproduced from Ref. [63]/with permission from American Chemical Society, (b) Reproduced from Ref. [64]/with permission from Elsevier.

spectroscopy was applied to study the chemical states of Fe in IrFe/SiO_2 [63]. Mössbauer spectra of the calcined Ir–Fe/SiO_2 were composed of a sextuplet and a doublet (Figure 5.6a and Table 5.2), ascribed to larger particles of α-Fe_2O_3 and superparamagnetic α-Fe_2O_3, respectively. After H_2 reduction for 2 h at 300 °C, the catalyst contained Fe^{n+} ($2 < n < 3$), Fe^{2+}, FeIr alloy, and Fe^0, indicating that both the superparamagnetic and larger particles of α-Fe_2O_3 in the calcined Ir–Fe/SiO_2 were reduced to low-valent Fe species [69–71]. Subsequently, the reduced Ir–Fe/SiO_2 catalyst was treated in 1 vol.% O_2 at 80 °C for 1 h. The spectrum of this reoxidized catalyst consisted of Fe^0 (24%) and Fe^{3+} (76%), indicating that Fe^{n+}, Fe^{2+}, FeIr alloy, and a portion of Fe^0 were oxidized to Fe^{3+} species. This remained portion of Fe^0 (24%) was in the core of the Fe^0 particles [72], whereas the shell of these Fe^0 particles was oxidized to Fe^{3+} during the reoxidation process (Figure 5.6a). The spectra of Fe/SiO_2 after H_2 reduction at 300 °C for 2 h consist of a Fe^{3+} doublet

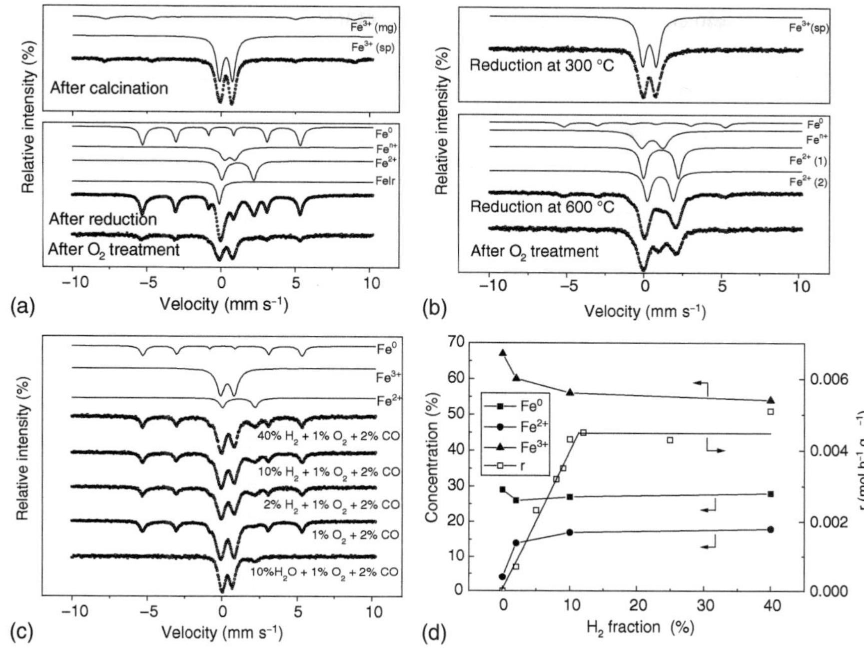

Figure 5.6 Quasi-in situ ^{57}Fe Mössbauer spectra of (a) Ir–Fe/SiO_2 (before and after 100% H_2 reduction at 300 °C for 2 h and followed by 1% O_2 treatment at 80 °C for 1 h), (b) Fe/SiO_2 (after H_2 reduction at 300 and 600 °C and followed by O_2 treatment after reduction at 600 °C), (c) Ir–Fe/SiO_2 after PROX with different H_2 contents in the gas feed, and (d) correlation between the reaction rate and amounts of various Fe species. Source: Reproduced from Ref. [63]/with permission from American Chemical Society.

(Figure 5.6b), indicating that it is more difficult to reduce the Fe species in Fe/SiO_2 than in Ir–Fe/SiO_2. Ir benefited the reduction of the Fe species. Fe species in Fe/SiO_2 reduced at 600 °C included Fe^{n+}, Fe^{2+} (1), Fe^{2+} (2), and Fe^0 (Figure 5.6b), and the subsequent 1 vol.% O_2 treatment at 80 °C for 1 h resulted in the disappearance of Fe^{n+} and Fe^0 and an almost unchanged overall amount of Fe^{2+} species. Since no Fe^{2+} was present in Ir–Fe/SiO_2 after reoxidation, it is more difficult to reoxidize the Fe^{2+} species in Fe/SiO_2 than in Ir–Fe/SiO_2. Ir facilitates redox ability of the Fe species in Ir–Fe/SiO_2.

Mössbauer spectra of Ir–Fe/SiO_2 during PROX at 80 °C were collected to clarify the active Fe species and to quantify different Fe species. As shown in Figure 5.6c, Fe^{3+}, Fe^{2+}, and Fe^0 were present in Ir–Fe/SiO_2 during PROX at 80 °C, but Fe^{n+} and FeIr alloy disappeared. The dramatic difference between these spectra is with the H_2 concentration increasing from 0% to 40%, the percentage of the Fe^{2+} species increased from 4% to 18%. The percentage of Fe^0 was almost constant (~28%) both in the presence and absence of H_2, and this portion of Fe^0 was encapsulated by ferric oxides and could not participate in the reactions. The changing trend of activity is quite similar to that of the percentage Fe^{2+} (Figure 5.6d), suggesting that Fe^{2+} in Ir–Fe/SiO_2 was essential for CO oxidation and PROX reaction. This conclusion was confirmed by the Mössbauer results of Ir–Fe/SiO_2 during CO oxidation in the presence of 10 vol.%

Table 5.2 Quasi-in situ ^{57}Fe Mössbauer parameters of $Ir-Fe/SiO_2$ and Fe/SiO_2 after different treatments.

Catalysts	Treatments	Chemical state	δ (mm s^{-1})	ΔE_Q (mm s^{-1})	MF (T)	RI (%)
$Ir-Fe/SiO_2$	Calcination at 300 °C for 5 h	Fe^{3+} (sp)	0.33	0.87	—	87
		Fe^{3+} (mg)	0.36	0.40	52.0	13
	H_2 reduction at 300 °C for 2 h	Fe^{n+} $(2<n<3)$	0.58	0.71	—	20
		Fe^0	0.00	0.00	32.9	43
		Fe^{2+}	1.13	2.15	—	26
		FeIr	−0.11	—	—	11
	1% O_2 at 80 °C for 1 h	Fe^{3+}	0.34	0.94	—	76
		Fe^0	0.00	0.00	33.0	24
Fe/SiO_2	Reduction at 300 °C for 2 h	Fe^{3+}	0.34	0.87	—	100
	Reduction at 600 °C for 2 h	Fe^{n+} $(2<n<3)$	0.53	1.38	—	29
		Fe^0	0.00	0.00	32.4	10
		Fe^{2+} (1)	1.10	2.25	—	30
		Fe^{2+} (2)	1.06	1.69	—	31
	1% O_2 at 80 °C for 1 h	Fe^{3+}	0.35	0.92	—	44
		Fe^{2+} (1)	0.98	2.40	—	40
		Fe^{2+} (2)	0.95	1.73	—	16

δ: isomer shift; ΔE_Q: electric quadrupole splitting; MF: magnetic field; RI: relative intensity; sp: superparamagnetic; mg: magnetic; uncertainty is ±5% of reported value.
Source: Ref. [63]/with permission from American Chemical Society.

water vapor. Seen from Figure 5.6c, the spectrum was composed of two doublets for Fe^{3+} in ferrihydrite ($\delta = 0.36$ mm s^{-1} and $\Delta E_Q = 0.66$ mm s^{-1}) and Fe^{2+}, but Fe^0 and FeIr alloy were absent due to their transformation to Fe^{3+} or Fe^{2+} during the reaction. Fe^{2+} was the only low-valence Fe species in $Ir-Fe/SiO_2$ and was the active site for activating O_2. It is widely accepted that Fe^{2+} species in Fe promoted PGM catalysts is active for PROX reaction based on different experimental measurements and theoretical calculations [56, 73–75], e.g. XPS [56, 74], XANES [73], and density functional theory (DFT) calculations [74, 75]. In comparison with the reports above, quasi-in situ Mössbauer studies provide quantified data for determining the roles of Fe^{2+} in CO oxidation and PROX reaction. Intimately contacted Ir and Fe had a strong metal–metal interaction between Ir and Fe, and the facilitation effect of Ir on the redox ability of Fe can take place. FeIr alloy disappeared during the reaction, forming active Fe^{2+} for CO oxidation and the PROX reaction. Besides the FeIr alloy, the shell of Fe^0 in intimate contact with Ir might also be oxidized to Fe^{2+} during the reaction [63, 64]. Fe^{n+} is more likely to be oxidized to Fe^{3+} instead of being reduced to Fe^{2+} during the reaction and cannot activate O_2 or participate in CO oxidation.

The impregnation sequence could influence the redox ability of the Fe species in the Al_2O_3 supported IrFe catalysts. Quasi-in situ Mössbauer studies were conducted on the catalysts after H_2 reduction at 300 °C for 2 h [64]. Similar to Ir–Fe/SiO_2, the Fe species in Ir–Fe/Al_2O_3 was also composed of Fe^{n+} ($2 < n < 3$), Fe^{2+}, Fe^0, and FeIr alloy. However, in Ir–Fe/Al_2O_3, Fe^{2+} (a) ($\delta = 1.00\,mm\,s^{-1}$, $\Delta E_Q = 2.38\,mm\,s^{-1}$) and Fe^{2+} (a) (IS = $1.04\,mm\,s^{-1}$, $\Delta E_Q = 1.55\,mm\,s^{-1}$) were observed. These two kinds of Fe^{2+} with similar δ values and different ΔE_Q values were assigned to the Fe^{2+} species in different positions in aluminate [76]. The percentage of FeIr alloy in Ir–Fe/Al_2O_3 (4%) was lower than that in Ir–Fe/SiO_2 (11%), possibly due to the stronger interaction between Al_2O_3 and Fe than that between SiO_2 and Fe. The reducibility follows the order of Ir/Fe/Al_2O_3 < Fe/Ir/Al_2O_3 < Ir–Fe/Al_2O_3, the trend of which is the same as that of the PROX activity. Ir could facilitate the reduction of Fe species [63] because of the Ir–Fe interaction, and thus, the strength of the Ir–Fe interaction in Ir–Fe/Al_2O_3 was the strongest while that in Ir/Fe/Al_2O_3 is the weakest. The reduced Ir–Fe/Al_2O_3 after reoxidation consisted of 80% Fe^{3+} and 20% Fe^0 (Table 5.3), and similar to Ir–Fe/SiO_2, this portion of Fe^0 was encapsulated by Fe^{3+} coming from reoxidation of the outer layer of the Fe^0 particles [64]. Interestingly, some Fe^{2+} (a) remained after 1 vol.% O_2 treatment in Fe/Ir/Al_2O_3 and Ir/Fe/Al_2O_3 (Table 5.3) due to the interaction between these Fe^{2+} (a) species and the aluminate support or the coverage of Fe^{2+} (a) by Fe^{3+}. These stable Fe^{2+} species cannot be reoxidized to Fe^{3+} by O_2 or activate O_2, and thus, might be spectators during CO oxidation. Consequently, 6% Fe^{2+} (a) in Fe/Ir/Al_2O_3 and 19% Fe^{2+} (a) in Ir/Fe/Al_2O_3 do not participate in PROX reaction.

Quasi-in situ Mössbauer spectra of Ir–Fe/Al_2O_3 and Fe/Ir/Al_2O_3 during PROX reaction show the presence of Fe^{3+}, Fe^{2+}, (a) and Fe^0, whereas the spectrum of Ir/Fe/Al_2O_3 shows Fe^{3+} and Fe^{2+} (a). The similar percentage of Fe^0 in Ir–Fe/Al_2O_3 after reoxidation (20%) and reaction (22%) indicates that Fe^0 is spectator and does not contribute to the PROX reaction. Similar conclusion can be obtained for Fe/Ir/Al_2O_3. The percentages of Fe^{2+} (a) in Ir–Fe/Al_2O_3, Fe/Ir/Al_2O_3, and Ir–Fe/Al_2O_3 were 17%, 15%, and 22%, respectively, and 6% Fe^{2+} (a) in Fe/Ir/Al_2O_3 and 19% Fe^{2+} (a) in Ir–Fe/Al_2O_3 were inert and spectators. Thus, the percentages of the active Fe^{2+} (a) species in Fe/Ir/Al_2O_3 and Ir/Fe/Al_2O_3 were $15\% - 6\% = 9\%$ and $22\% - 19\% = 3\%$, respectively. The CO conversion increased almost linearly with the increasing percentage of the active Fe^{2+} (a) species (Figure 5.7), indicating that Fe^{2+} (a) in the Ir–Fe catalysts plays the role of activating O_2 and is essential for the PROX reaction. The stronger interaction between the Fe species and Ir in Ir–Fe/Al_2O_3 led to more Fe^{2+} (a) maintained during PROX reaction. The percentage of FeIr alloy in Ir–Fe/Al_2O_3 was 4%, much less than that of the active Fe^{2+} (a) (17%), indicating that most of the active Fe^{2+} (a) did not come from FeIr alloy.

OH groups were reported to exert a positive effect on CO oxidation [77–79]. Seen from Figure 5.6c, a doublet for Fe^{3+} in ferrihydrite ($\delta = 0.36\,mm\,s^{-1}$ and $\Delta E_Q = 0.66\,mm\,s^{-1}$) was observed in Ir–Fe/$SiO_2$ during CO oxidation in 10 vol.% H_2O, indicating that OH groups were present under this reaction condition, possibly improving the PROX activity. Ir/Fe$(OH)_x$ can convert the CO completely at room temperature in PROX [80]. Also, Rh supported on Fe$(OH)_x$ and an

Table 5.3 ^{57}Fe Mössbauer parameters of Al_2O_3 supported IrFe catalysts after different treatments.

Catalysts	Treatments	Chemical state	δ (mm s^{-1})	ΔE_Q (mm s^{-1})	MF (T)	RI (%)
Ir–Fe/Al_2O_3	Reduction at 300 °C for 2 h	Fe^{n+}	0.43	1.00		15
		Fe^0	0.00	0.00	33.4	31
		Fe^{2+}(a)	1.00	2.38		15
		FeIr	0.09	—		4
		Fe^{2+}(b)	1.04	1.55		35
	1% O_2 at 80 °C for 1 h	Fe^{3+}	0.33	1.08		80
		Fe^0	0.00	0.00	33.4	20
Fe/Ir/Al_2O_3	Reduction at 300 °C for 2 h	Fe^{n+}	0.47	0.85		27
		Fe^0	0.00	0.00	33.3	18
		Fe^{2+}(a)	1.02	2.4		20
		Fe^{2+}(b)	0.99	1.67		35
	1% O_2 at 80 °C for 1 h	Fe^{3+}	0.32	1.09		80
		Fe^0	0.00	0.00	33.3	13
		Fe^{2+}(a)	1.07	2.14		6
Ir/Fe/Al_2O_3	Reduction at 300 °C for 2 h	Fe^{3+}	0.36	1.14		24
		Fe^{2+}(a)	1.09	2.22		29
		Fe^{2+}(b)	1.02	1.56		47
	1% O_2 at 80 °C for 1 h	Fe^{3+}	0.32	1.08		81
		Fe^{2+}(a)	1.09	2.15		19

δ: isomer shift; ΔE_Q: electric quadrupole splitting; MF: magnetic field; RI: relative intensity; uncertainty is ±5% of reported value.
Source: Adapted from Ref. [64].

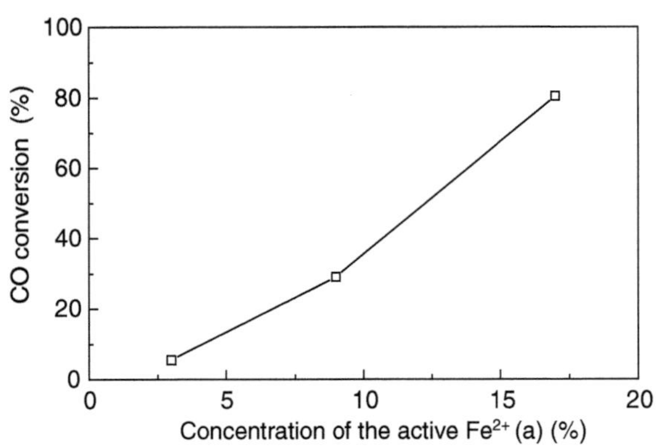

Figure 5.7 Correlation between the CO conversions and the relative amounts of the active Fe^{2+} (Fe^{2+} (a)). Source: Reproduced from Ref. [64]/with permission from Elsevier.

atomically dispersed iron hydroxide anchored on silica supported Pt nanoparticles could achieve 100% CO conversion in a wide temperature range of 20–70 °C and 198–380 K, respectively [81, 82]. All the reports above demonstrate that the supports containing OH groups are beneficial for PROX, leading to a higher PROX activity even at room temperature.

5.3.4 CuO/CeO_2 with Fe_2O_3 Additive

CuO/CeO_2 modified by the Fe or Ni species prepared using impregnation method exhibited a higher PROX activity than CeO_2, CuO/CeO_2, and Fe_2O_3/CeO_2, which could remove CO completely at 150 °C in an atmosphere containing 1 vol.% CO, 1 vol.% O_2, and 98 vol.% H_2 with 20 mg catalyst at a flow rate of 40 cm^3/min [83]. The impregnation sequence affected the PROX activity. $CuO/(Fe_2O_3/CeO_2)$ was prepared by supporting iron oxide on CeO_2 first and then supporting CuO. $Fe_2O_3/(CuO/CeO_2)$ was prepared by first supporting CuO on CeO_2, followed by supporting Fe_2O_3. 2.5% CuO/(2/5% Fe_2O_3/CeO_2) exhibited a higher PROX activity than 2.5% Fe_2O_3/(2/5% CuO/CeO_2). ^{57}Fe Mössbauer spectra of both 2.5% Fe_2O_3/(2/5% CuO/CeO_2) and 2.5% CuO/(2/5% Fe_2O_3/CeO_2) consisted of "paramagnetic" doublet of Fe^{3+} ion ascribed to iron oxide nanocluster with the particle size about 3 ∼ 7 nm. The parameters of Mössbauer spectra were almost the same before and after PROX reaction on 2.5% Fe_2O_3/(2/5% CuO/CeO_2), indicating that Fe state did not change after reaction. On the other hand, paramagnetic 31% Fe^{2+} ions were also observed in 2.5% CuO/(2/5% Fe_2O_3/CeO_2) after reaction. A ternary oxide compound in 2.5% CuO/(2/5% Fe_2O_3/CeO_2) was formed, in which oxygen transfer from CeO_2 via the lattice of Fe_2O_3 to the active centers CuO was promoted.

5.4 Concluding Remarks

By using ^{57}Fe Mössbauer spectroscopy, the state of Fe species in different Fe-containing PROX and CO oxidation catalysts (e.g. PtFe, IrFe, and Au/HMS-Fe catalysts) before reduction can be identified. On Au/Fe_2O_3, ferrihydrite was suggested to be important for accelerating CO oxidation, while hematite increased with the increasing calcination temperature. The transformation of ferrihydrite to α-Fe_2O_3 at higher calcination temperature might be responsible for the decrease of stability. Fe^{3+} species is present in the Fe-containing catalysts after calcination. When Fe_2O_3 was the support for Pt, the amount of Fe_3O_4 after reduction decreased with the increasing amount of ZrO_2 added, whereas the PROX activity decreased with the increasing ZrO_2 loading. Biogenic iron oxides produced from the metabolic activities of the iron-oxidizing bacteria contained mainly Fe^{3+} species and different FeOOH species. These materials show potential in the application as catalyst supports in CO oxidation. In situ ^{119}Sn and quasi-in situ ^{57}Fe Mössbauer spectroscopy can determine the active sites and the distribution of different Sn and Fe species after reduction and reaction quantitatively. By correlating the reaction rates and the amounts of different Fe species in Ir–Fe/SiO_2 and Ir–Fe/Al_2O_3 after PROX reaction, the active sites for O_2 activation were found to be Fe^{2+}. The PtSn alloy was important for CO oxidation, and the FeIr and PtFe alloys were suggested to be important for PROX, since they would segregate to provide intimately contacted noble metal sites for activating CO and metal oxides for activating O_2. Fe^{3+} in ferrihydrite was present in Ir–Fe/SiO_2 during CO oxidation in 10 vol.% H_2O, demonstrating that the OH groups benefit the PROX activity.

^{197}Au Mössbauer spectroscopy shows that Au(III) oxyhydroxide species ($AuOOH{\cdot}xH_2O$) in Au/Fe_2O_3 is beneficial for CO oxidation. However, Au^I atoms in Au/$Mg(OH)_2$ or Au–titania–zirconia were suggested to be important for CO oxidation. It seems that the active Au species for CO oxidation differs depending on different supports. During the PROX reaction, some researchers pointed out that metallic Au species in Au/Fe_2O_3 was active for CO oxidation, while cationic Au species was active for the reverse water gas shift reaction. ^{193}Ir Mössbauer spectroscopy shows that Ir species provide atomic oxygen that will react with CO molecules adsorbed on the adjacent metallic Pt site during CO oxidation in the SDB granules supported Pt and Ir catalyst.

For PGM catalysts, the strength of CO adsorption decreased when the size of the noble metal decreased from nano to sub-nano, promoting CO oxidation [84]. Recently, single-atom catalysts (SACs) have even been developed, which show excellent PROX performance [8, 85]. However, Mössbauer studies on these catalysts are still lacking. Future studies could be conducted on clarifying the Fe species on different Fe-containing single-atom metal PROX catalysts, and new light may be shed on the change of the state of Fe species in SACs.

Acknowledgments

Financial supports from the National Natural Science Foundation of China (Nos. 21476232, 21607029, 21961142006), the International Partnership Program of Chinese Academy of Sciences (No. 121421KYSB20170020), and the State Key Laboratory of Catalysis in Dalian Institute of Chemical Physics (N-16-07) are greatly acknowledged.

References

1 Liu, Z.M., Ma, L.L., Zhang, J. et al. (2013). *Catal. Rev. Sci. Eng.* 55: 255.
2 Si, R. and Flytzani-Stephanopoulos, M. (2008). *Angew. Chem. Int. Ed.* 47: 2884–2887.
3 O'Connell, M., Kolb, G., Schelhaas, K. et al. (2010). *Int. J. Hydrogen Energy* 35: 2317–2327.
4 Liu, K., Wang, A., and Zhang, T. (2012). *ACS Catal.* 2: 1165.
5 Zhang, W.S., Wang, A.Q., Li, L. et al. (2008). *Catal. Today* 131: 457.
6 Zhang, W.S., Wang, A.Q., Li, L. et al. (2008). *Catal. Lett.* 121: 319.
7 Zhang, W.S., Huang, Y.Q., Wang, J.H. et al. (2010). *Int. J. Hydrogen Energy* 35: 3065.

8 Qiao, B.T., Wang, A.Q., Yang, X.F. et al. (2011). *Nat. Chem.* 3: 634.

9 Liu, H.H., Jia, A., Wang, Y. et al. (2015). *Chin. J. Catal.* 36 (11): 1976–1986.

10 Allian, A.D., Takanabe, K., Fujdaja, K.L. et al. (2011). *J. Am. Chem. Soc.* 133: 4498.

11 Aragao, I.B., Bueno, J.M.C., and Zanchet, D. (2018). *Appl. Catal. A* 568: 86.

12 Navas-Cárdenas, C., Benito, N., Wolf, E.E., and Gracia, F. (2019). *Appl. Catal. A* 576: 11–19.

13 Moscu, A., Veyre, L., Thieuleux, C. et al. (2015). *Catal. Today* 258: 241.

14 Millet, J.-M.M. (2007). *Adv. Catal.* 51: 309.

15 Gutlich, P. (2012). *Z. Anorg. Allg. Chem.* 638: 15.

16 Stievano, L. and Wagner, F.E. (2012). Mößbauer spectroscopy. In: *Characterisation of Solid Materials and Heterogeneous Catalysts* (ed. M. Che and J. Vedrine), 407. Wiley-VCH.

17 Liu, K., Rykov, A.I., Wang, J., and Zhang, T. (2015). *Adv. Catal.* 58: 1.

18 Li, X., Zhu, K., Pang, J. et al. (2018). *Appl. Catal. B* 224: 518.

19 Zhang, W., Wang, J., Liu, K. et al. (2013). Mössbauer spectroscopic investigations of novel bimetal catalysts for preferential CO oxidation in H_2. In: *Mössbauer Spectroscopy, Applications in Chemistry, Biology and Nanotechnology* (ed. V.K. Sharma, G. Klingelhöfer and T. Nishida), 564–575. Wiley.

20 Wagner, F.E., Galvagno, S., Milone, C. et al. (1997). *J. Chem. Soc. Faraday Trans.* 93 (18): 3403.

21 Šmit, G., Strukan, N., Crajé, M.W.J., and Lázár, K. (2006). *J. Mol. Catal. A Chem.* 252: 163.

22 Šmit, G., Zrnčević, S., and Lázár, K. (2006). *J. Mol. Catal. A Chem.* 252: 103.

23 Wang, G.Y., Lian, H.L., Zhang, W.X. et al. (2002). *Kinet. Catal.* 43: 433.

24 Šmit, G., Lázár, K., and Crajé, M.W.J. (2007). *Croat. Chem. Acta* 80: 141.

25 Mizera, J., Spiridis, N., Socha, R. et al. (2012). *Catal. Today* 187: 20.

26 Zhao, K.F., Qiao, B.T., Zhang, Y.J., and Wang, J. (2013). *Chin. J. Catal.* 34: 1386.

27 Li, Y., Guan, Y., van Santen, R.A. et al. (2009). *J. Phys. Chem. C* 113: 21831.

28 Kocemba, I., Rynkowski, J., Gurgul, J. et al. (2016). *Appl. Catal. A* 519: 16.

29 Schoch, R., Huang, H., Schünemann, V., and Bauer, M. (2014). *ChemPhysChem* 15: 3768.

30 Carriazo, J.G., Centeno, M.A., Odriozola, J.A. et al. (2007). *Appl. Catal. A* 317: 120.

31 Oliveira, L.C.A., Fabris, J.D., Rios, R.R.V.A. et al. (2004). *Appl. Catal. A* 259: 253.

32 Kulshreshtha, S.K., Sharma, S., Vijayalakshmi, R., and Sasikala, R. (2004). *Indian J. Chem. Technol.* 11: 427.

33 Maksimov, Y.V., Tsodikov, M.V., Trusova, E.A. et al. (2001). *Catal. Lett.* 72 (1–2): 11.

34 Lelis, M.F.F., Rios, R.V.R.A., Lago, R.M. et al. (2002). *Hyperfine Interact. (C)* 345.

35 Shopska, M., Cherkezova-Zheleva, Z.P., Paneva, D.G. et al. (2013). *Cent. Eur. J. Chem.* 11 (2): 215.

36 Shopska, M., Paneva, D., Kadinov, G. et al. (2016). *Reac. Kinet. Mech. Catal.* 118: 179.

37 Shopska, M.G., Paneva, D.G., Kolev, H.G. et al. (2017). *Helv. Chim. Acta* 100: e1700172.

38 Shopska, M., Paneva, D., Kolev, H. et al. (2019). *React. Kinet. Mech. Catal.* 127: 617.
39 Margitfalvi, J.L., Borbáth, I., Hegedűs, M. et al. (2000). *J. Catal.* 196: 200.
40 Margitfalvi, J.L., Borbáth, I., Lázár, K. et al. (2001). *J. Catal.* 203: 94.
41 Margitfalvi, J.L., Borbáth, I., Hegedűs, M. et al. (2002). *Catal. Today* 73: 343.
42 Lázár, K., Rhodes, W.D., Borbáth, I. et al. (2002). *Hyperfine Interact.* 139/140: 87.
43 Somodi, F., Borbáth, I., Hegedűs, M. et al. (2009). *Appl. Surf. Sci.* 256: 726.
44 Somodi, F., Borbáth, I., Margitfalvi, J.L. et al. (2009). *Hyperfine Interact.* 192: 13.
45 Stievano, L. and Wagner, F.E. (2005). Investigation of catalytic materials using ^{197}Au Mössbauer spectroscopy. In: *Industrial Applications of Mössbauer Effect* (ed. M. Gracia, J.F. Marco and F. Plazaola), *AIP Conf. Proc*, 765, 3.
46 Daniells, S.T., Overweg, A.R., Makkee, M., and Moulijn, J.A. (2005). *J. Catal.* 230: 52.
47 Hutchings, G.J., Hall, M.S., Carley, A.F. et al. (2006). *J. Catal.* 242: 71.
48 Finch, R.M., Hodge, N.A., Hutchings, G.J. et al. (1999). *Phys. Chem. Chem. Phys.* 1: 485.
49 Lee, S.-J., Gavriilidis, A., Pankhurst, Q.A. et al. (2001). *J. Catal.* 200: 298.
50 Kobayashi, Y., Nasu, S., Tsubota, S., and Haruta, M. (2000). *Hyperfine Interact.* 126: 95.
51 Boyd, D., Golunski, S., Hearne, G.R. et al. (2005). *Appl. Catal. A* 292: 76.
52 Magadzu, T., Hearne, G.R., and Scurrell, M.S. (2007). Characteristics of gold-zeolite Y catalysts in CO oxidation and ethylene hydrogenation. In: *From Zeolites to Porous MOF Materials – the 40th Anniversary of International Zeolite Conference* (ed. R. Xu, Z. Gao, J. Chen and W. Yan), 1059. Elsevier.
53 Sawicki, J.A., Marcinkowska, K., and Wagner, F.E. (2010). *Nucl. Instrum. Methods Phys. Res. B* 268: 2544.
54 Yin, J., Wang, J., Zhang, T., and Wang, X.D. (2008). *Catal. Lett.* 125: 76.
55 Kotobuki, M., Shido, T., Tada, M. et al. (2003). *Catal. Lett.* 103: 263.
56 Zhang, H., Liu, X., Zhang, N. et al. (2016). *Appl. Catal. B* 180: 237.
57 Cherkezova-Zheleva, Z., Paneva, D., Todorova, S. et al. (2014). *Hyperfine Interact.* 226: 529.
58 Schmal, M., Scheunemann, R., Ribeiro, N.F.P. et al. (2011). *Appl. Catal. A* 392: 1.
59 Vandenberghe, R.E., De Grave, E., Landuydt, C., and Bowen, L.H. (1990). *Hyperfine Interact.* 53: 175.
60 Zepeda, T.A., Martinez-Hernández, A., Guil-López, R., and Pawelec, B. (2010). *Appl. Catal. B* 100: 450.
61 Landon, P., Ferguson, J., Solsona, B.E. et al. (2006). *J. Mater. Chem.* 16: 199.
62 Zhao, X., Huang, Y., Zhang, W. et al. (2014). *J. Nanosci. Nanotechnol.* 14: 1.
63 Liu, K., Wang, A., Zhang, W. et al. (2010). *J. Phys. Chem. C* 114: 8533.
64 Liu, K., Wang, A., Zhang, W. et al. (2010). *Chin. J. Catal.* 31: 1335.
65 Xu, J., Xu, X.-C., Ouyang, L. et al. (2012). *J. Catal.* 287: 114.
66 Potemkin, D.I., Filatov, E.Y., Zadesenets, A.V., and Sobyanin, V.A. (2017). *Catal. Commun.* 100: 232.
67 Kotobuki, M., Watanabe, A., Uchida, H. et al. (2005). *J. Catal.* 236: 262.

68 Hu, X., Gao, Y., Wang, W., and Chen, C. (2016). *Inter. J. Hydrogen Energy* 41: 14079.

69 Berry, F.J. and Jobson, S. (1988). *Hyperfine Interact.* 41: 745.

70 Niemantsverdriet, J.W. and Van Der Kraan, A.M. (1986). *Surf. Interface Anal.* 9: 221.

71 Van Gruijthuijsen, L.M.P., Howsmon, G.J., Delgass, W.N. et al. (1997). *J. Catal.* 170: 331.

72 Wang, C.M., Baer, D.R., Thomas, L.E. et al. (2005). *J. Appl. Phys.* 98: 094308.

73 Kotobuki, M., Watanabe, A., Uchida, H. et al. (2005). *Catal. Lett.* 103: 263.

74 Lou, Y. and Liu, J. (2017). *Chem. Commun.* 53: 9020.

75 Fu, Q., Li, W.X., Yao, Y.X. et al. (2010). *Science* 328: 1141–1144.

76 Larsson, L., O'Neill, H.S., and Annersten, H. (1994). *Eur. J. Mineral* 6: 39.

77 Liu, K., Wang, A., Zhang, W. et al. (2011). *Ind. Eng. Chem. Res.* 50: 758.

78 Lin, J., Qiao, B., Li, L. et al. (2014). *J. Catal.* 319: 142.

79 Saavedra, J., Whittaker, T., Chen, Z. et al. (2016). *Nat. Chem.* 8: 584.

80 Lin, J., Qiao, B., Liu, J. et al. (2012). *Angew. Chem. Int. Ed.* 51: 2920.

81 Guan, H., Lin, J., Li, L. et al. (2016). *Appl. Catal. B* 184: 299.

82 Cao, L., Liu, W., Luo, Q. et al. (2019). *Nature* 565: 631.

83 Firsova, A.A., Il'ichev, A.N., Khomenko, T.I. et al. (2007). *Kinet. Catal.* 48 (2): 282.

84 Lin, J., Wang, X., and Zhang, T. (2016). *Chin. J. Catal.* 37 (11): 1805.

85 Wang, A., Li, J., and Zhang, T. (2018). *Nat. Rev. Chem.* 2: 65.

6

Application of ^{57}Fe Mössbauer Spectroscopy in Studying Fe–N–C Catalysts for Oxygen Reduction Reaction in Proton Exchange Membrane Fuel Cells

Xinlong Xu[1], Junhu Wang[2], Suli Wang[1], and Gongquan Sun[1]

[1] *Dalian Institute of Chemical Physics, Chinese Academy of Sciences, Division of Fuel Cells and Battery, Dalian National Laboratory for Clean Energy, No. 457 Zhongshan Road, Dalian 116023, P. R. China*
[2] *Dalian Institute of Chemical Physics, Chinese Academy of Sciences, Center for Advanced Mössbauer Spectroscopy, Mössbauer Effect Data Center, No. 457 Zhongshan Road, Dalian 116023, P. R. China*

6.1 Introduction

With the increasing concerns of energy and environmental issues, proton exchange membrane fuel cells (PEMFCs) featured with high capacity, high energy conversion efficiency, and low emissions receive extensive attention, while the large-scale application is still limited by the high cost [1, 2]. One reason is the use of expensive and scarce platinum-group metal (PGM) as electrocatalysts for both hydrogen oxidation reaction (HOR) in the anode and oxygen reduction reaction (ORR) in the cathode in most existing commercial PEMFCs [3, 4]. In particular, the higher Pt loading in the cathode (about eight times as much as the anode) is more critical due to the far more sluggish kinetics of ORR than HOR [5]. In the last decades, tremendous efforts have been devoted to the development of efficient nonprecious metal catalysts (NPMCs) for ORR in acidic and alkaline environments, including metal oxides, metal nitrides, metal sulfides, heteroatomic doped carbon, and so on [6–10]. There unto, iron and nitrogen co-doped carbon (Fe–N–C) aroused extensive interest in the study of PEMFCs due to the low cost, convenient synthesis, and relatively high activity in the acidic environment [11–14].

Studies on Fe–N–C catalysts could be traced back to the 1960s when Jasinski discovered the ORR activity of cobalt phthalocyanine in alkaline conditions, which was later extended to Fe macrocycles such as Fe phthalocyanine (FePc) and Fe porphyrin [15]. Then, the enhanced activity and stability were achieved through the heat treatment of Fe macrocycles (or Fe macrocycles adsorbed on carbon supports), deriving a new category of catalysts which was composed of carbon supports, N dopants and various Fe species, and labeled as Fe–N–C [16]. In a later pioneer work, it was found that the expensive Fe macrocycles could be replaced by a mixture of individual Fe salts, nitrogen sources, and carbon precursors. This low-cost, convenient and flexible method led to the rapid growth of Fe–N–C [17–19].

Mössbauer Spectroscopy: Applications in Chemistry and Materials Science,
First Edition. Edited by Yann Garcia, Junhu Wang, and Tao Zhang.

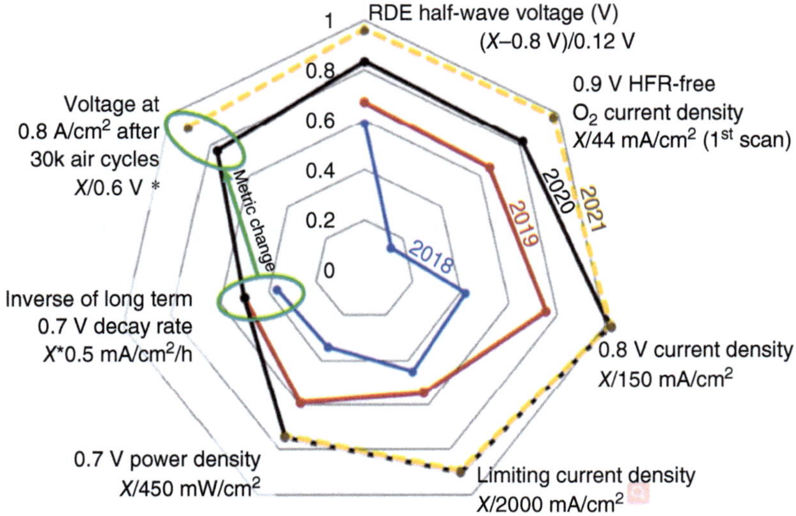

Figure 6.1 The spider chart representing the technical target and status of PGM-free catalysts for ORR from the Department of Energy (DOE) hydrogen program 2021 annual merit review (AMR). X = spider chart values. Source: Adapted from Litster [20].

In recent progress, Fe–N–C exhibited the half-wave potential ($E_{1/2}$) over 0.85 V vs. RHE in the acidic electrolyte tested by rotating ring-disk electrode (RRDE) and peak power density (P_{max}) over 1 W cm^{-2} in the PEMFC single cell, exceeding other types of NPMCs [21, 22]. However, as shown in Figure 6.1, the state-of-the-art activity is still struggling to match the requirement of the practical application according to the target proposed by the US Department of Energy (DOE). Worse still, PEMFCs assembled with Fe–N–C usually lose 40–80% of the initial performance after a short-term aging test (<100 h), which is far behind the stability target [20, 23–26].

By using current density at applied potential ($j_{(E)}$) as a descriptor, the performance of Fe–N–C in PEMFCs could be simply described as follows:

$$j_{(E)} = \mathrm{TOF}^{*}_{(E)}\mathrm{SD}^{*}e^{*}d$$

where SD is the site density (sites $g_{cat}{}^{-1}$), TOF_E is the turnover frequency at applied potential (z_{e-} site^{-1} s^{-1}), e is the electric charge of one electron, and d is the mass loading of the catalyst on the electrode (g_{cat} cm^{-2}) [27]. It's easily concluded that the improvement of performance could be achieved either by increasing the instinct activity (TOF) or the number of active sites (SD). Also, the loss of performance could be correlated to the decrease of TOF or SD with respect to catalysts, though some problems at the level of membrane electrode assemble (MEA), such as the degradation of the membrane, the loss of triple phase boundary, water flooding and so on are also critical [28, 29]. Under this premise, the approach to highly reactive and stable Fe–N–C relies heavily on the interpretation of several crucial issues:

(1). As mentioned earlier, the preparation of Fe–N–C depends on a pyrolysis routine, which results in Fe–N–C complexities with various Fe phases including metallic Fe, Fe carbides, Fe nitrides, Fe coordinated with N (FeN_x), etc. Besides, the activity is significantly impacted by the adsorption behavior according to Sabatier's principle. Thus, it is strongly related to the local environment and the electron density of Fe atoms [30]. Therefore, the recognition of Fe species with high TOF and the understanding of the structure–activity relationship is an effective guide for the design of Fe–N–C.

(2). The parallel formation of active and inactive Fe species during pyrolysis is inevitable and the formation of some inactive species is thermodynamically favorable, which means the Fe–N–C with high SD could not be obtained by simply increasing the Fe loading [13]. Therefore, investigation of the formation of active species is essential to the design of synthesis for higher SD.

(3). The degradation mechanism of Fe–N–C still remains further exploration up to now, which is related to both materials and conditions. The structural evolution and destruction of active sites may lead to the decrease of TOF and SD, respectively. It occurs continuously during working and significantly depends on the operation conditions (e.g. temperature, pH, and potential) and intermediate processes (e.g. the formation of H_2O_2) [31–34]. However, the characterization technique is still limited, especially for the in situ and *operando* observations.

To solve these problems, ^{57}Fe Mössbauer spectroscopy was widely introduced as a powerful tool for its sensitivity to the valence, spin state, and coordination environment of Fe [27, 35]. Different from X-ray absorption spectroscopy (XAS), which gives the "average" structure information of selected elements in the bulk catalyst, ^{57}Fe Mössbauer spectroscopy allows structural identification and quantitative analysis of different Fe species [36]. In combination with electrochemical measurements, physical characterizations and theoretical calculations, more insights into the fine structure of Fe–N–C and the structure–activity/stability relationship are able to be obtained. In this chapter, we introduced the advanced techniques of room temperature (RT), low temperature (LT), and electrochemical in situ/*operando* ^{57}Fe Mössbauer spectroscopy in studying Fe–N–C catalysts for ORR. The advancement in the identification of active sites, investigation of degradation mechanism, and rational design of synthesis with the assistance of ^{57}Fe Mössbauer spectroscopy was summarized. Finally, future trends for the ^{57}Fe Mössbauer spectroscopy in studying Fe–N–C were discussed.

6.2 Advanced ^{57}Fe Mössbauer Spectroscopy Technique

6.2.1 Room Temperature ^{57}Fe Mössbauer Spectroscopy

^{57}Fe Mössbauer spectroscopy conducted at RT by using bulk Fe–N–C power or electrode is a common method in the traditional study. In the later part, most studies were conducted in this way without an external magnetic field unless otherwise stated. The structural analysis depends on three important parameters: isomer shift (δ), quadrupole splitting (Δ), and magnetic hyperfine field (B). The isomer shift arises from the interaction between the Fe nucleus and extranuclear electrons, which is typically given relative to a standard reference (e.g. alpha iron) and provides the information of electronic structure such as valence and spin state; the quadrupole splitting arises from the interaction between electric field gradient (EFG) and electric quadrupole moment, which provides the information of electronic symmetry; the magnetic field arises from the magnetic interaction, which provides the information of effective magnetic field. According to various δ, Δ, and B, the original spectrum could be fitted into several subspectra including singlet, doublet, and sextet, and the absorption area represents the relative content with the hypothesis of identical Lamb-Mössbauer factors (f_{LM}) for all Fe species. The assignments are usually based on the parameters of Fe macrocycles and inorganic Fe species.

Also, the combination of RT ^{57}Fe Mössbauer spectroscopy with other characterization techniques could lead to a more comprehensive understanding of Fe–N–C. For example, based on the total Fe loading determined by induced coupled plasma-mass spectroscopy (ICP-MS) or neutron activation analysis (NAA) and the content of different Fe species determined by Mössbauer spectroscopy, the density of the specific site in the bulk Fe–N–C (SD_{bulk}) could be calculated. This method was widely used for the investigation of SD_{bulk} and TOF of Fe–N–C [40–42]. However, such an evaluation suffered the limitation because not all active sites in the bulk contributed to ORR. To address the puzzle, probe molecules such as carbon monoxide and nitrite (regarded as an O_2 analog) were further introduced to determine sites that were accessible and reactive to ORR [43–45]. Leonard et al. used CO *cyro* chemisorption in combination with Mössbauer spectroscopy to unravel the SD of Fe–N–C, in which the Mössbauer spectroscopy represented the SD_{bulk} and the CO *cyro* adsorption represented the quantification of surface-sensitive sites ($SD_{surface}$). The monotonic correlation between the $SD_{surface}$ and mass activity of Fe–N–C tended strong evidence for the claim of probing catalytical active sites. And a descriptor of Φ, which referred to $SD_{surface}/SD_{bulk}$, was proposed to estimate the site utilization in Fe–N–C [44].

6.2.2 Low Temperature and Computational ^{57}Fe Mössbauer Spectroscopy

However, it's still difficult to give unambiguous identification of those subspectra to complex Fe species in Fe–N–C just by Mössbauer spectroscopy at RT. As a recoilless nuclear resonance absorption spectroscopy, the lower temperature means better measurement conditions, which allows the minimized recoil effect and maximized resonant absorption. In recent years, LT ^{57}Fe Mössbauer spectroscopy was developed as the powerful tool for the study on Fe–N–C, which enabled the more precise identification and quantification of the active sites. At RT, nanosized particles of α-Fe and of most iron carbides may undergo superparamagnetic relaxation, leading to singlets or quadrupole doublets. But these structures could be well distinguished at LT through magnetic sextets. Moreover, the temperature-dependent test could be used to determine the absolute values of the Debye temperature and further calculate the temperature-dependent f_{LM} values for quantification [46].

As shown in Figure 6.2a, the temperature-dependent ^{57}Fe Mössbauer spectroscopy study of the Fe–N–C comprising quasi-exclusively FeN_x ($Fe_{0.5d}$) revealed that only two doublets could be measured at as low as 50 K, while an additional sextet was observed and assigned to γ-Fe_2O_3 at 5 K. Such low content of γ-Fe_2O_3 in $Fe_{0.5d}$ (14% in 1.5 wt% Fe) could not even be detected in extended X-ray absorption fine structure (EXAFS), indicating the irreplaceable role of LT ^{57}Fe Mössbauer spectroscopy. Moreover, the f_{LM} values were found significantly smaller than those of α-Fe and γ-Fe at RT [37].

Because the size of Fe macrocycle molecules for reference is smaller than the average size of graphitic crystallites embedded with FeN_x in pyrolyzed Fe–N–C, the computational Mössbauer spectroscopy was also introduced and the δ and Δ could be predicted with good accuracy using hybrid functionals and suitable basis, which enabled the more accurate assignment of doublets. Recently, computational Mössbauer spectroscopy was calibrated for plausible active sites in Fe–N–C and the data was made publicly available in an interactive notebook. TPSSh, B3LYP, and PBE0 density functionals were found able to predict δ and Δ values accurately and the trust regions of B3LYP were 0.13 mm s^{-1} for the δ and 0.36 mm s^{-1} for the Δ, respectively (Figure 6.2b–c). That made it possible to identify the hyperfine structure of active sites for ORR in Fe–N–C [38].

6.2.3 In Situ Electrochemical ^{57}Fe Mössbauer Spectroscopy

By definition, "in situ" characterization describes testing the structure of catalysts under conditions relevant to the practical catalytic operation. There are two types of equipment developed for in situ electrochemical ^{57}Fe Mössbauer spectroscopy: one is conducted by the three-electrode system in the half-cell, the Fe–N–C electrode is immersed in the electrolyte and the potential is controlled by an electrochemical workstation. The other is the in situ PEMFC test, where Fe–N–C catalysts are assembled into MEA. Through in situ characterization, the "true" active sites for ORR among various Fe species could be identified and the stability could be investigated by the structural evolution under working conditions. As soon as 1984, in situ ^{57}Fe Mössbauer spectroscopy was conducted to examine an operating oxygen-fed cathode fabricated with FePc in an alkaline solution (Figure 6.2d). When ORR occurred, the decreased Δ was observed and attributed the interaction in the axial position with O–C and OH groups (Figure 6.2e). Moreover, the degradation of the catalyst related to demetallation was confirmed by the decrease (15%) of the total resonant absorption area [39]. The demetallation of FePc was also revealed by electrochemical tip-enhanced Raman spectroscopy (EC-TERS) recently, which was considered the possible reason for the degradation of Fe–N–C [49]. The in situ test in 0.5 M H_2SO_4 by Bouwkamp-Wijnoltz et al. showed the valence change of Fe ions in heat-treated FeTPPCl/carbon but definitive assignments were difficult to make due to the considerable site heterogeneity of catalysts [50]. Recently, in situ electrochemical ^{57}Fe Mössbauer spectroscopy becomes booming in the study of Fe–N–C, and the progress in the identification of active sites and demonstration of degradation mechanism will be introduced in subsequent sections [30, 51–54].

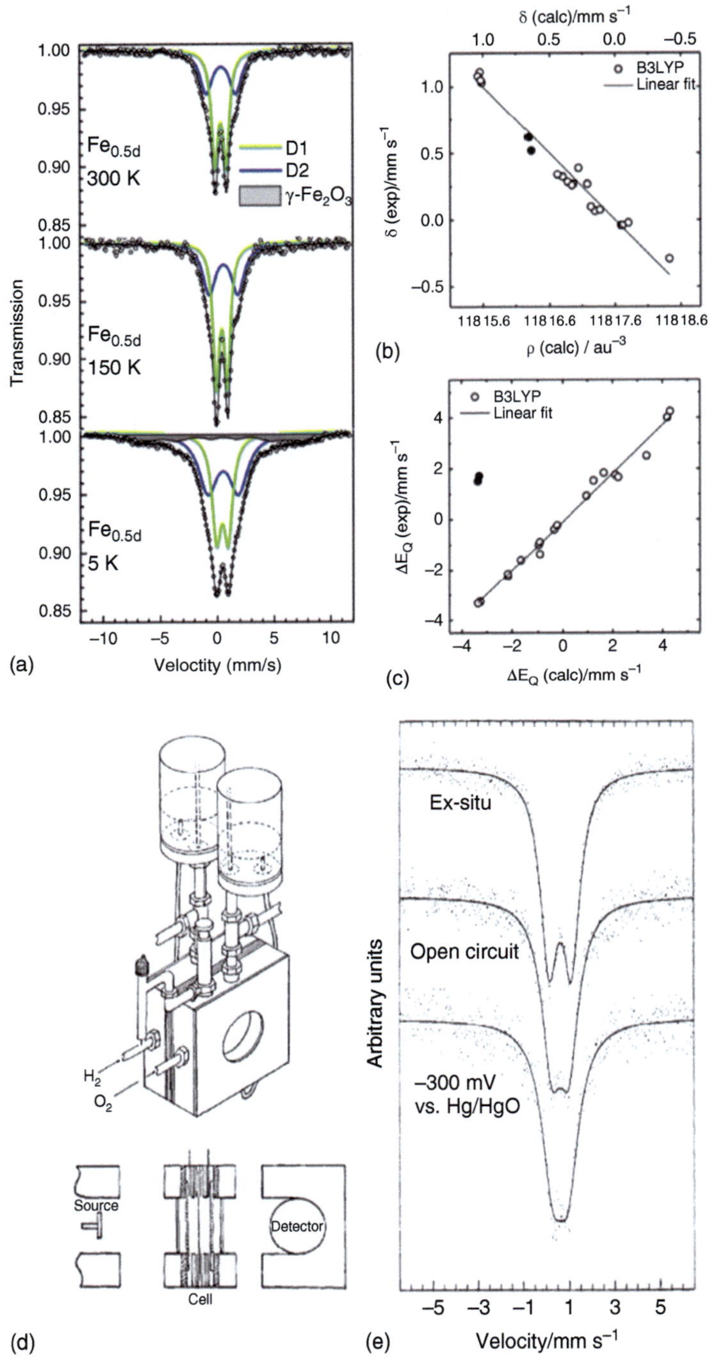

Figure 6.2 (a) The temperature-dependent Mössbauer spectra of Fe–N–C denoted as $Fe_{0.5d}$ recorded at 300, 150, and 5 K. Source: Sougrati et al. [37]/with permission of Elsevier; (b–c) Correlation of experimental and calculated δ and Δ for the density functionals of B3LYP. Source: Gallenkamp et al. [38]/John Wiley & Sons/Public Domain; (d–e) The schematic diagram of the in situ Mössbauer spectroscopy designed by Scherson et al. and the collected spectra of ^{57}FePc adsorbed on Vulcan XC-72 at ex situ (dry powder), open circuit and −300 mV vs. Hg/HgO. Source: Reproduced with permission from ref. [39], Elsevier.

6.3 Characterization of Fe–N–C Using ^{57}Fe Mössbauer Spectroscopy

6.3.1 Identification of Active Sites

Although tremendous achievements have been made over the past decades, the understanding of active sites in terms of the Fe phase, chemical state, and coordination environment remains uncertain. One of the key reasons is the complex components of pyrolyzed Fe–N–C including different Fe phases (Fe metal, oxides, carbides, nitrides, and FeN_x) and various FeN_x configurations.

In early studies, researchers attempted to establish the relationship between ORR activity and the content of a particular type of Fe species, which was regarded as powerful evidence for the identification of active sites. For example, Zhong et al. prepared Fe–N–C catalysts denoted as G@FeNPs containing α-Fe, γ-Fe, γ-Fe_2O_3, Fe_3C, and FeN_x and found the ORR activity was positively correlated with the Fe_3C content, indicating its involvement as the active site in ORR [55]. Varnell et al. also gave powerful evidence for Fe_3C as the active site by utilizing gas purification (Cl_2 and H_2 treatments) to control the components of Fe–N–C. As shown in Figure 6.3a–c, Fe_3C and α-Fe along with FeN_4 were observed in the as-prepared catalyst while FeN_4 was absent after the H_2 treatment. The H_2-treated catalyst exhibited great ORR activity and it was explained that reduced Fe species protected by carbon were ORR-active species while FeN_4 sites were not essential. However, the control sample without N displayed minimal ORR activity and the author discussed that the function of N was to promote the formation of the carbon-encapsulated Fe nanoparticles [47]. By using ex situ ^{57}Fe Mössbauer spectroscopy and in situ XAS under simulated conditions of operating PEMFCs, Strickland et al. revealed that ORR occurs primarily on the N-doped carbon in a system without direct Fe–N coordination. Fe was considered a catalyst for the graphitization of N-doped carbon during the heat treatment and the formed Fe/Fe_xC nanoparticles allowed the stabilization of the peroxide intermediate and the full 4e^- ORR through the synergistic effect [56].

Among different insights, it is more widely accepted that the ORR activity of Fe–N–C origins from FeN_x, which was proposed from the very beginning because of its structural similarity to ORR-active Fe macrocycles. Kneebone et al. introduced electrochemical reduction and nitric oxide (NO) as the probe in the Mössbauer spectroscopy study (80 K with a 0.07 T applied magnetic field) to identify ORR active sites of PANI-^{57}Fe–C (Figure 6.3d–f). After the electrochemical reduction at −0.1 V vs. RHE, no changes occurred to bulk Fe species (sextet and singlet) and Doubl 1 ($\delta = 0.87\ \text{mm s}^{-1}$, $\Delta = 2.05\ \text{mm s}^{-1}$), while Doubl 2, 3, 4 converted into three new doublets (Doubl 2_{red}, 3_{red}, 4_{red}). The subsequent NO treatment revealed that only Doubl 4_{red} ($\delta = 1.54\ \text{mm s}^{-1}$, $\Delta = 2.72\ \text{mm s}^{-1}$) was perturbed upon NO treatment to generate a new Doubl 4_{NO} ($\delta = 0.60\ \text{mm s}^{-1}$, $\Delta = 0.56\ \text{mm s}^{-1}$), and the decreased δ and Δ were consistent with nonheme Fe–NO adducts. It is demonstrated that only specific FeN_x were gas-phase reactive to the electrochemical and molecular probe and possibly serving as the ORR active sites, while the bulk-Fe species were not [48].

During the development, the relationship between the ORR activity and fine structures of FeN_x were widely investigated. Li et al. systematically investigated the effect

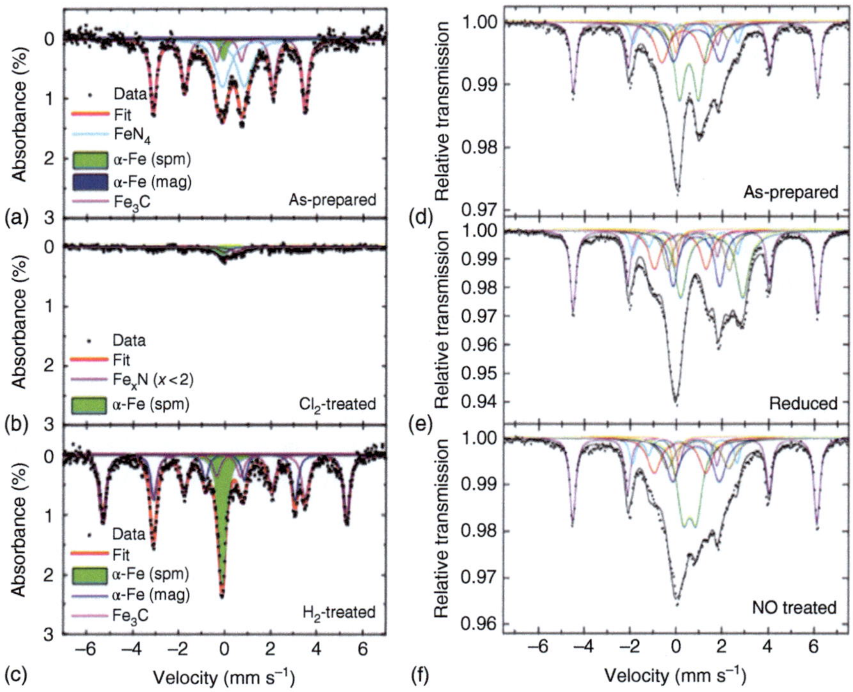

Figure 6.3 (a–c) Mössbauer spectra of as-prepared (a), Cl_2-treated (b) and H_2-treated (c) catalysts collected at 300 K. Source: Varnell et al. [47]/Springer Nature/CC BY 4.0; (d–f) Mössbauer spectra of as-prepared (d), reduced (e) and NO treated (f) PANI-^{57}Fe–C collected at 80 K with a 0.07 T applied magnetic field. Source: Reproduced with permission from ref. [48], American Chemical Society.

of coordination number by tuning the annealing temperature and it was found the ORR activity and PEMFC performance followed the order $FeN_4 > FeN_3 > FeN_1$ [57]. Shen et al. introduced a template casting strategy to produce Fe–N–C and a new doublet ($\delta = 0.60\,mm\,s^{-1}$, $\Delta = 3.13\,mm\,s^{-1}$) was found in ^{57}Fe Mössbauer spectroscopy, which was further identified as FeN_2 moieties through EXAFS. The tendency of $E_{1/2}$ with FeN_2 concentrations was also demonstrated in this work but it is limited in the alkaline environment. [58]. Besides N coordination, the coordination of other heteroatoms with Fe (instead of doping in carbon matrix), such as O, F, P, and so on also played key roles. Chen et al. reported an axial Fe–O coordination of O–FeN_4 (D2) and O–FeN_4–O_2 (D3) detected by Mössbauer spectroscopy at 300 K. The effect of axial O coordination was revealed in combination with various techniques but the Mössbauer parameters were absent and the ORR activity was studied in 0.1 M KOH solution [59]. Also, replacing N with P could significantly change the coordination bond length and electropositivity because P had a larger radius and lower electronegativity than N and C (N (3.04) > C (2.55) > P (2.19)) [60]. Recently, a high-performance Fe–N/P–C catalyst was developed by Yuan et al. using cutting edge techniques and a unique singlet ($\delta = 0.16\,mm\,s^{-1}$) was observed in Mössbauer spectroscopy, which was distinct from the commonly observed doublets of FeN_x and

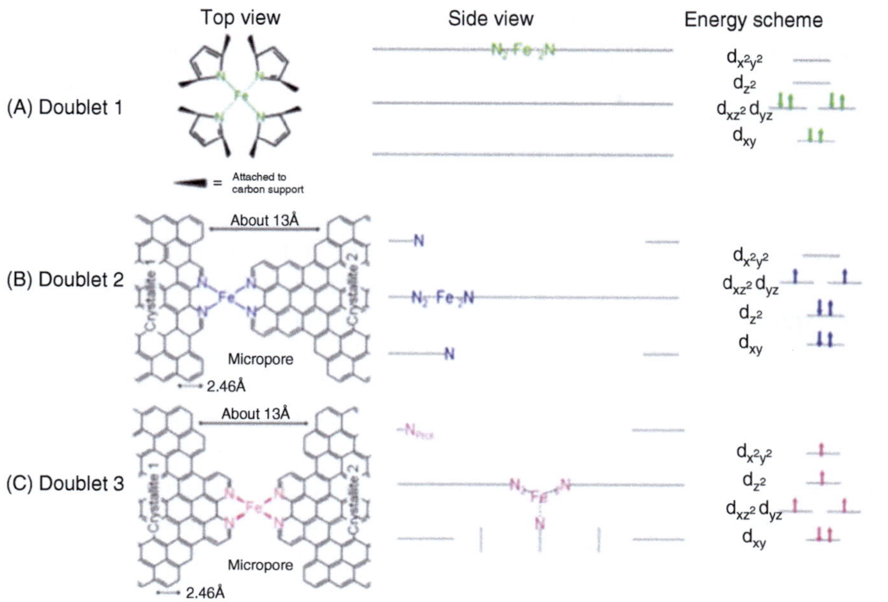

Figure 6.4 The top view, side view, and energy scheme of D1 (FeN_4/C), D2 (FeN_{2+2}), and D3 (N–FeN_{2+2}). Source: Reproduced with permission from ref. [40], Royal Society of Chemistry.

identified as Fe–N_3P sites. Compared with FeN_4, Fe–N_3P was more thermodynamically conducive to the adsorption of O_2, cleavage of the O–O bond, and reduction of *OH [61].

When it comes to the state of central Fe, complicated factors including valence, electron density, and spin state are involved in the structure–activity relationship. In the Mössbauer spectroscopy study by Kramm et al. at RT, five doublets and an additional broad singlet were deconvoluted in the fitted spectra of a highly active Fe–N–C catalyst. The proposed structure of D1, D2, and D3 were shown in Figure 6.4. The low-spin (LS) D1 and high-spin (HS) D3 were suggested active for ORR but medium-spin (MS) D2 was inactive due to the completely filled $3d_z^2$ orbital, which prevented the end-on adsorption of O_2 [40]. However, things may be different under alkaline conditions as the high ORR activity of an MS Fe(III)N_4 system activated by Mn–N moieties suggested by Yang et al., which possessed one e_g electron ($t_{2g}4e_g1$) readily penetrating the antibonding π-orbital of O_2 [62]. Wang et al. summarized the δ for six Fe macrocycles and related it to the $E_{1/2}$ of ORR. The enhanced catalytic activity was observed along with the increased δ, which could be attributed to the growing 3d electron density and higher spin state of Fe atoms [63]. This behavior was similar to the one observed for pyrolyzed Fe–N–C that a higher 3d-electron density of Fe enabled a higher TOF [64].

Recently, more attention was paid to the coordination environment around FeN_x, which was expressed by FeN_xC_y, in which y was the number of carbon atoms around the FeN_x site. Some typical FeN_xC_y sites with different local carbon structures were suggested such as the FeN_4 site bridging two adjacent armchair-like graphitic edges

(FeN_4C_8), the FeN_4 site embedded in an otherwise intact graphitic layer (FeN_4C_{10}), and the FeN_4 site bridging two adjacent zigzag graphitic edges with a porphyrin-like architecture (FeN_4C_{12}) [65]. An important breakthrough was the confirmation of porphyrin-like FeN_4C_{12} moieties as the active site by Zitolo et al. The experimental X-ray absorption near edge structure (XANES) spectra was well identical with the simulative structure for both Fe–N–C suffered Ar or NH_3 pyrolysis, and the higher activity after NH_3 treatment was attributed to the adjustment of the basicity of the N-doped carbon [11]. However, though the XANES of the model catalyst was fitted with only a porphyrin-like FeN_4C_{12} structure, the fitting of its Mössbauer spectrum still required at least two distinct doublets (D1 and D2).

Later, the study of computational and LT Mössbauer spectroscopy was conducted on the same model catalyst for the more explicit fine structure [46]. As shown in Figure 6.5a–b, D1 and D2 had a similar δ while their Δ varied in distribution (~0.9 mm s^{-1} for D1 and ~2.3 mm s^{-1} for D2). D1 and D2 were confirmed as HS Fe(III)N_4C_{12} and LS or MS Fe(II)N_4C_{10}, respectively. It indicated that D1 corresponded to surface-exposed sites, whereas D2 might be assigned to either bulk sites that were inaccessible to O_2 or surface sites that bound O_2 weaker than D1. This was further implied by the in situ Mössbauer spectroscopy in PEMFCs at RT (Figure 6.5c–d). The δ and Δ of both D1 (labeled D1H) and D2 were similar to ex situ measured values at 0.8 V while the substantial increase of δ and Δ for D1 (labeled D1L) was observed at 0.2 V. This reversible change with potential corresponded to the switch from HS OH–Fe(III)N_4C_{12} (D1H) to HS Fe(II)N_4C_{12} (D1L) [53]. At the same time, three other in situ Mössbauer spectroscopy studies of Fe–N–C catalysts were published. Xu et al. conducted in situ tests in 0.1 M $HClO_4$ at RT and found that D1 (also assigned as FeN_4C_8 besides FeN_4C_{12} [21]) mainly contributed to ORR at high potential while D2 (assigned as FeN_4C_{10}) only exhibited catalytic activity as the overpotential increased [30]. Ni et al. concluded two different FeN_x moieties from in situ data in 0.1 M H_2SO_4 including a ferrous HS site with pyrrolic-N_4 coordination (site 1a and site 1b) and a ferric MS site with pyridinic-N_4 coordination and anion bounding (site 2). Site 1 was not only considered to catalyze ORR in a direct pathway but also contributed to a small fraction of H_2O_2. However, Site 2 was able to reabsorb H_2O_2, thus contributing to the overall reduction via the 2*2 pathway [52]. In their latest work, it is found a direct structural correspondence between the newly emerging Mössbauer signal under *operando* conditions and the loss of other Mössbauer signatures. The thermodynamically and spectroscopically consistent catalytic cycle for ORR on Fe–N–C was given based on the pyrrolic N-coordination active site [54]. Li et al. also showed the structural change of Fe–N–C under working conditions, but limited in an alkaline environment [51]. It should be noted that the few in situ studies ever conducted were different in material, experimental practice and emphasis, and further study is still urgent for a more comprehensive understanding.

6.3.2 Investigation of Degradation Mechanism

In addition to the unsatisfactory activity, poor stability is another obstacle that hinders the application of Fe–N–C in PEMFCs. In the early study, Mössbauer

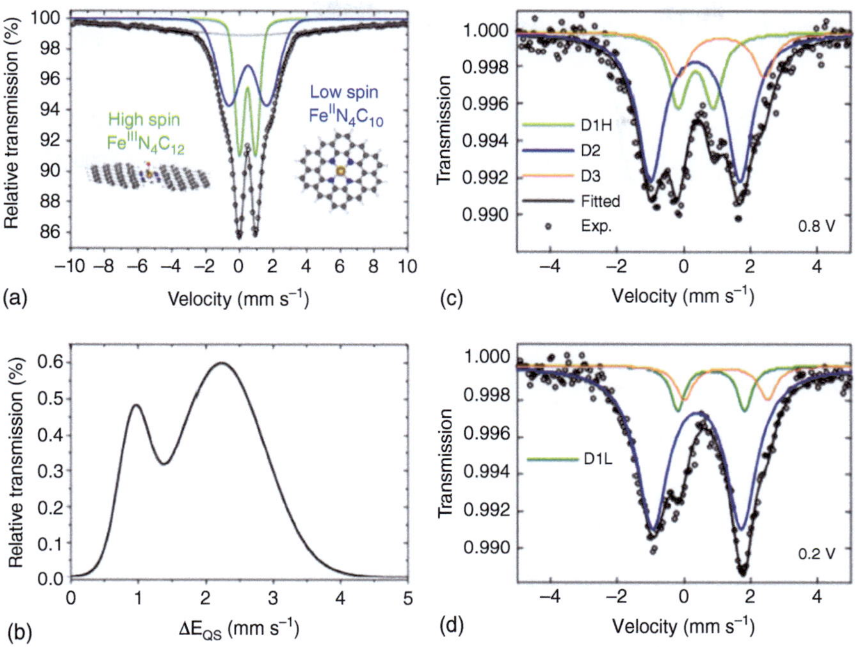

Figure 6.5 (a–b) ^{57}Fe Mössbauer spectra (D1: green, D2: blue) of $Fe_{0.5}$–NC measured at 80 K in the air in the absence of external magnetic field (a) and the corresponding distribution of Δ (b) Source: Mineva et al. [46]/American Chemical Society; (c–d) In situ ^{57}Fe Mössbauer spectra in PEMFCs at 0.8 V (c) and 0.2 V (d), the cell was at RT and fed with H_2 (anode) and Ar (cathode) without back pressure. Source: Reproduced with permission from ref. [53], Springer Nature.

spectroscopy provided insights into the deactivation behavior of Fe–N–C suffered external acid treatment, which suggested FeN_4 as the possible active site and its leaching was responsible for the decrease of activity [67]. Through the postmortem study after aging in fuel cells, the decreasing current along with the loss of active sites was confirmed by the decreased ^{57}Fe Mössbauer resonant absorbance (Figure 6.6a–c) [66]. This was also supported by the Mössbauer spectroscopy analysis that the demetallation of FeN_4 sites in micropores appeared proportional to the decay [29]. The formation of Fe and Fe oxide clusters as decomposition products associated with the disintegration of FeN_4 sites was further confirmed as illustrated in Figure 6.6d [32, 68].

Recently, the development of in situ Mössbauer spectroscopy was well implemented in the investigation of the degradation mechanism. Li et al. found that S1 (HS FeN_4C_{12}) was not durable in PEMFC and rapidly converted to ferric oxide. In contrast, S2 (LS or MS FeN_4C_{10}) was more durable, with no measurable decrease of active sites after 50 h of operation at 0.5 V. Both the direct and indirect pathways were attributed to the degradation of S1, and the indirect pathway might be caused by the localized surface oxidation of carbon or the protonation of highly basic N involved in S1. However, the more graphitic local structure, lower productivity of ROS, and subsurface activation resulted in the high stability of S2 [53]. By bridging

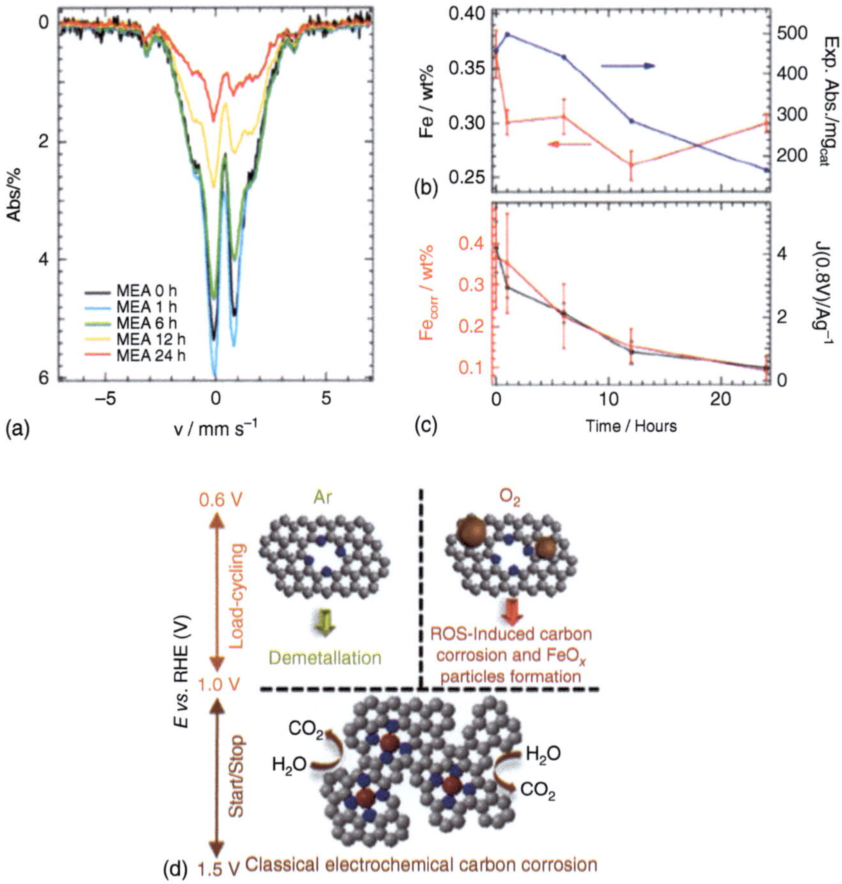

Figure 6.6 (a) ^{57}Fe Mössbauer spectra of MEA suffered the oxidizing treatment from 0 to 24 h; (b) Change of the Fe concentration and integrated Mössbauer absorption as a function of treatment time; (c) Change of the total corrected Fe concentration and mass activity at 0.8 V as a function of treatment time. Source: Kramm et al. [66]/American Chemical Society; (d) Illustration of the loss of FeN$_x$ sites and formation of FeO$_x$ induced by reactive oxygen species (ROS) produced from H_2O_2. Source: Reproduced with permission from ref. [32], John Wiley & Sons.

Mössbauer spectroscopy and theoretical study, the demetallation of Fe–N–C under working conditions was investigated by Xu et al. at the molecular level. As shown in Figure 6.7a–b, at higher potential along with more severe degradation, the more significant losses of D1 (assigned to FeN_4C_8 or FeN_4C_{12}) and D2 (assigned to FeN_4C_{10}) were observed in the end-of-state Mössbauer spectra, which was attributed to the demetallation of FeN_4 sites. Moreover, in situ Mössbauer spectra (Figure 6.7c–d) in O_2-saturated 0.1 M $HClO_4$ indicated that D1 was more active to ORR according to the transformation from D1 and D2 to D4 (O_x–FeN_4 intermediates). A density functional theory (DFT) study revealed that the weak Fe–N coordination in FeN_4C_8 was responsible for the rapid demetallation and ab initio molecular dynamic (AIMD)

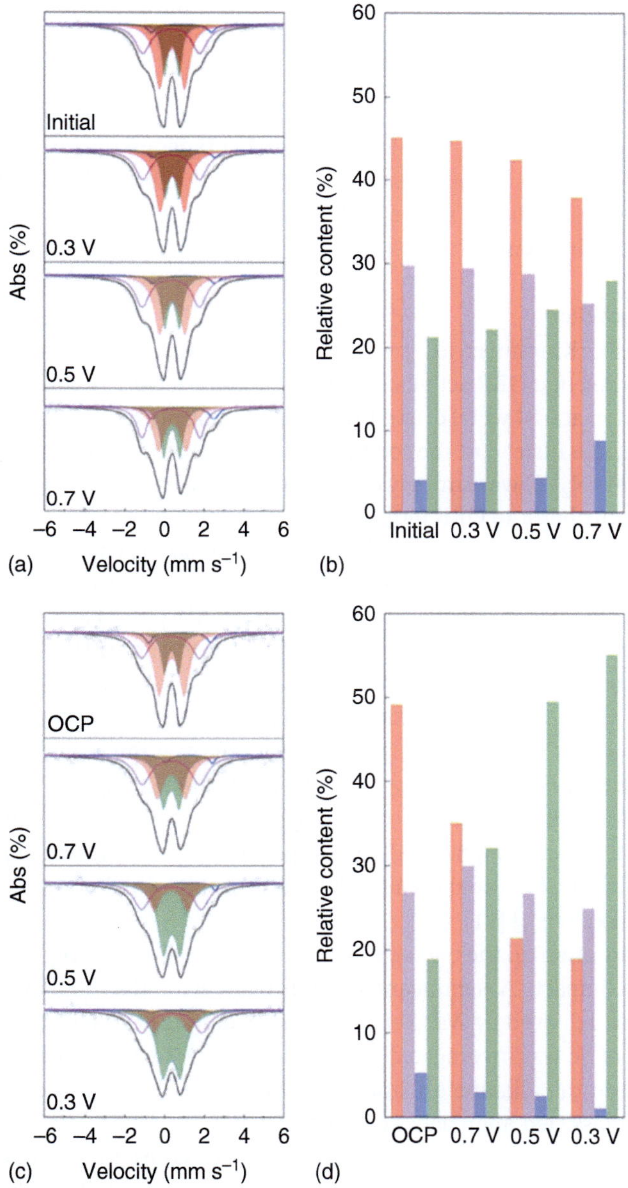

Figure 6.7 (a) The end-of-state Mössbauer spectra at RT of the ^{57}Fe–N–C at the initial state and after degradation at different potentials for 150 h and (b) the corresponding relative content of different types of FeN_4; (c) The in situ Mössbauer spectra at RT of the ^{57}Fe–N–C under different potentials and (d) the corresponding relative content of different types of FeN_4 (Red: D1; Purple: D2; Blue: D3; Green: D4). Source: Reproduced with permission from ref. [30], Elsevier.

simulations further revealed the accelerated demetallation at higher potential due to the adsorption of oxygenated intermediate and activation electric field [30].

6.3.3 Optimization for Synthesis of Fe–N–C

The goal of all fundamental research on activity and stability is to improve the performance in the practical application, and the solution somewhat relies on the rational synthesis of Fe–N–C finally. As mentioned earlier, the preparation of Fe–N–C still mostly depends on a pyrolysis routine, where both the precursor composition and heat-treatment strategy are influential.

6.3.3.1 Precursor Composition

As shown in Figure 6.8a, Kramm et al. investigated the impact of Fe content (0.03–1.55 wt%) and showed that ORR activity increased from the lowest Fe concentration and reached the top at 0.84 wt% Fe, and then dropped sharply for the higher Fe concentration (1.55 wt%). Instead of the total Fe content, only the formation of D2 (FeN_{2+2}) and D3 (N–FeN_{2+2}) was strongly correlated to the activity and D3 was identified as the main contributor to ORR [40].

Kübler et al. showed that surface functionalization with indazole enabled the direct attachment of FeN_4 centers to CNTs and no Fe agglomeration was detected in Mössbauer spectroscopy, which resulted in an improved SD [70]. The effect of ligands was also investigated by Tian et al. in a the Fe(II)/ligand/ZIF-8 system. The competition for Fe^{2+} cations was observed between 2-methylimidazole (2-MeIm) in ZIF-8 and 2,4,6-tris(2-pyridyl)-striazine (TPTZ) ligands according to the doublets assigned to [Fe(II)(2-MeIm)] in the precursor after wet impregnation. Such a cation exchange was considered disadvantageous to the activity as FeN_x derived from [Fe(II)(2-MeIm)] was situated on average further to the surface of the catalytic particles. Therefore, a higher activity was generally observed for using 1,10-phenanthroline (phen)as the ligand because the cation exchange did not occur for the more stable [Fe(II)(phen)$_3$] [71].

Moreover, the addition of auxiliaries, which acts as either dopant or template, also made a significant impact. A significant breakthrough with the P_{max} over 1 W cm^{-2} in PEMFCs was achieved by Wang et al. through S-doping [72]. The effect of S-doping on the state of Fe was investigated by *operando* ^{57}Fe Mössbauer spectra but the study is limited to the alkaline environment [73]. Fu et al. developed an ammonium chloride salt-assisted approach, and Mössbauer spectroscopy revealed that the enhanced nitrogen content would anchor more Fe atoms to generate FeN_x sites in a graphene-like porous scaffold [74]. As shown in Figure 6.8b, in the silica-coating-mediated synthesis developed by Woo et al., the interaction between Fe and Si was identified by Mössbauer spectroscopy at 295 K with the emergence of an Fe–Si peak ($\delta = 0.21$ mm s^{-1}, $\Delta = 0.55$ mm s^{-1}) in the precursor with silica coating (CNT/PC(SiO_2)). The Fe–Si interaction could mitigate the aggregation into large Fe-based particles, thereby preferentially generating FeN_x sites (doublet) in Fe–N–C (CNT/PC), while numerous inorganic Fe species (singlet and sextet) were observed in the sample without silica coating (CNT/PC w/o SiO_2) (Figure 6.8c–d) [69].

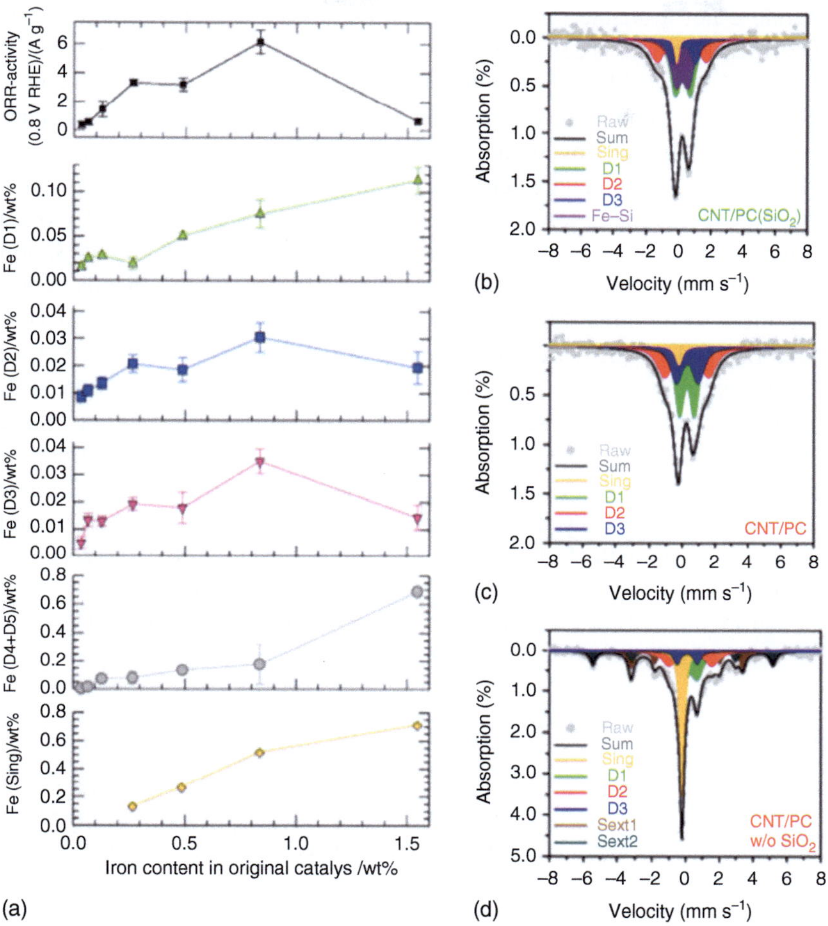

Figure 6.8 (a) The ORR activity and relative content of each Fe species as a function of total Fe loading. Source: Kramm et al. [40]/Royal Society of Chemistry; (b–d) ^{57}Fe Mössbauer spectra for CNT/PC(SiO$_2$) (b), CNT/PC (c) and CNT/PC w/o SiO$_2$ (d) Source: Reproduced with permission from ref. [69], American Chemical Society.

6.3.3.2 Heat Treatment

The Mössbauer spectroscopy and X-ray photoelectron spectroscopy (XPS) study on the formation of FeN_4 sites as a function of the pyrolysis temperature revealed that higher pyrolysis temperatures led to the shift of electron density from N to coordinated Fe, which resulted in higher activity toward ORR [64]. However, the higher temperature and pyrolysis duration might also result in the decomposition of FeN_4. Besides the inert atmosphere such as N_2 and Ar, NH_3 atmosphere was also commonly introduced for the preparation of Fe–N–C, which could facilitate the formation of FeN_x, tune the basicity of the N-doped carbon and optimize the porous structure. It was considered that most of the active sites in Fe–N–C treated by NH_3 were located at the surface and accessible to O_2 and H^+, while pyrolysis only in Ar tended to generate active sites that were secluded inside the catalytic material and therefore not available for ORR [75]. Moreover, Fe nitrides formed during NH_3 treatment were considered to affect either the carbon environment or the constitution of catalytically active centers, thus enabling an increase in TOF [76].

In the latest development, the preparation of Fe–N–C with the exclusive presence of FeN_x became a subject of intense interest for higher site density and atomic utilization. Through exchanging Zn with Fe in a ZnN_x-riched matrix, a highly active Fe–N–C (Fe–NC$^{\Delta\text{-DCDA}}$) was developed, in which 7 wt% Fe exclusively formed FeN_4 sites as determined by ^{57}Fe Mössbauer spectroscopy at 5 K, ICP-MS, and XAS. SD values measured by in situ nitrite stripping and ex situ CO chemisorption methods are $4.7*10^{19}$ and $7.8*10^{19}$ sites g^{-1}, respectively (Figure 6.9a–b) [12]. Likewise, Li et al. implemented chemical vapor deposition (CVD) to prepare Fe–N–C through the high-temperature *trans*-metalation of ZnN_4 sites into FeN_4 sites. Only two doublets labeled as D1 ($\delta = 0.50\,\text{mm s}^{-1}$, $\Delta = 1.02\,\text{mm s}^{-1}$) and D3 ($\delta = 1.0\,\text{mm s}^{-1}$, $\Delta = 3.5\,\text{mm s}^{-1}$) were found in the Mössbauer spectroscopy collected at 5 K (Figure 6.9c), which were assigned to O_2–Fe(III)–N_4–C_{12} moiety and $FeCl_2{\cdot}4H_2O$, respectively. The absence of D2 (Fe(III)–N_4 with oxidation and spin state unchanged with electrochemical potential) suggested that all FeN_4 sites formed via the CVD approach were gas-phase accessible. As a result, the Fe–N–C catalyst delivered the SD of $1.92*10^{20}$ sites g^{-1} with 100% site utilization [13].

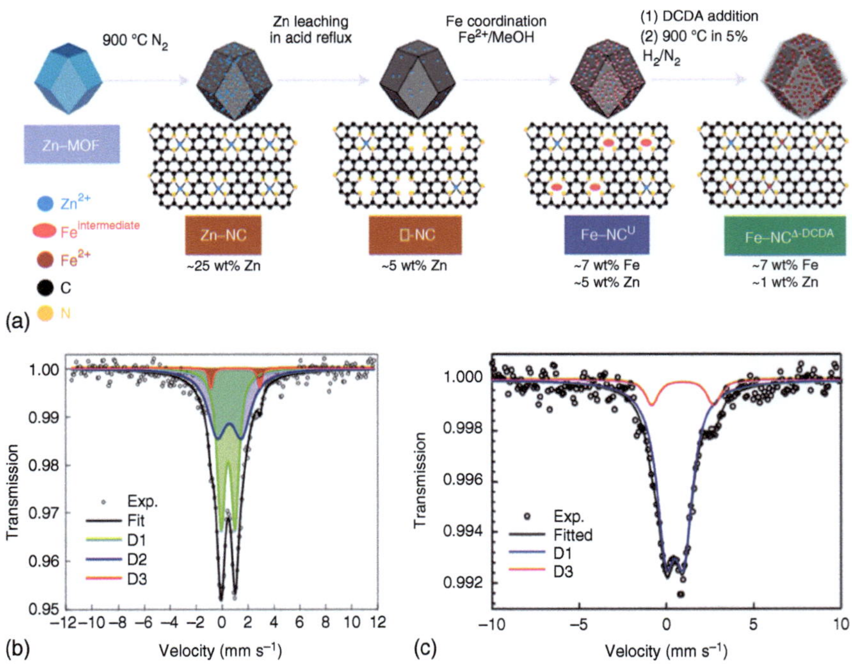

Figure 6.9 (a) The synthesis of Fe–N–C with high active site density by exchanging Zn with Fe from a ZnN_x-enriched matrix and (b) ^{57}Fe Mössbauer spectroscopy of as-prepared catalyst (Fe–NC$^{\Delta\text{-DCDA}}$) collected at 5 K. Source: Adapted from Mehmood et al. [12]; (c) ^{57}Fe Mössbauer spectroscopy collected at 5 K of FeNC-CVD-750 prepared via a CVD procedure. Source: Reproduced with permission from ref. [13], Springer Nature.

6.4 Summary and Perspective

In summary, this chapter introduced the important role of ^{57}Fe Mössbauer spectroscopy in studying Fe–N–C catalysts for ORR from the following aspects:

(1). Determining the structure and density of active sites and establishing the structure–activity relationship.
(2). Revealing the structure and content changes of Fe species upon catalyst deactivation.
(3). Guiding the synthesis of Fe–N–C with high TOF, density, and utilization of active sites.

Benefitting from the above progress, high performances of Fe–N–C in PEMFCs close to the DOE 2025 target have been achieved, but there are still some tough challenges such as:

(1). The low performance at practical pressure or H_2-air conditions due to the high mass transport resistance of O_2.
(2). The rapid degradation with the ambiguous mechanism.

To solve these problems, the deep understanding of structure and mechanism is essential, and Mössbauer spectroscopy will play an important role. For example, to achieve better performance at practical pressure or H_2-air conditions, the SD and TOF of active sites remain improving to reduce the thickness of the catalyst layer. To improve stability, the rational design of active sites is the promising pathway as evidenced by Wu et al. recently [77]. In the future, LT Mössbauer spectroscopy should get more common to avoid the blind area at RT. The computational Mössbauer spectroscopy is expected to give accurate assignments as some Fe species are not matched with those well-established Fe crystalline and macrocycles. Besides the sustained attention on the configuration of FeN_xC_y sites, more attention should be paid on the adsorption of reactants and intermediates to understand the mechanism, which demands the cooperation of Mössbauer spectroscopy and soft XAS at the C, N, or O edge to complete the weakness of each other.

Moreover, in situ/*operando* Mössbauer spectroscopy is becoming the power tool to investigate Fe–N–C. More further works are urgent for the comprehensive evaluation, especially for the in situ/*operando* observation in PEMFCs, as the behavior of Fe–N–C in RDE and PEMFC measurements is significantly different. In addition, the time resolution requires elevation and time-dependent data are imperative. This is particularly important for the study on stability because Fe–N–C usually suffers rapid degradation in the first several hours, while the developed in situ/*operando* Mössbauer spectroscopy also demands a time scale of several hours.

With the progress of technology, it is convinced that Mössbauer spectroscopy will provide sharper insights into problems in practical application, and facilitate the development of low-cost PEMFC technologies based on Fe–N–C.

Acknowledgments

Financial supports from the National Natural Science Foundation of China (22109161) and the Postdoctoral Science Foundation of China (2021M693130) are greatly acknowledged.

References

1 Banham, D. and Ye, S. (2017). Current status and future development of catalyst materials and catalyst layers for proton exchange membrane fuel cells: an industrial perspective. *ACS Energy Lett.* 2 (3): 629–638.
2 Kodama, K., Nagai, T., Kuwaki, A. et al. (2021). Challenges in applying highly active Pt-based nanostructured catalysts for oxygen reduction reactions to fuel cell vehicles. *Nat. Nanotechnol.* 16 (2): 140–147.
3 Chen, C., Kang, Y., Huo, Z. et al. (2014). Highly crystalline multimetallic nanoframes with three-dimensional electrocatalytic surfaces. *Science* 343 (6177): 1339–1343.
4 Chung, H.T., Cullen, D.A., Higgins, D. et al. (2017). Direct atomic-level insight into the active sites of a high-performance PGM-free ORR catalyst. *Science* 357 (6350): 479–484.
5 Shao, M., Chang, Q., Dodelet, J.P., and Chenitz, R. (2016). Recent advances in electrocatalysts for oxygen reduction reaction. *Chem. Rev.* 116 (6): 3594–3657.
6 Thompson, S.T. and Papageorgopoulos, D. (2019). Platinum group metal-free catalysts boost cost competitiveness of fuel cell vehicles. *Nat. Catal.* 2 (7): 558–561.
7 Gong, K., Du, F., Xia, Z. et al. (2009). Nitrogen-doped carbon nanotube arrays with high electrocatalytic activity for oxygen reduction. *Science* 323 (5915): 760–764.
8 Lefevre, M., Proietti, E., Jaouen, F., and Dodelet, J.P. (2009). Iron-based catalysts with improved oxygen reduction activity in polymer electrolyte fuel cells. *Science* 324 (5923): 71–74.
9 Suntivich, J., Gasteiger, H.A., Yabuuchi, N. et al. (2011). Design principles for oxygen-reduction activity on perovskite oxide catalysts for fuel cells and metal-air batteries. *Nat. Chem.* 3 (7): 546–550.
10 Yan, B., Krishnamurthy, D., Hendon, C.H. et al. (2017). Surface restructuring of nickel sulfide generates optimally coordinated active sites for oxygen reduction catalysis. *Joule* 1 (3): 600–612.
11 Zitolo, A., Goellner, V., Armel, V. et al. (2015). Identification of catalytic sites for oxygen reduction in iron- and nitrogen-doped graphene materials. *Nat. Mater.* 14 (9): 937–942.
12 Mehmood, A., Gong, M., Jaouen, F. et al. (2022). High loading of single atomic iron sites in Fe–NC oxygen reduction catalysts for proton exchange membrane fuel cells. *Nat. Catal.* 5 (4): 311–323.

13 Jiao, L., Li, J., Richard, L.L. et al. (2021). Chemical vapour deposition of Fe-N-C oxygen reduction catalysts with full utilization of dense Fe-N4 sites. *Nat. Mater.* 20 (10): 1385–1391.

14 Xu, X., Xia, Z., Zhang, X. et al. (2019). Atomically dispersed Fe-N-C derived from dual metal-organic frameworks as efficient oxygen reduction electrocatalysts in direct methanol fuel cells. *Appl. Catal. B Environ.* 259 (15): 118042.

15 Jasinski, R. (1964). A new fuel cell cathode catalyst. *Nature* 201 (4925): 1212–1213.

16 Jahnke, H., Schonborn, M., and Zimmermann, G. (1976). Organic dyestuffs as catalysts for fuel cells. *Top. Curr. Chem.* 61: 133–181.

17 Vanderputten, A., Elzing, A., Visscher, W., and Barendrecht, E. (1986). Oxygen reduction on pyrolyzed carbon-supported transition-metal chelates. *J. Electroanal. Chem.* 205 (1–2): 233–244.

18 Gupta, S., Tryk, D., Bae, I. et al. (1989). Heat-treated polyacrylonitrile-based catalysts for oxygen electroreduction. *J. Appl. Electrochem.* 19 (1): 19–27.

19 Wang, W., Jia, Q., Mukerjee, S., and Chen, S. (2019). Recent insights into the oxygen-reduction electrocatalysis of Fe/N/C materials. *ACS Catal.* 9 (11): 10126–10141.

20 Litster, S. (2021). ElectroCat: Advanced PGM-free cathode engineering for high power density and durability. *Annual Merit Review Presentation of DOE Hydrogen and Fuel Cells Program*. Pittsburgh, PA: Carnegie Mellon University.

21 Zhang, H., Chung, H.T., Cullen, D.A. et al. (2019). High-performance fuel cell cathodes exclusively containing atomically dispersed iron active sites. *Energy Environ. Sci.* 12 (8): 2548–2558.

22 Wan, X., Liu, X., Li, Y. et al. (2019). Fe–N–C electrocatalyst with dense active sites and efficient mass transport for high-performance proton exchange membrane fuel cells. *Nat. Catal.* 2 (3): 259–268.

23 Choi, C.H., Baldizzone, C., Polymeros, G. et al. (2016). Minimizing operando demetallation of Fe-N-C electrocatalysts in acidic medium. *ACS Catal.* 6 (5): 3136–3146.

24 Choi, C.H., Baldizzone, C., Grote, J.P. et al. (2015). Stability of Fe-N-C catalysts in acidic medium studied by operando spectroscopy. *Angew. Chem. Int. Ed.* 54 (43): 12753–12757.

25 Martinez, U., Komini Babu, S., Holby, E.F., and Zelenay, P. (2018). Durability challenges and perspective in the development of PGM-free electrocatalysts for the oxygen reduction reaction. *Curr. Opin. Electrochem.* 9: 224–232.

26 Choi, C.H., Lim, H.-K., Chung, M.W. et al. (2018). The Achilles' heel of iron-based catalysts during oxygen reduction in an acidic medium. *Energy Environ. Sci.* 11 (11): 3176–3182.

27 Kramm, U.I., Ni, L., and Wagner, S. (2019). ^{57}Fe Mossbauer spectroscopy characterization of electrocatalysts. *Adv. Mater.* 31 (31): e1805623.

28 Wang, Y.-C., Huang, L., Zhang, P. et al. (2017). Constructing a triple-phase interface in micropores to boost performance of Fe/N/C catalysts for direct methanol fuel cells. *ACS Energy Lett.* 2 (3): 645–650.

29 Chenitz, R., Kramm, U.I., Lefèvre, M. et al. (2018). A specific demetalation of Fe–N4 catalytic sites in the micropores of NC_Ar + NH3 is at the origin of the initial activity loss of the highly active Fe/N/C catalyst used for the reduction of oxygen in PEM fuel cells. *Energy Environ. Sci.* 11: 365–382.

30 Xu, X., Zhang, X., Kuang, Z. et al. (2022). Investigation on the demetallation of Fe-N-C for oxygen reduction reaction: the influence of structure and structural evolution of active site. *Appl. Catal. B Environ.* 309: 121290.

31 Gao, Y., Hou, M., Qi, M. et al. (2021). New insight into effect of potential on degradation of Fe-N-C catalyst for ORR. *Front. Energy* 15: 421–430.

32 Kumar, K., Dubau, L., Mermoux, M. et al. (2020). On the influence of oxygen on the degradation of Fe-N-C catalysts. *Angew. Chem. Int. Ed.* 59 (8): 3235–3243.

33 Santori, P.G., Speck, F.D., Li, J. et al. (2019). Effect of pyrolysis atmosphere and electrolyte pH on the oxygen reduction activity, stability and spectroscopic signature of FeN_x moieties in Fe-N-C catalysts. *J. Electrochem. Soc.* 166 (7): F3311–F3320.

34 Goellner, V., Baldizzone, C., Schuppert, A. et al. (2014). Degradation of Fe/N/C catalysts upon high polarization in acid medium. *Phys. Chem. Chem. Phys.* 16 (34): 18454–18462.

35 Xu, X., Zhang, X., Xia, Z. et al. (2021). Fe-N-C with intensified exposure of active sites for highly efficient and stable direct methanol fuel cells. *ACS Appl. Mater. Interfaces* 13 (14): 16279–16288.

36 Zeng, Y., Li, X., Wang, J. et al. (2021). In situ/operando Mössbauer spectroscopy for probing heterogeneous catalysis. *Chem Catal.* 1 (6): 1215–1233.

37 Sougrati, M.T., Goellner, V., Schuppert, A.K. et al. (2016). Probing active sites in iron-based catalysts for oxygen electro-reduction: a temperature-dependent ^{57}Fe Mössbauer spectroscopy study. *Catal. Today* 262: 110–120.

38 Gallenkamp, C., Kramm, U.I., Proppe, J., and Krewald, V. (2020). Calibration of computational Mössbauer spectroscopy to unravel active sites in FeNC catalysts for the oxygen reduction reaction. *Int. J. Quantum Chem.* 121 (3): e26394.

39 Scherson, D.A., Fierro, C., Yeager, E.B. et al. (1984). In situ Mössbauer spectroscopy on an operating fuel cell. *J. Electroanal. Chem. Interfacial Electrochem.* 169 (1–2): 287–302.

40 Kramm, U.I., Herranz, J., Larouche, N. et al. (2012). Structure of the catalytic sites in Fe/N/C-catalysts for O_2-reduction in PEM fuel cells. *Phys. Chem. Chem. Phys.* 14 (33): 11673–11688.

41 Xiao, M., Zhu, J., Ma, L. et al. (2018). Microporous framework induced synthesis of single-atom dispersed Fe-N-C acidic ORR catalyst and its in situ reduced Fe-N4 active site identification revealed by X-ray absorption spectroscopy. *ACS Catal.* 8 (4): 2824–2832.

42 Koslowski, U.I., Abs-Wurmbach, I., Fiechter, S., and Bogdanoff, P. (2008). Nature of the catalytic centers of porphyrin-based electrocatalysts for the ORR: a correlation of kinetic current density with the site density of Fe–N4 centers. *J. Phys. Chem. C* 112 (39): 15356–15366.

43 Kozhushner, A., Zion, N., and Elbaz, L. (2021). Methods for assessment and measurement of the active site density in platinum group metal–free oxygen reduction reaction catalysts. *Curr. Opin. Electrochem.* 25: 100620.

44 Leonard, N.D., Wagner, S., Luo, F. et al. (2018). Deconvolution of utilization, site density, and turnover frequency of Fe–nitrogen–carbon oxygen reduction reaction catalysts prepared with secondary N-precursors. *ACS Catal.* 8 (3): 1640–1647.

45 Sahraie, N.R., Kramm, U.I., Steinberg, J. et al. (2015). Quantifying the density and utilization of active sites in non-precious metal oxygen electroreduction catalysts. *Nat. Commun.* 6: 8618.

46 Mineva, T., Matanovic, I., Atanassov, P. et al. (2019). Understanding active sites in pyrolyzed Fe–N–C catalysts for fuel cell cathodes by bridging density functional theory calculations and ^{57}Fe Mössbauer spectroscopy. *ACS Catal.* 9 (10): 9359–9371.

47 Varnell, J.A., Tse, E.C., Schulz, C.E. et al. (2016). Identification of carbon-encapsulated iron nanoparticles as active species in non-precious metal oxygen reduction catalysts. *Nat. Commun.* 7: 12582.

48 Kneebone, J.L., Daifuku, S.L., Kehl, J.A. et al. (2017). A combined probe-molecule, Mössbauer, nuclear resonance vibrational spectroscopy, and density functional theory approach for evaluation of potential iron active sites in an oxygen reduction reaction catalyst. *J. Phys. Chem. C* 121 (30): 16283–16290.

49 Chen, Z., Jiang, S., Kang, G. et al. (2019). Operando characterization of iron phthalocyanine deactivation during oxygen reduction reaction using electrochemical tip-enhanced Raman spectroscopy. *J. Am. Chem. Soc.* 141 (39): 15684–15692.

50 Bouwkamp-Wijnoltz, A.L., Visscher, W., van Veen, J.A.R. et al. (2002). On active-site heterogeneity in pyrolyzed carbon-supported iron porphyrin catalysts for the electrochemical reduction of oxygen: an in situ Mössbauer study. *J. Phys. Chem. B* 106 (50): 12993–13001.

51 Li, X., Cao, C.-S., Hung, S.-F. et al. (2020). Identification of the electronic and structural dynamics of catalytic centers in single-Fe-atom material. *Chem* 6 (12): 3440–3454.

52 Ni, L., Gallenkamp, C., Paul, S. et al. (2021). Active site identification in FeNC catalysts and their assignment to the oxygen reduction reaction pathway by in situ ^{57}Fe Mössbauer spectroscopy. *Adv. Energy Sustain. Res.* 2 (2): 2000064.

53 Li, J., Sougrati, M.T., Zitolo, A. et al. (2020). Identification of durable and non-durable FeN_x sites in Fe–N–C materials for proton exchange membrane fuel cells. *Nat. Catal.* 4 (1): 10–19.

54 Ni, L., Gallenkamp, C., Wagner, S. et al. (2022). Identification of the catalytically dominant iron environment in iron- and nitrogen-doped carbon catalysts for the oxygen reduction reaction. *J. Am. Chem. Soc.* 144 (37): 16827–16840.

55 Zhong, L., Frandsen, C., Mørup, S. et al. (2018). ^{57}Fe-Mössbauer spectroscopy and electrochemical activities of graphitic layer encapsulated iron electrocatalysts for the oxygen reduction reaction. *Appl. Catal. B Environ.* 221: 406–412.

56 Strickland, K., Miner, E., Jia, Q. et al. (2015). Highly active oxygen reduction non-platinum group metal electrocatalyst without direct metal-nitrogen coordination. *Nat. Commun.* 6: 7343.

57 Li, Y., Liu, X., Zheng, L. et al. (2019). Preparation of Fe–N–C catalysts with FeNx (x = 1, 3, 4) active sites and comparison of their activities for the oxygen reduction reaction and performances in proton exchange membrane fuel cells. *J. Mater. Chem. A* 7 (45): 26147–26153.

58 Shen, H., Gracia-Espino, E., Ma, J. et al. (2017). Atomically FeN_2 moieties dispersed on mesoporous carbon: a new atomic catalyst for efficient oxygen reduction catalysis. *Nano Energy* 35: 9–16.

59 Chen, K., Liu, K., An, P. et al. (2020). Iron phthalocyanine with coordination induced electronic localization to boost oxygen reduction reaction. *Nat. Commun.* 11 (1): 4173.

60 Yan, Y., Cheng, H., Qu, Z. et al. (2021). Recent progress on the synthesis and oxygen reduction applications of Fe-based single-atom and double-atom catalysts. *J. Mater. Chem. A* 9 (35): 19489–19507.

61 Yuan, K., Lutzenkirchen-Hecht, D., Li, L. et al. (2020). Boosting oxygen reduction of single iron active sites via geometric and electronic engineering: nitrogen and phosphorus dual coordination. *J. Am. Chem. Soc.* 142 (5): 2404–2412.

62 Yang, G., Zhu, J., Yuan, P. et al. (2021). Regulating Fe-spin state by atomically dispersed Mn-N in Fe-N-C catalysts with high oxygen reduction activity. *Nat. Commun.* 12 (1): 1734.

63 Guo, L., Hwang, S., Li, B. et al. (2021). Promoting atomically dispersed MnN_4 sites via sulfur doping for oxygen reduction: unveiling intrinsic activity and degradation in fuel cells. *ACS Nano* 15 (4): 6886–6899.

64 Kramm, U.I., Abs-Wurmbach, I., Herrmann-Geppert, I. et al. (2011). Influence of the electron-density of FeN[sub 4]-centers towards the catalytic activity of pyrolyzed FeTMPPCl-based ORR-electrocatalysts. *J. Electrochem. Soc.* 158 (1): https://doi.org/10.1149/1.3499621.

65 Liu, K., Wu, G., and Wang, G. (2017). Role of local carbon structure surrounding FeN_4 sites in boosting the catalytic activity for oxygen reduction. *J. Phys. Chem. C* 121 (21): 11319–11324.

66 Kramm, U.I., Lefèvre, M., Bogdanoff, P. et al. (2014). Analyzing structural changes of Fe–N–C cathode catalysts in PEM fuel cell by Mößbauer spectroscopy of complete membrane electrode assemblies. *J. Phys. Chem. Lett.* 5 (21): 3750–3756.

67 Mamtani, K., Singh, D., Tian, J. et al. (2016). Evolution of N-coordinated iron–carbon (FeNC) catalysts and their oxygen reduction (ORR) performance in acidic media at various stages of catalyst synthesis: an attempt at benchmarking. *Catal. Lett.* 146 (9): 1749–1770.

68 Martinaiou, I., Monteverde Videla, A.H.A., Weidler, N. et al. (2020). Activity and degradation study of an Fe-N-C catalyst for ORR in direct methanol fuel cell (DMFC). *Appl. Catal. B Environ.* 262: 118217.

69 Woo, J., Choi, H., Sa, Y.J. et al. (2021). Structural evolution of atomically dispersed Fe species in Fe–N/C catalysts probed by X-ray absorption and ^{57}Fe Mössbauer spectroscopies. *J. Phys. Chem. C* 125 (22): 11928–11938.

70 Kübler, M., Wagner, S., Jurzinsky, T. et al. (2020). Impact of surface functionalization on the intrinsic properties of the resulting Fe–N–C catalysts for fuel cell applications. *Energy Technol.* 8 (9): 2000433.

71 Tian, J., Morozan, A., Sougrati, M.T. et al. (2013). Optimized synthesis of Fe/N/C cathode catalysts for PEM fuel cells: a matter of iron-ligand coordination strength. *Angew. Chem. Int. Ed.* 52 (27): 6867–6870.

72 Wang, Y.C., Lai, Y.J., Song, L. et al. (2015). S-doping of an Fe/N/C ORR catalyst for polymer electrolyte membrane fuel cells with high power density. *Angew. Chem. Int. Ed.* 54 (34): 9907–9910.

73 Chen, Z., Niu, H., Ding, J. et al. (2021). Unraveling the origin of sulfur-doped Fe-N-C single-atom catalyst for enhanced oxygen reduction activity: effect of iron spin-state tuning. *Angew. Chem. Int. Ed.* 60 (48): 25404–25410.

74 Fu, X., Li, N., Ren, B. et al. (2019). Tailoring FeN_4 sites with edge enrichment for boosted oxygen reduction performance in proton exchange membrane fuel cell. *Adv. Energy Mater.* 9 (11): 1803737.

75 Kramm, U.I., Lefevre, M., Larouche, N. et al. (2014). Correlations between mass activity and physicochemical properties of Fe/N/C catalysts for the ORR in PEM fuel cell via ^{57}Fe Mossbauer spectroscopy and other techniques. *J. Am. Chem. Soc.* 136 (3): 978–985.

76 Kramm, U.I., Herrmann-Geppert, I., Bogdanoff, P., and Fiechter, S. (2011). Effect of an ammonia treatment on structure, composition, and oxygen reduction reaction activity of Fe–N–C catalysts. *J. Phys. Chem. C* 115 (47): 23417–23427.

77 Liu, S., Li, C., Zachman, M.J. et al. (2022). Atomically dispersed iron sites with a nitrogen–carbon coating as highly active and durable oxygen reduction catalysts for fuel cells. *Nat. Energy* 7: 652–663.

7

^{197}Au Mössbauer Spectroscopy of Thiolate-protected Gold Clusters

Norimichi Kojima[1], Yasuhiro Kobaqyashi[2], and Makoto Seto[2]

[1] *The University of Tokyo, Graduate School of Arts and Sciences, Department of Basic Science, Komaba 3-8-1, Meguroku, Tokyo, Japan*
[2] *Kyoto University, Institute for Integrated Radiation and Nuclear Science, Division of Nuclear Radiation Physics, Asashiro-Nishi 2, Kumatori-cho, Sennan-gun, Osaka, Japan*

List of Abbreviations

DFT	density functional theory
ESI	electrospray ionization
EXAFS	extended X-ray absorption fine structure
FWHM	full width at half maximum
GSH	glutathione
HPLC	high-performance liquid chromatography
PAGE	electrophoresis
Ph	phenyl group (C_6H_5)
SG	glutathionate
δ	isomer shift
Δ	quadrupole splitting

7.1 Introduction

Nano-clusters consisting of 10–10^3 atoms have been considered to form a new phase of materials showing both macroscopic and microscopic features such as the coexistence of solid-like and liquid-like states, or fluctuating states with permutation isomers [1]. For example, Au nanoclusters consisting of about 460 atoms move on the surface of a Si substrate covered with SiO_2 film as if they are alive, which was observed by means of transmission electron microscopy (TEM) [2]. As for the development of Au nano-clusters, D. M. P. Mingos, et al. have synthesized various kinds of Au clusters by means of the reduction of mononuclear Au(I) phosphine complexes,

Mössbauer Spectroscopy: Applications in Chemistry and Materials Science, First Edition. Edited by Yann Garcia, Junhu Wang, and Tao Zhang.

$AuX(PR_3)$ (X = halogen, SCN, etc.: R = C_6H_5, etc.), and systematically investigated their geometric structures and the electronic properties since the 1980s [3].

In relation to the ^{197}Au Mössbauer spectroscopic research on Au nano-clusters, A. Vollenbroek, et al. synthesized $Au_{11}(PPh_3)_7X_3$ (Ph = C_6H_5, p-ClC_6H_4; X = CN, SCN, I), and investigated the ^{197}Au Mössbauer spectra in 1978 [4]. According to the molecular structure [5], there are electronically different five Au sites in the Au_{11} cluster as follows: (i) central core Au atom, (ii) peripheral Au atoms coordinated by X ligands, and (iii) three different sites of peripheral Au atoms coordinated by PPh_3 ligands, which reflects on the ^{197}Au Mössbauer spectra having one singlet and four doublet peaks. On the other hand, in 1981, G. Schmid, et al. synthesized $Au_{55}(PPh_3)_{12}Cl_6$ and investigated its ^{197}Au Mössbauer spectra [6]. In the case of $Au_{55}(PPh_3)_{12}Cl_6$, there are 13 inner core Au atoms and 42 Au atoms on the surface shell having 24 nonbonding bare Au atoms, 12 Au atoms coordinated by PPh_3 ligands, and 6 Au atoms coordinated by Cl ligands, which reflects on the ^{197}Au Mössbauer spectra with two singlet peaks corresponding to the inner core Au atoms and nonbonding bare surface Au atoms, with two doublet peaks corresponding to the surface Au atoms coordinated by PPh_3 and Cl ligands. After these pioneering ^{197}Au Mössbauer spectroscopic research on Au nano-clusters, L. J. de Jongh, et al. systematically analyzed the ^{197}Au Mössbauer spectra for $[Au_9(PPh_3)_8](NO_3)_3$, $Au_{11}(PPh_3)_7(SCN)_3$, and $Au_{55}(PPh_3)_{12}Cl_6$ [7]. Through these analyses, ^{197}Au Mössbauer spectroscopy has been recognized as the most powerful method to distinguish the geometrically different sites of Au atoms and investigate their electronic states in Au nano-clusters. Therefore, from the viewpoint of ^{197}Au Mössbauer spectroscopy, the synthesis of Au nano-clusters filling in the missing link between Au_{11} and Au_{55} clusters had been desired.

In 2004–2005, Y. Negishi, T. Tsukuda, et al. succeeded in the isolation of size-controlled gold nano-clusters protected by glutathionate (SG), $Au_n(SG)_m$ with n = 10–39, using polyacrylamide gel electrophoresis (PAGE) [8, 9]. In 2007, we proposed the structure of $Au_{25}(SG)_{18}$ by means of ^{197}Au Mössbauer spectroscopy [10], based on the "core-in-cage" model theoretically predicted by T. Iwasa and K. Nobusada; the planar Au_7 core is sandwiched between two –[Au–SR–]$_3$ cyclic oligomers and surrounded by one –[Au–SR–]$_{12}$ cyclic oligomer [11]. Then, the geometrical structure of $Au_{25}(SC_2H_4Ph)_{18}$ was determined by single-crystal X-ray diffraction [12, 13], which was supported by density functional theory (DFT) calculation [14]. According to the single-crystal X-ray analysis, $Au_{25}(SC_2H_4Ph)_{18}$ consists of Au_{13} icosahedron and six staples of –SR–[Au–SR–]$_2$. The surface Au atoms of Au_{13} icosahedron are completely protected by the S atoms in –SR–[Au–SR–]$_2$. Based on the molecular structure of $Au_{25}(SC_2H_4Ph)_{18}$, we estimated the relative ratio of recoil-free fractions for the central core and surface Au atoms of Au_{13}, and the Au atoms in the –SR–[Au–SR–]$_2$ staple for $Au_{25}(SG)_{18}$ [15], and investigated the structural evolution of $Au_n(SG)_m$ from $n = 10$ to $n \approx 55$ [16, 17]. Since the first report of precisely size-controlled thiolate-protected Au nano-clusters [8, 9], a huge number of thiolate-protected Au clusters have been developed, and their molecular structures and electronic states have extensively been investigated [18].

In this chapter, we focus on the structural evolution and the electronic states of Au nano-clusters, $Au_n(SG)_m$, based on the analysis of ^{197}Au Mössbauer spectra for $Au_{10}(SG)_{10}$ and $Au_{25}(SG)_{18}$. Moreover, we describe the analysis of ^{197}Au Mössbauer spectra for dodecanethiolate ($SC_{12}H_{25}$)-protected Au clusters, $Au_n(SC_{12}H_{25})_m$, with the average diameters of 2 and 4 nm, and $Au_{24}Pd(SC_{12}H_{25})_{18}$.

7.2 Synthesis of Thiolate Protected Gold Clusters

The Au nano-clusters $Au_n(SG)_m$ with $(n, m) = (10, 10), (15, 13), (18, 14), (22, 16), (22, 17), (25, 18), (29, 20), (33, 22), (39, 24), (45, 28)$, and $(\sim 55, m)$ were fractionated from nearly monodisperse Au:SG clusters ($d_{av}\sim 1$ nm) using polyacrylamide gel electrophoresis (PAGE) [8, 9]. The complete isolation of $Au_n(SG)_m$ was confirmed by electrospray ionization (ESI) mass spectroscopy [8, 9]. In order to obtain the ^{197}Au Mössbauer spectra with sufficient S/N ratio, each $Au_n(SG)_m$ cluster was accumulated up to 50~100 mg by repeating the PAGE procedure.

On the other hand, $Au_n(SC_{12}H_{25})_m$ with average diameters of 2 and 4 nm were prepared by the direct chemisorption method and the ligand-exchange method, respectively [19]. The distribution of particle size and the average diameter were analyzed by small-angle X-ray scattering as well as transmission electron microscopy (TEM) [19]. The average diameter of $Au_n(SC_{12}H_{25})_m$ prepared by the direct chemisorption method was 1.9 nm, while the average diameter of $Au_n(SC_{12}H_{25})_m$ prepared by the ligand-exchange method was 4.0 nm.

The synthetic processes of $Au_n(SG)_m$, $PdAu_{24}(SC_{12}H_{25})_{18}$, and $Au_n(SC_{12}H_{25})_m$ are shown in Scheme 7.1. In the case of $PdAu_{24}(SC_{12}H_{25})_{18}$, the starting material was the mixture of $H[Au^{III}Cl_4]$ and $Na_2[Pd^{II}Cl_4]$ in water, and a reverse-high-performance liquid chromatography (HPLC) was used to fractionate $PdAu_{24}(SC_{12}H_{25})_{18}$ [20].

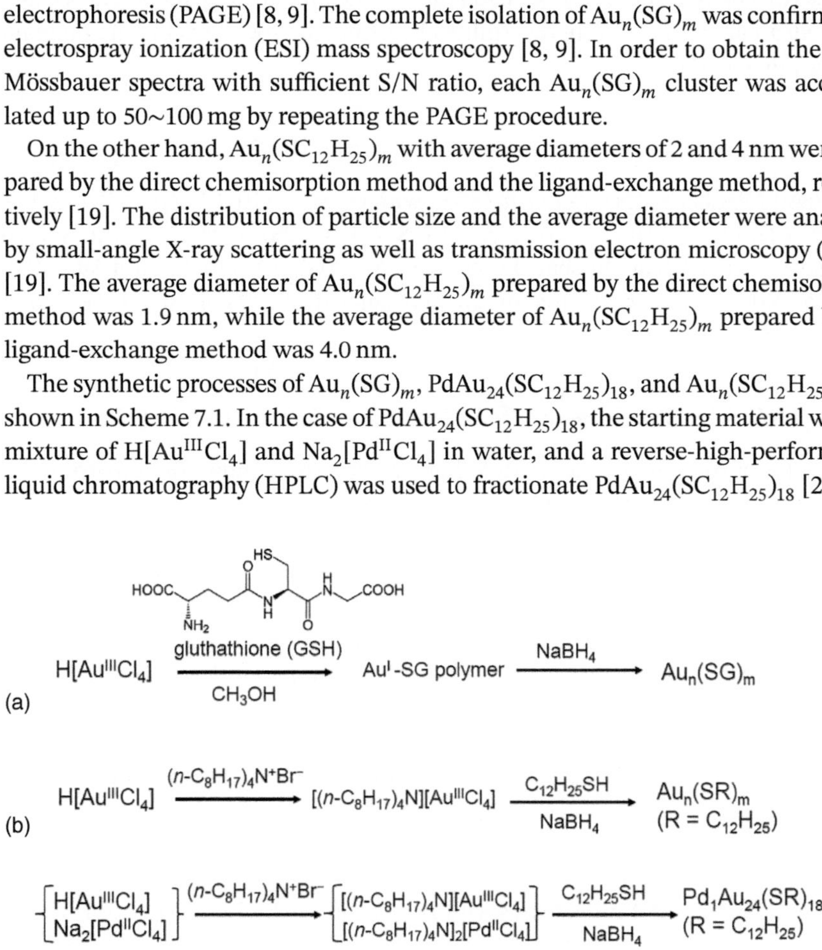

Scheme 7.1 Synthetic processes of (a) $Au_n(SG)_m$, (b) $Au_n(SC_{12}H_{25})_m$, and (c) $PdAu_{24}(SC_{12}H_{25})_{18}$.

7.3 ^{197}Au Mössbauer Spectroscopy of Gold Nano-clusters

7.3.1 Experimental Procedure of ^{197}Au Mössbauer Spectroscopy

The γ-ray source (77.3 keV) of ^{197}Au Mössbauer spectroscopy, ^{197}Pt, was obtained by the neutron irradiation to 98%-enriched ^{196}Pt metal foil, following the nuclear reaction of ^{196}Pt [n, γ] ^{197}Pt. We carried out ^{197}Au Mössbauer spectroscopy at the research reactor of the Institute for Integrated Radiation and Nuclear Science, Kyoto University. The γ-ray source and samples were cooled down to 16 K. The transmitted γ-ray was detected by a NaI(Tl) scintillation counter. The isomer shift (δ) of Au foil was referenced to 0 mm/s. The spectra were calibrated and referenced by using the six lines of a body-centered cubic Fe foil (α-Fe). The Mössbauer parameters of δ and Δ (quadrupole splitting) were determined by fitting the spectra using the program MossWinn 3.0 [21].

7.3.2 ^{197}Au Mössbauer Spectra of $Au_n(SG)_m$ ($n = 10{\sim}55$)

The ^{197}Au Mössbauer spectra of $Au_n(SG)_m$ with $(n, m) = (10, 10), (15, 13), (18, 14), (22, 16), (22, 17), (25, 18), (29, 20), (33, 22), (38, 24), (45, 28)$, and $({\sim}55, m)$ are shown in Figure 7.1. The ^{197}Au Mössbauer spectra for $Au_n(SG)_m$ ($n = 10$–22) and $Au_n(SG)_m$ ($n = 25{\sim}55$) were analyzed based on the molecular structures of $Au_{10}(SCH_3)_{10}$ and $Au_{25}(SC_2H_4Ph)_{18}$, respectively.

7.3.3 Molecular Structure and ^{197}Au Mössbauer Spectra of $Au_{10}(SG)_{10}$

In the case of $Au_{10}(SG)_{10}$, the ^{197}Au Mössbauer spectrum shows the superposition of two sets of doublets. The δ and Δ values obtained are 3.06 and 6.68 mm/s, respectively, for the major component (**C1**), whereas those for the minor component (**C1′**) are 2.48 and 5.71 mm/s, respectively. According to the correlation between the δ and Δ values established for various kinds of Au^{I} and Au^{III} compounds [22–25], both components fall into a category of Au^{I} coordinated by two SG ligands. The presence of two doublets (**C1** and **C1′**) suggests that there are two different environments for Au atoms coordinated by two SG ligands. There are two possible scenarios to explain the two doublets, based on DFT calculation for $Au_{10}(SCH_3)_{10}$ [26]. The first scenario is the formation of a catenane structure consisting of two interlocked pentamers (Figure 7.2(a)). In the case of two interlocked pentamers, $2[Au_5(SCH_3)_5]$, there are two Au–S distances: 2.40 Å for the Au atom at the center of the ring and 2.35 Å for the others [26]. In the case of ^{197}Au Mössbauer spectrum, δ shifts to the lower velocity side, and Δ decreases with increasing Au–S distance [22–25]. Therefore, **C1′** and **C1** should be attributed to the center Au site of the ring and the other Au sites, respectively, where the numbers of Au atoms for **C1′** and **C1** are estimated to be 2.92 and 7.08, respectively, on condition that the recoil-free fractions for the center Au site of the ring and the other Au sites are same. These numbers are almost consistent with the numbers of the center Au site of the ring and the others (2 and 8).

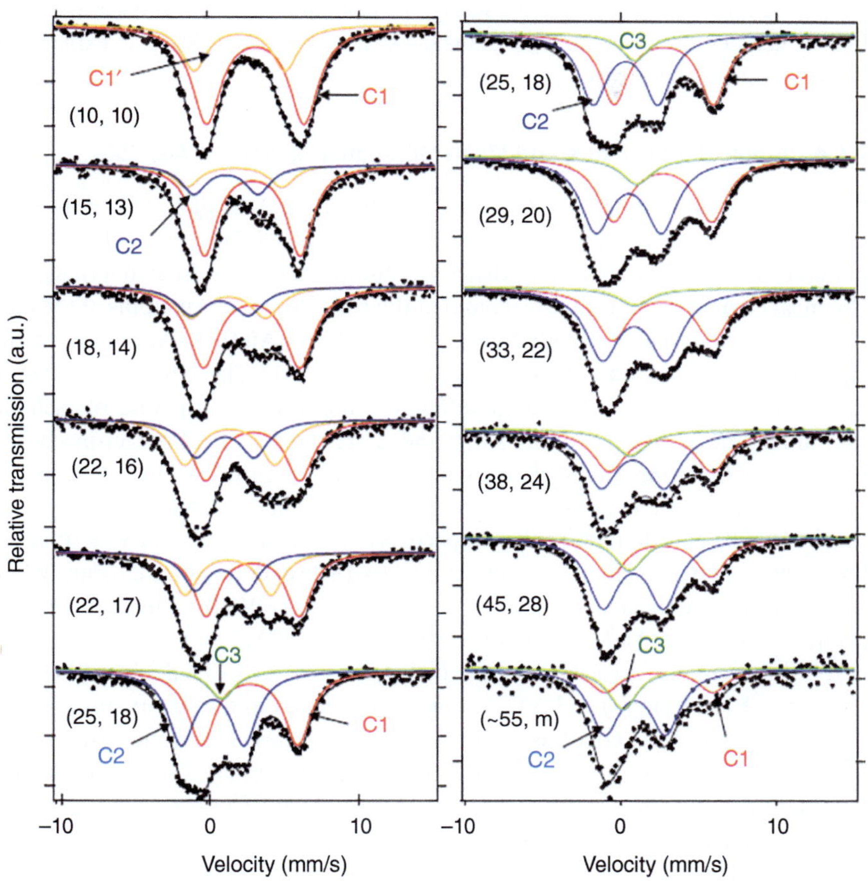

Figure 7.1 ^{197}Au Mössbauer spectra of $Au_n(SG)_m$ ($n = 10{\sim}55$) at 16 K. $Au_n(SG)_m$ is represented as (n, m). Source: Ref. [17]/with permission from Springer Nature.

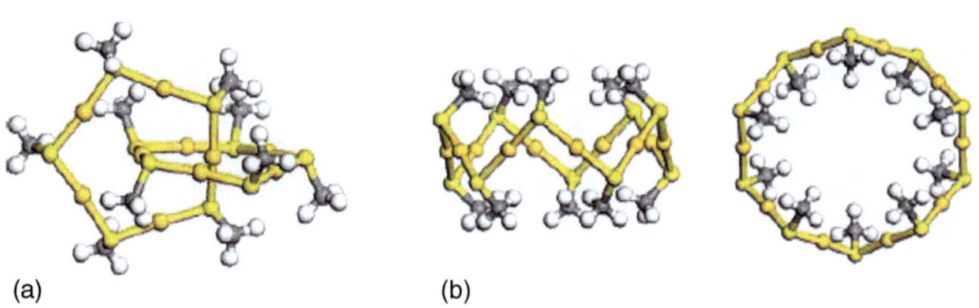

Figure 7.2 Predicted molecular structure of $Au_{10}(SCH_3)_{10}$ based on DFT calculation. (a) Catenane structure, (b) cyclic structure. Source: Ref. [26]/with permission from American Chemical Society.

The second scenario is a formation of cyclic oligomers (Figure 7.2(b)) with different lengths as minor fractions. Consequently, the catenane structure consisting of two interlocked pentamers is the most plausible structure for $Au_{10}(SG)_{10}$ to explain the ^{197}Au Mössbauer spectrum.

7.3.4 Molecular Structure and ^{197}Au Mössbauer Spectra of $Au_{25}(SG)_{18}$

Figure 7.3(a) shows the skeletal molecular structure of $Au_{25}(SR)_{18}$ (SR = thiolate) predicted by the single crystal X-ray analysis of $Au_{25}(SC_2H_4Ph)_{18}$ [12, 13]. As shown in Figure 7.3(a), the Au atoms in $Au_{25}(SR)_{18}$ are classified into three in terms of the chemical environment; central Au atom of the core (**Au3**), 12 Au atoms at the core surface each of which are bound by a single thiolate (**Au2**), 12 Au atoms on the outermost layer which are bound by two thiolates (**Au1**). These three sites should exhibit different δ and Δ due to the difference in electronic structures of each Au site. In Au, the main contributions to δ and Δ are the density of s-electrons at the nucleus and the symmetry of charge distribution around the nucleus, respectively. On the basis of consideration for symmetry, the **Au3** and **Au1/Au2** atoms produce singlet (Δ = 0) and doublet peaks, respectively. Thus, the spectrum was fitted by the superposition of a singlet Lorentzian and two sets of doublet Lorentzians while assuming the identical full width at half maximum (FWHM) values for all the components. The ^{197}Au Mössbauer spectrum of $Au_{25}(SG)_{18}^-$ in Figure 7.2(b) is the best fitting result taking account of the molecular structure. The doublet (**C1**) with the largest δ (2.78 mm/s) and Δ (6.35 mm/s) is assigned to the 12 Au atoms (**Au1**) on the outermost layer coordinated by two thiolates. These Mössbauer parameters, δ and Δ, are typical of Au^I coordinated by two sulfur atoms [27]. The second doublet (**C2**) with δ = 0.34 mm/s and Δ = 4.14 mm/s is assigned to the 12 Au atoms at the core surface, each of which is bound by single thiolate (**Au2**). The singlet (**C3**) with δ = 0.94 mm/s and Δ = 0.00 mm/s is assigned to the central Au atom (**Au3**) of

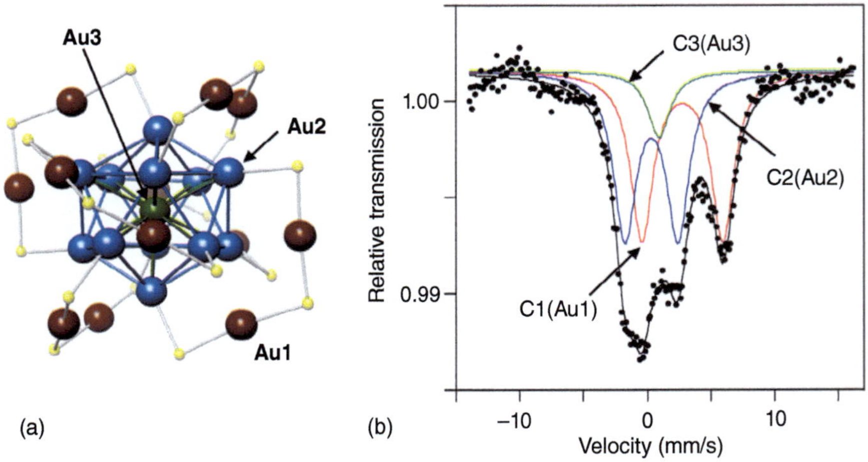

Figure 7.3 (a) Skeletal molecular structure of $Au_{25}(SR)_{18}$ (SR = thiolate). (b) ^{197}Au Mössbauer spectrum of $Au_{25}(SG)_{18}$ at 16 K. Source: Ref. [15]/with permission from The Chemical Society of Japan.

the icosahedron. In this manner, the line profile of ^{197}Au Mössbauer spectrum for $Au_{25}(SG)_{18}$ can be reproduced well by the superposition of a singlet (**C3**) and two doublets, **C1** and **C2**, The relative ratio of recoil-free fractions was determined to be f(**Au1**) : f(**Au2**) : f(**Au3**) = 1.03 : 1.00 : 2.44 by dividing the spectral areas of **C1**, **C2**, and **C3** (46.1%, 44.8%, and 9.1%) with the numbers of Au atoms (12, 12, and 1) for the **Au1**, **Au2**, and **Au3** sites.

7.3.5 Structural Evolution of $Au_n(SG)_m$ ($n = 10{\sim}55$) Based on ^{197}Au Mössbauer Spectroscopy

As shown in Figure 7.1, the ^{197}Au Mössbauer spectra of a series of $Au_n(SG)_m$ evolve drastically as a function of the number of Au atoms. The spectra of $Au_n(SG)_m$ ($15 \leq n \leq 22$) were fitted by three sets of doublets. Two components are assigned to Au^I coordinated by two SG ligands since the δ and Δ values are smoothly correlated to those of $Au_{10}(SG)_{10}$. The third component having the smallest δ and Δ values is assigned to Au^I coordinated by a single SG ligand which is smoothly correlated to those of the **Au2** atoms coordinated by a single SG ligand in $Au_{25}(SG)_{18}$. The spectral profile abruptly changes on going from $Au_{22}(SG)_{16}$ to $Au_{25}(SG)_{18}$, where a core Au atom (**Au3**) appears for the first time. On the other hand, on going from $Au_{25}(SG)_{18}$ to $\sim Au_{55}(SG)_m$, the line profile smoothly converges to that of $\sim Au_{55}(SG)_m$ whose average diameter is about 1.4 nm. In this region, the spectrum was fitted by the superposition of a singlet Lorentzian (core Au atoms) and two sets of doublet Lorentzians. With increasing n, the number of the Au atoms at the core surface coordinated by single SG ligand increases compared to that of the Au atoms coordinated by two SG ligands. The δ and Δ for the ^{197}Au Mössbauer spectra of $Au_n(SG)_m$ ($n = 10{\sim}55$) as a function of the number of Au atoms are shown in Figure 7.4. As shown in Figure 7.4, both the δ and Δ for C2 discontinuously change between $n = 22$ and $n = 25$, which is attributed to the formation of Au icosahedron for $Au_{25}(SG)_{18}$.

Figure 7.5 shows the correlation between δ and Δ of the ^{197}Au Mössbauer spectra of $Au_n(SG)_m$ ($n = 10 \sim 55$). Since almost all the δ and Δ of Au^I compounds are located in the gray-colored band in Figure 7.5 [25], the valence states of **Au1** (corresponding to **C1**) and **Au1′** (corresponding to **C1′**) coordinated by two SG can be regarded as Au^I. As shown in Section 7.3.2, in the case of ^{197}Au Mössbauer spectrum, δ shifts to the lower velocity side, and Δ decreases with increasing Au–S distance [22–25]. Therefore, the Au–S distance for **Au1′** is considered to be longer than that for **Au1**. In the case of **Au2**-coordinated single SG ligand, the covalency between Au and S atoms becomes weaker than that for **Au1** and **Au1′**, which decreases the δ and Δ of **Au2** compared with those of **Au1** and **Au1′**. **Au3** (corresponding to **C3**) without SG ligand appears at $Au_{25}(SG)_{18}$ for the first time, and the δ decreases with increasing the number of Au atoms from $n = 25$ to $n \approx 55$, and converges to that of bulk Au foil.

In order to estimate the number of Au atoms in the **Au1**, **Au2**, and **Au3** sites for $Au_n(SG)_m$, the following relative ratio of recoil-free fractions, f(**Au1**) : f(**Au2**) : f(**Au3**) = 1.03 : 1.00 : 2.44 for $Au_{25}(SG)_{18}$ was used. The numbers of Au atoms for the **Au1**, **Au2**, and **Au3** sites for other $Au_n(SG)_m$ were estimated on the basis of these recoil-free fractions, which are summarized in Table 7.1.

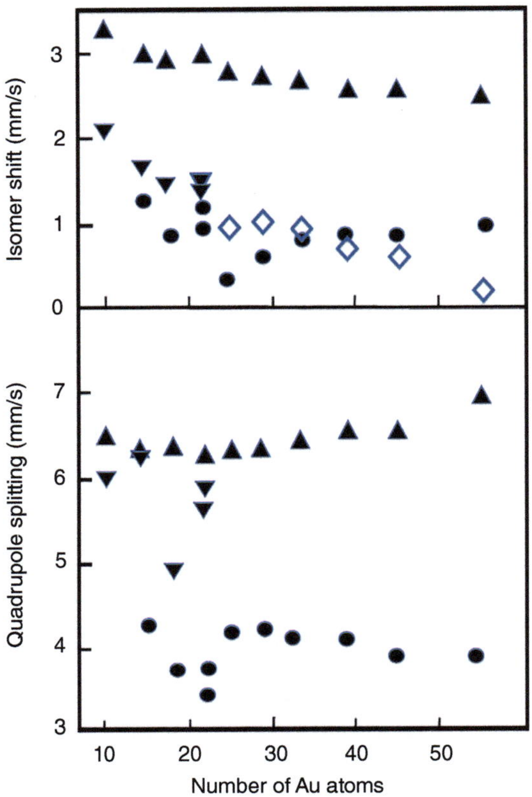

Figure 7.4 Isomer shift and quadrupole splitting for the ^{197}Au Mössbauer spectra of $Au_n(SG)_m$ (n = 10~55) as a function of the number of Au atoms at 16 K. ▲: **C1** corresponding to the Au atom coordinated by two SG ligands, ▼: **C1′** corresponding to the Au atom coordinated by two SG ligands, •: **C2** corresponding to the Au atom coordinated by single SG ligand, ◇: **C3** corresponding to the center core Au atom.

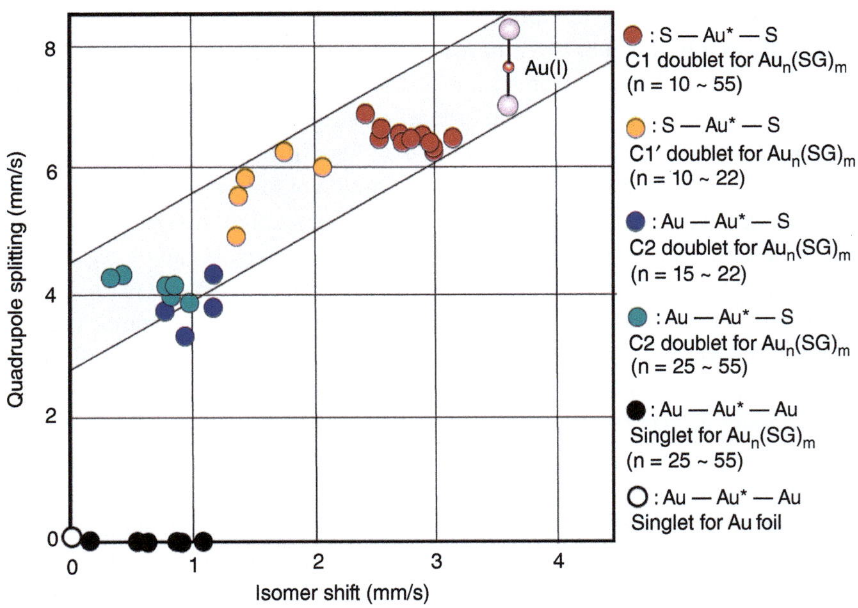

Figure 7.5 Correlation between the isomer shift and quadrupole splitting for Au* atoms in the ^{197}Au Mössbauer spectra of $Au_n(SG)_m$ (n = 10~55).

Table 7.1 Estimated numbers of Au atoms for the **Au1**, **Au2**, and **Au3** sites in $Au_n(SG)_m$ ($n = 10{\sim}55$).

	Number of Au atoms		
(n, m)	**N (Au1)**	**N (Au2)**	**N (Au3)**
(10, 10)	10	0	0
(15, 13)	12	3	0
(18, 14)	14	4	0
(22, 16)	17	5	0
(22, 17)	17	5	0
(25, 18)	12	12	1
(29, 20)	13	15	1
(33, 22)	14	18	1
(38, 24)	15	21	2
(45, 28)	15	27	3
(~55, *m*)	~13	~37	5

From the analysis of the number of Au atoms in the **Au1**, **Au2**, and **Au3** sites for $Au_n(SG)_m$, the structural evolution of $Au_n(SG)_m$ is summarized as follows:

(1) $\boldsymbol{n = 15{\sim}22}$: The first appearance of **Au2** at $Au_{15}(SG)_{13}$ indicates that the formation of small Au aggregates starts at this critical size. There are two notable features in Table 7.1. First, the numbers of **Au2** for $Au_{15}(SG)_{13}$, $Au_{22}(SG)_{16}$, and $Au_{22}(SG)_{17}$ are not even, but odd. This indicates that at least one of the –SG–Au oligomers attached to the Au core has an open structure. The second feature is that the numbers of the **Au1** and thiolates are nearly equivalent: (N(**Au1**), m) = (12, 13), (14, 14), (17, 16), and (17, 17) for $Au_{15}(SG)_{13}$, $Au_{18}(SG)_{14}$, $Au_{22}(SG)_{16}$, and $Au_{22}(SG)_{17}$, respectively. This result excludes the possibility that Au cores are protected by –SG–(Au–SG–)$_{1,2,3}$ oligomers in a bidentate fashion but indicates the attachment of –SR–Au oligomers to the Au end. Instability of $Au_n(SG)_m$ with $n = 15{\sim}22$ against etching by GSH [28] is probably associated with such structures. A. Tlahuice and I. L. Garzón predicted that the most stable structure of $Au_{18}(SR)_{14}$ (R = CH_3) consists of a prolate Au_8 core covered with two-dimer (–SR–Au–SR–Au–SR–) and two-trimer (–SR–Au–SR–Au–SR–Au–SR–) motifs [29]. This calculation indicates that 10 and 8 Au atoms belong to **Au1** and **Au2**, respectively. Although this model explains high stability, it is not consistent with our ^{197}Au Mössbauer results (Table 7.1).

(2) $\boldsymbol{n = 25{\sim}55}$: At $Au_{25}(SG)_{18}$, two notable transitions are observed: an **Au3** atom appears for the first time, and the ratio of N(**Au1**) with respect to m decreases significantly from ~1. As already established by X-ray crystallography [12, 13], these changes are related to the formation of a 3-dimensional Au_{13} icosahedral

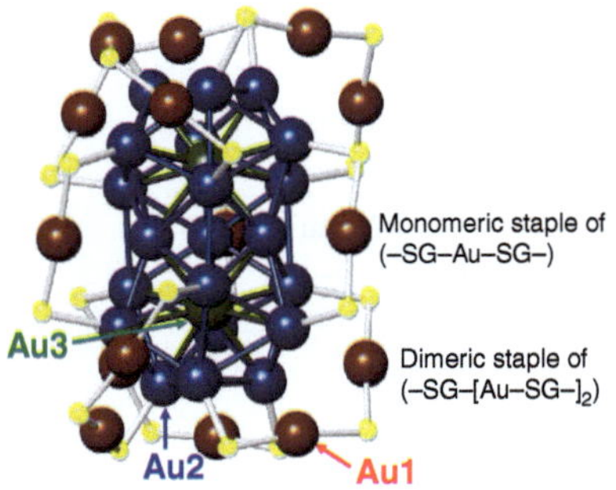

Figure 7.6 Predicted molecular structure of $Au_{38}(SG)_{24}$.

core and its protection by –SG–[Au–SG–]$_2$ in a bidentate fashion. For $Au_{29}(SG)_{20}$ and $Au_{33}(SG)_{22}$, bidentate oligomers are probably bonded on the Au_{16} and Au_1 core, respectively because the ratios of *N*(**Au1**) with respect to *m* for $Au_{29}(SG)_{20}$ and $Au_{33}(SG)_{22}$ are similar to that of $Au_{25}(SG)_{18}$. However, at least one surface atom of the Au_{16} core for $Au_{29}(SG)_{20}$ is bonded by a linear oligomer in a monodentate fashion, as the *N*(**Au2**) value is odd. For $Au_{33}(SG)_{22}$, we cannot construct a satisfactory model in which all the core surface atoms are completely protected by bidentate –SG–(Au–SG–)$_{1,2,3}$ oligomers. The imperfectness of these structures is the origin of the chemical instability of $Au_{29}(SG)_{20}$ and $Au_{33}(SG)_{22}$ against etching by an excess amount of GSH [28].

At $Au_{38}(SG)_{24}$, the number of **Au3** increases from 1 to 2, suggesting that a structural transition occurs at this size. The estimated numbers (15, 21, 2) of Au atoms for the **Au1**, **Au2**, and **Au3** sites for $Au_{38}(SG)_{24}$ are almost consistent with those for $Au_{38}(SC_2H_4Ph)_{24}$, whose structure was determined by the single-crystal X-ray diffraction analysis [30]. The geometrical structure of $Au_{38}(SC_2H_4Ph)_{24}$ is based on a face-fused bi-icosahedral Au_{23} core, which is protected by three monomeric staples (–SR–Au–SR–) and six dimeric staples (–SR–[Au–SR–]$_2$). In this geometrical structure, the numbers of Au atoms for the **Au1**, **Au2**, and **Au3** sites are 15, 21, and 2, respectively. The predicted molecular structure of $Au_{38}(SG)_{24}$ is schematically shown in Figure 7.6.

In this way, we can estimate the structural evolution of $Au_n(SG)_m$ ($n = 10{\sim}55$) as shown in Figure 7.7.

7.3.6 ^{197}Au Mössbauer Spectra of $Au_{24}Pd_1(SC_{12}H_{25})_{18}$

Since the synthesis of thiolate-protected intermetallic clusters, $Au_{25-x}Pd_x(SC_2H_4Ph)_{18}$ by C. A. Fields-Zinna, et al. in 2009 [31], a lot of intermetallic nanoclusters such as $Au_nM_{n'}(SR)_m$ (M = Cu, Ag, Pd, Pt), $Ag_nM_{n'}(SR)_m$ (M = Ni, Pt) have been developed [32]. In 2010, Y. Negishi, et al. synthesized a dodecanethiolate-protected bimetallic cluster, $Au_{24}Pd_1(SC_{12}H_{25})_{18}$, and proposed its possible molecular structure by means of DFT calculation [20]. Table 7.2 shows the possible position of Pd in $Au_{24}Pd_1(SC_{12}H_{25})_{18}$.

In order to elucidate the position of Pd site in $Au_{24}Pd_1(SC_{12}H_{25})_{18}$, we measured the ^{197}Au Mössbauer spectra of $Au_{24}Pd_1(SC_{12}H_{25})_{18}$ and compared the line profile

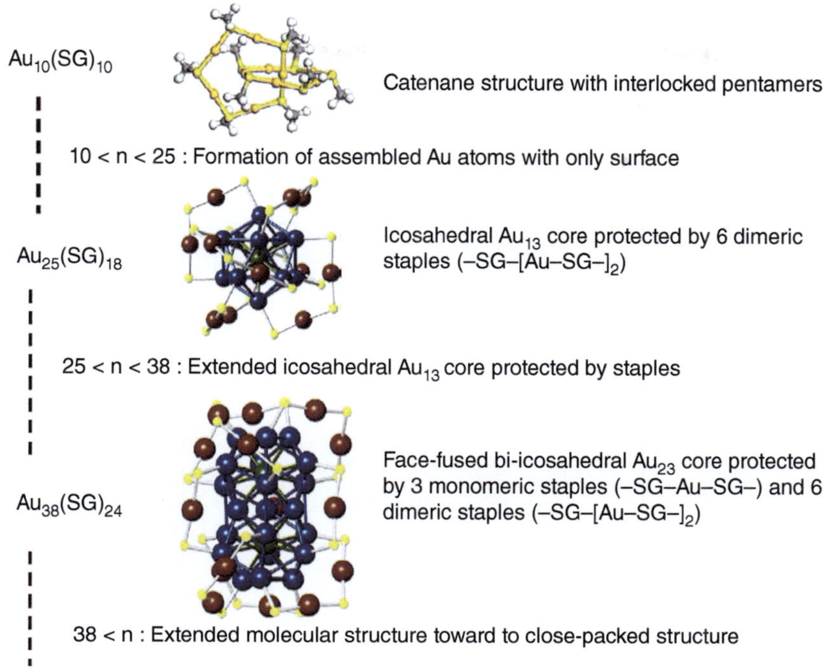

Figure 7.7 Structural evolution of $Au_n(SG)_m$ ($n = 10{\sim}55$).

with that of $Au_{25}(SC_{12}H_{25})_{18}$ [33], which is shown in Figure 7.8. As shown in Figure 7.8, the contribution from the central Au atom observed in the monometallic $Au_{25}(SC_{12}H_{25})_{18}$ counterpart obviously disappeared in the ^{197}Au Mössbauer spectrum of $Au_{24}Pd_1(SC_{12}H_{25})_{18}$. The coordination number of the Pd atom with the neighboring Au atoms determined by EXAFS measurement indicates that the Pd dopant is fully bonded by Au atoms [33]. Based on these results, it is obvious that the single Pd atom doped in $Au_{24}Pd_1(SC_{12}H_{25})_{18}$ is located at the center of the icosahedral $Au_{12}Pd$ core.

7.3.7 ^{197}Au Mössbauer Spectra of $Au_n(SC_{12}H_{25})_m$

Figure 7.9 shows the ^{197}Au Mössbauer spectra of $Au_n(SC_{12}H_{25})_m$ with average diameters of 2 and 4 nm [16]. These spectra are fitted by the superposition of a singlet Lorentzian (core Au atoms) and two sets of doublet Lorentzians, whose components correspond to those of $Au_n(SG)_m$ ($n \sim 55$), **C1**, **C2**, and **C3**. The Mössbauer parameters (δ, Δ, area) of **C1**, **C2**, and **C3** for the Au cluster with an average diameter of 2 nm are estimated at (2.60 mm/s, 6.90 mm/s, 17.3%), (0.88 mm/s, 3.47 mm/s, 43.7%), (0.73 mm/s, 0.0 mm/s, 39.0%), respectively. On the other hand, in the case of the Au cluster with an average diameter of 4 nm, those of **C1**, **C2**, and **C3** are estimated at (2.23 mm/s, 6.51 mm/s, 5.60%), (0.51 mm/s, 3.07 mm/s, 37.7%), (0.35 mm/s, 0.0 mm/s, 56.8%), respectively. Taking account of the relative ratio of recoil-free fractions, f(**Au1**) : f(**Au2**) : f(**Au3**) = 1.03 : 1.00 : 2.44, the fractions of

Table 7.2 Possible position of Pd in $Au_{24}Pd_1(SC_{12}H_{25})_{18}$. Large-size ball: Pd atom, middle-size ball: Au atom, small-size ball: S atom.

Pd position		Case A	Case B	Case C
Molecular structure		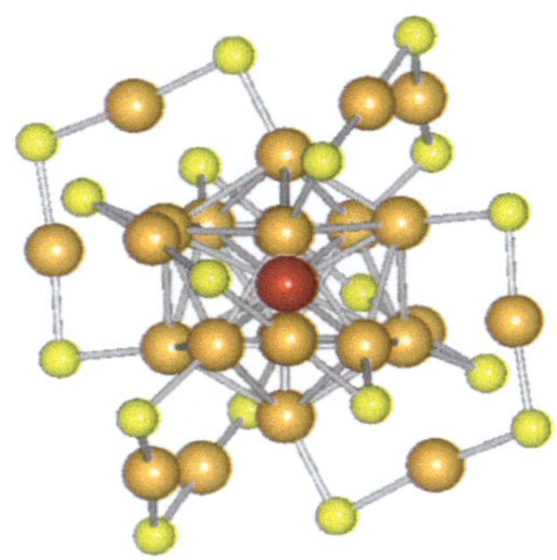	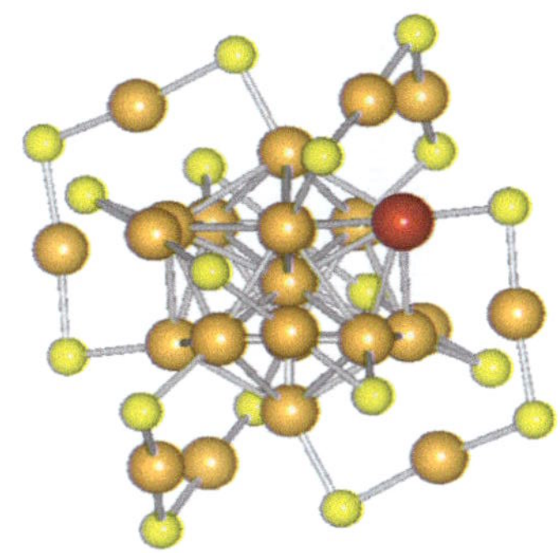	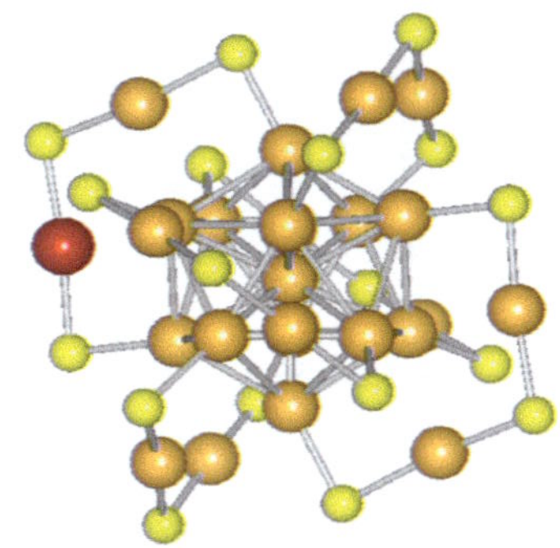
Coordination number	Pd–Au	12	6	0
	Pd–S	0	1	2

Source: Adapted from Ref. [20].

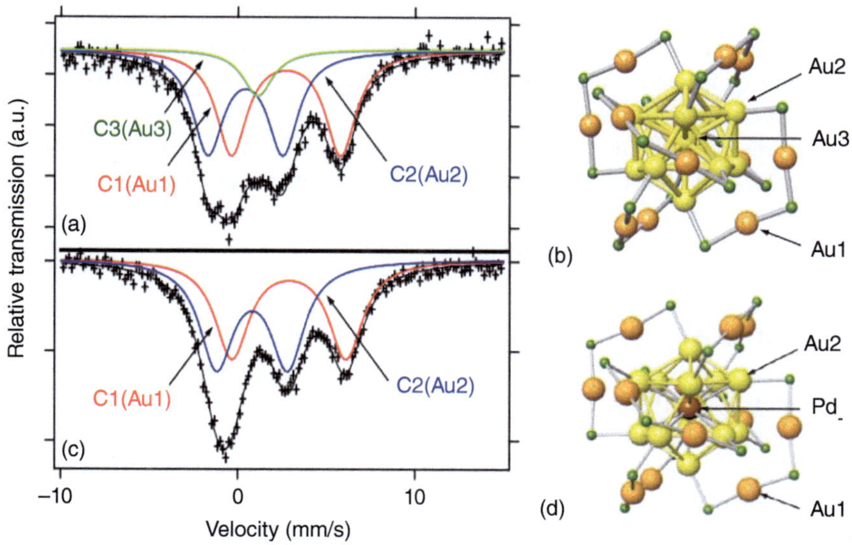

Figure 7.8 (a) ^{197}Au Mössbauer spectrum and (b) molecular structure of $Au_{25}(SC_{12}H_{25})_{18}$, (c) ^{197}Au Mössbauer spectrum and (d) molecular structure of $Au_{24}Pd_1(SC_{12}H_{25})_{18}$ at 16 K. Source: Ref. [33]/with permission from American Chemical Society.

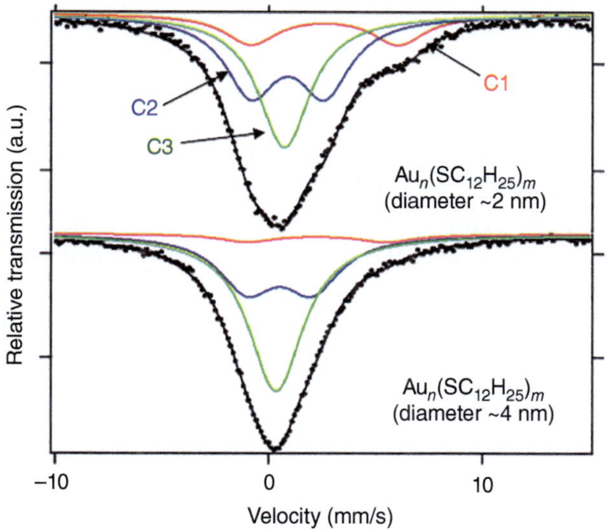

Figure 7.9 ^{197}Au Mössbauer spectra of $Au_n(SC_{12}H_{25})_m$ with the average diameters of 2 and 4 nm at 16 K. Source: Adapted from Ref. [16].

surface Au atoms for 2 and 4 nm Au clusters are estimated at 61.0% and 43.3%, respectively. In this way, the ^{197}Au Mössbauer spectra of $Au_n(SC_{12}H_{25})_m$ with average diameters of 2 and 4 nm could be analyzed on the basis of the structure and electronic state of $Au_n(SG)_m$ ($n \sim 55$).

In connection with the ferromagnetism for thiolate-protected Au nano-clusters, the following should be mentioned. In 2008, J. S. Garitaonandia, et al. reported the existence of a ferromagnetic phase in $Au_n(SC_{12}H_{25})_m$ with an average diameter of 1.9 ± 0.2 nm from the analysis of ^{197}Au Mössbauer spectroscopy at 4.2 K [34]. However, we could not detect the existence of a ferromagnetic phase for $Au_n(SC_{12}H_{25})_m$ with average diameters of 2 and 4 nm at 16 K as shown in Figure 7.9. The systematic study on the existence of ferromagnetism for thiolate-protected Au nano-clusters is an important issue in the future.

7.4 Conclusion

The evolution of the geometrical SG-protected gold clusters, $Au_n(SG)_m$, with $n = 10 \sim 55$, was studied by using ^{197}Au Mössbauer spectroscopy. Successful analysis of the $Au_{25}(SG)_{18}$ spectrum, based on the crystallographically determined structure, enabled us to estimate quantitatively the numbers of Au atoms coordinated by different numbers (0, 1, and 2) of SG ligands for other $Au_n(SG)_m$ clusters. In $Au_{10}(SG)_{10}$, all the Au atoms are bonded to SG ligands, indicating –Au–SG– cyclic structures. A catenane structure was proposed for $Au_{10}(SG)_{10}$. At $n = 15$, Au atoms coordinated by a single SG ligand appeared, suggesting the formation of small Au clusters. At $n = 25$, a single Au atom without the SG ligation appeared, consistent with the formation of an icosahedral Au_{13} core protected by six staples, –SG–[Au–SG–]$_2$. At $n = 39$, the number of core Au atoms without the SG ligation increased from one to two, which suggests that $Au_{39}(SG)_{24}$ has a similar structure to that of $Au_{38}(SC_2H_4Ph)_{24}$ with face-fused bi-icosahedral Au_{23} core. These results demonstrate that ^{197}Au Mössbauer spectroscopy will provide detailed information on the structures of thiolate-protected Au nanoclusters. In fact, we elucidated the location of Pd atom in $Au_{24}Pd_1(SC_{12}H_{25})_{18}$ by means of ^{197}Au Mössbauer spectroscopy. Moreover, we could successfully analyze the ^{197}Au Mössbauer spectra of $Au_n(SC_{12}H_{25})_m$ with the average diameters of 2 and 4 nm, based on the ^{197}Au Mössbauer spectrum of $Au_n(SG)_m$ ($n \sim 55$). However, we could not detect the existence of a ferromagnetic phase for $Au_n(SC_{12}H_{25})_m$ with average diameters of 2 and 4 nm, and $Au_n(SG)_m$. The systematic study on the existence of ferromagnetism for thiolate-protected Au nanoclusters is an important issue in the future.

Acknowledgments

We wish to thank Prof. Y. Negishi (Tokyo University of Science) and Prof. T. Tsukuda (University of Tokyo) for preparing sufficient amounts of $Au_n(SG)_m$ ($n = 10 \sim 55$) by means of polyacrylamide gel electrophoresis (PAGE) in order to measure ^{197}Au Mössbauer spectra, and fruitful discussions on their molecular structures and electronic states. We also wish to thank Emeritus Prof. T. Sugawara and Dr. G. Harada (University of Tokyo) for preparing dodecanethiolate-protected Au clusters with average diameters of 2 and 4 nm.

References

1 Sugano, S. (1991). *Microcluster Physics*, 11–36. Springer-Verlag, Chapter 2.
2 Sugano, S. (1987). Cluster reaction from the viewpoint of nuclear reaction. In: *Microclusters* (ed. S. Sugano, Y. Nishina, and S. Ohnishi), 226–231. Springer.
3 Mingos, D.M.P. (2014). Structural and bonding issues in clusters and nano-clusters. In: *Gold Clusters, Colloids and Nanoparticles II*, Structure and Bonding, vol. 162 (ed. D.M.P. Mingos), 1–65. Springer, Chapter 1.
4 Vollenbroek, F.A., Bouten, P.C.P., Trooster, J.M. et al. (1978). Mössbauer investigation and novel synthesis of gold cluster compounds. *Inorg. Chem.* 17: 1345–1347.
5 Albano, V.G., Bellon, P.L., Manassero, M., and Sansoni, M. (1970). Intermetallic pattern in metal-atom clusters. Structural studies on $Au_{11}X_3(PR_3)_7$ species. *Chem. Commun.* 1210–1211. https://doi.org/10.1039/C29700001210.
6 Schmid, G., Pfeil, R., Boese, R. et al. (1981). $Au_{55}[P(C_6H_5)_3]_{12}Cl_6$ – ein Goldcluster ungewöhnlicher Größe. *Chem. Ber.* 114: 3634–3642.
7 Smit, H.H.A., Nugteren, P.R., Thiel, R.C., and de Jongh, L.J. (1988). Mössbauer and specific heat studies of vibrations of metal core atoms in polynuclear gold cluster compounds. *Physica, B* 153: 33–52.
8 Negishi, Y., Takasugi, Y., Sato, S. et al. (2004). Magic-numbered Au_n clusters protected by glutathione monolayers (n = 18, 21, 25, 28, 32, 39): isolation and spectroscopic characterization. *J. Am. Chem. Soc.* 126: 6518–6519.
9 Negishi, Y., Nobusada, K., and Tsukuda, T. (2005). Glutathione-protected gold clusters revisited: bridging the gap between gold(I)-thiolate complexes and thiolate-protected gold nanocrystals. *J. Am. Chem. Soc.* 127: 5261–5270.
10 Ikeda, K., Kobayashi, Y., Negishi, Y. et al. (2007). Thiolate-induced structural reconstruction of gold clusters probed by ^{197}Au Mössbauer spectroscopy. *J. Am. Chem. Soc.* 129: 7230–7231.
11 Iwasa, T. and Nobusada, K. (2007). Theoretical investigation of optimized structures of thiolated gold cluster $[Au_{25}(SCH_3)_{18}]^+$. *J. Phys. Chem. C* 111: 45–49.
12 Heaven, M.W., Dass, A., White, P.S. et al. (2008). Crystal structure of the gold nanoparticle $[N(C_8H_{17})_4][Au_{25}(SCH_2CH_2Ph)_{18}]$. *J. Am. Chem. Soc.* 130: 3754–3755.
13 Zhu, M., Lanni, E., Garg, N. et al. (2008). Kinetically controlled, high-yield synthesis of Au_{25} clusters. *J. Am. Chem. Soc.* 130: 1138–1139.
14 Akola, J., Walter, M., Whetten, R.L. et al. (2008). On the structure of thiolate-protected Au_{25}. *J. Am. Chem. Soc.* 130: 3756–3757.
15 Tsukuda, T., Negishi, Y., Kobayashi, Y., and Kojima, N. (2011). ^{197}Au Mössbauer spectroscopy of $Au_{25}(SG)_{18}^-$ revisited. *Chem. Lett.* 40: 1292–1293.
16 Kojima, N., Ikeda, K., Kobayashi, Y. et al. (2012). Study of the structure and electronic state of thiolate-protected gold clusters by means of ^{197}Au Mössbauer spectroscopy. *Hyperfine Interact.* 207: 127–131.

17 Kojima, N., Kobayashi, Y., Negishi, Y. et al. (2013). Structural evolution of glutathionate-protected gold clusters studied by means of ^{197}Au Mössbauer spectroscopy. *Hyperfine Interact.* 217: 91–98.

18 Takano, S. and Tsukuda, T. (2015). Controlled systhesis: size control. In: *Protected Metal Clusters: From Fundamentals to Applications*, Frontiers of Nanoscience Series, vol. 9 (ed. T. Tsukuda and H. Häkkinen). Elsevier, Chapter 2.

19 Nagao, O., Harada, G., Sugawara, T. et al. (2004). Small-angle X-Ray scattering method to determine the size distribution of gold nanoparticles chemisorbed by thiol ligands. *Jpn. J. Appl. Phys.* 43: 7742–7746.

20 Negishi, Y., Kurashige, W., Niihori, Y. et al. (2010). Isolation, structure, and stability of a dodecanethiolate-protected Pd_1Au_{24} cluster. *Phys. Chem. Chem. Phys.* 12: 6219–6225.

21 MossWin 3.0 (2011). Dr. Zoltan Klencsar. http://www.netx.hu/mosswinn/index.html (accessed 2011).

22 Barutnik, H.D., Potzel, W., Mössbauer, R.L., and Kaindl, G. (1970). Resonance spectroscopy of γ-radiation on Au (I) and Au (III) compounds. *Z. Phys.* 240: 1–16.

23 Charlton, J.S. and Nichols, D.I. (1970). ^{197}Au Mössbauer spectroscopy on some aurous and auric complexes. *J. Chem. Soc. A* 1970: 1484–1488.

24 Faltens, M.O. and Shirley, D.A. (1970). Mössbauer spectroscopy of gold compounds. *J. Chem. Phys.* 53: 4249–4264.

25 Parish, R.V. (1984). Gold-197 Mössbauer spectroscopy in the characterization of gold compounds. In: *Mössbauer Spectroscopy Applied to Inorganic Chemistry*, vol. 1 (ed. G.J. Long), 577–617. Plenum.

26 Grönbeck, H., Walter, M., and Häkkinen, H. (2006). Theoretical characterization of cyclic thiolated gold clusters. *J. Am. Chem. Soc.* 128: 10268–10275.

27 Stevens, J.G., Boss, C.R., Goforth, M.A. et al. (ed.) (1993). ^{197}Au Mössbauer spectroscopy 1993. In: *Mössbauer Handbooks*. Mössbauer Effect Data Center, The University of North Carolina, U.S.A.

28 Shichibu, Y., Negishi, Y., Tsunoyama, H. et al. (2007). Extremely high stability of glutathionate protected Au_{25} clusters against core etching. *Small* 3: 835–839.

29 Tlahuice, A. and Garzón, I.L. (2012). Structural, electronic, optical, and chiroptical properties of small thiolated gold clusters: the case of Au_6 and Au_8 cores protected with dimer $[Au_2(SR)_3]$ and trimer $[Au_3(SR)_4]$ motifs. *Phys. Chem. Chem. Phys.* 14: 7321–7329.

30 Qian, H., Eckenhoff, W.T., Zhu, Y. et al. (2010). Total structure determination of thiolate-protected Au_{38} nanoparticles. *J. Am. Chem. Soc.* 132: 8280–8281.

31 Fields-Zinna, C.A., Crowe, M.C., Dass, A. et al. (2009). Mass spectrometry of small bimetal monolayer-protected clusters. *Langmuir* 25: 7704–7710.

32 Negishi, Y., Niihori, Y., and Kurashige, W. (2015), Chapter 3). Controlled synthesis: composition and interface control. In: *Protected Metal Clusters: From Fundamentals to Applications*, Frontiers of Nanoscience Series, vol. 9 (ed. T. Tsukuda and H. Häkkinen), 39–71. Elsevier.

33 Negishi, Y., Kurashige, W., Kobayashi, Y. et al. (2013). Formation of a Pd@Au_{12} superatomic core in $Au_{24}Pd_1(SC_{12}H_{25})_{18}$ probed by ^{197}Au Mössbauer and Pd K-Edge EXAFS spectroscopy. *J. Phys. Chem. Lett.* 4: 3579–3583.

34 Garitaonandia, J.S., Insausti, M., Goikolea, E. et al. (2008). Chemically induced permanent magnetism in Au, Ag, and Cu nanoparticles: localization of the magnetism by element selective techniques. *Nano Lett.* 8: 661–667.

8

^{197}Au Mössbauer Spectroscopy of Gold Mixed-Valence Complexes, $Cs_2[Au^{I}X_2][Au^{III}Y_4]$(X, Y = Cl, Br, I) and $[NH_3(CH_2)_nNH_3]_2[(Au^{I}I_2)(Au^{III}I_4)(I_3)_2]$ (n = 7, 8)

Norimichi Kojima[1], Yasuhiro Kobaqyashi[2], and Makoto Seto[2]

[1] *The University of Tokyo, Graduate School of Arts and Sciences, Department of Basic Science, Komaba 3-8-1, Meguroku, Tokyo, Japan*
[2] *Kyoto University, Institute for Integrated Radiation and Nuclear Science, Division of Nuclear Radiation Physics, Asashiro-Nishi 2, Kumatori-cho, Sennan-gun, Osaka, Japan*

List of Abbreviations

CT	charge transfer
EFG	electric field gradient
E_R	recoil energy
IVCT	inter-valence charge transfer
Q	nuclear quadrupole moment
q	electric field gradient tensor
SCF–MS–Xα	self consistent field–multiple scattering–Xα
Δ	quadrupole splitting
δ	isomer shift
η	asymmetry parameter, $(V_{xx} - V_{yy})/V_{zz}$
Θ_D	Debye temperature

8.1 Introduction

Mixed-valence compounds, defined as compounds containing an element with more than one oxidation state, are classified into three groups according to the Robin-Day classification [1]:

Class I: Compounds whose valence states occupy different crystallographic sites and are too distant from each other to allow the electronic interaction between the mixed-valence states.

Class II: Compounds in which an electron exchange between the different valence states takes place as a thermally activated process. The inter-valence charge

Mössbauer Spectroscopy: Applications in Chemistry and Materials Science, First Edition. Edited by Yann Garcia, Junhu Wang, and Tao Zhang.

transfer transition (IVCT) between different valence states is observable in optical absorption spectra.

Class III(A): Compounds in which valence electrons are delocalized between atoms within discrete polynuclear ions.

Class III(B): Compounds in which valence electrons are delocalized between identically crystallographic sites.

Among them, the mixed-valence compounds whose electronic states are situated between class II and class III have attracted much attention from the viewpoint of their characteristic physical properties such as ferromagnetism caused by double exchange interaction [2–5], superconductivity [6–8], photo-generated soliton and polaron as the mismatch of valence alternation [9–12], photo-induced valence transition [13–15], and so forth.

In general, the mixed-valence system is classified into two types of (Z, Z + 1: ΔZ = 1) and (Z, Z + 2: ΔZ = 2), in accordance with the difference number between the lower and the higher oxidation numbers, where Z is the oxidation number. In the case of (Z, Z + 1: ΔZ = 1), the ground wave function of the mixed valence state is expressed as follows,

$$\Psi = (1 - \alpha^2)^{1/2} \left\{\varphi\left(A_i{}^{Z}\right) \varphi\left(A_j{}^{Z+1}\right)\right\} + \alpha \left\{\varphi\left(A_i{}^{Z+!}\right) \varphi\left(A_j{}^{Z}\right)\right\}, \tag{8.1}$$

where, A_i and A_j are the nearest neighbor metal sites to each other and α is the valence delocalization coefficient which represents the strength of charge transfer interaction. The physical property of class II is characterized by the IVCT transition induced by the charge transfer interaction and the intensity of IVCT transition increases with increasing α. For example, in the case of Prussian blue ($Fe^{III}{}_4[Fe^{II}(CN)_6]_3 \cdot 15H_2O$), the typical mixed-valence compound of class II, the charge transfer interaction induces the ferromagnetism with $T_C = 5.5$ K and α^2 was estimated to be less than 0.05 from the analysis of polarized neutron diffraction [16].

In the case of (Z, Z + 2: ΔZ = 2), on the other hand, the one-electron transfer state of $\varphi(A_i{}^{Z+1})\varphi(A_j{}^{Z+1})$ is generally unstable, and the charge disproportionation from $\varphi(A_i{}^{Z+1})\varphi(A_j{}^{Z+1})$ to $\varphi(A_i{}^{Z+2})\varphi(A_j{}^{Z})$ accompanied with lattice distortion occurs caused by strong electron-phonon interaction has systematically been studied [17–20]. The typical examples are $BaBiO_3$ ($Bi^{III, V}$), Cs_2SbX_6 ($Sb^{III, V}$; X = Cl, Br), $ATlX_3$ (A = Rb, Cs; $Tl^{I, III}$; X = F, Cl, Br), $Cs_2[Au^{I}X_2][Au^{III}X_4]$ (X = Cl, Br, I), halogen-bridged Pt mixed-valence complexes such as Wolfram's red salt, $[Pt^{II}(C_2H_5NH_2)_4][Pt^{IV}(C_2H_5NH_2)_4Cl_2]Cl_4 \cdot 4H_2O$, and so forth. The stability of $\varphi(A_i{}^{Z})\varphi(A_j{}^{Z+2})$ and $\varphi(A_i{}^{Z+2})\varphi(A_j{}^{Z})$ is considered to be realized by the so-called "negative U" effect [21], and the $\varphi(A_i{}^{Z})/\varphi(A_j{}^{Z})$ state is regarded as onsite bipolaron [22–25]. The two-electron transfer system between $\varphi(A_i{}^{Z})\varphi(A_j{}^{Z+2})$ and $\varphi(A_i{}^{Z+2})\varphi(A_j{}^{Z})$ is called "valence skipping" [26, 27], and the dynamics of "valence skipping" is regarded as electron-hole Bose liquid. The ground wave function of the mixed valence system for (Z, Z + 2: ΔZ = 2) is expressed as follows:

$$\Psi = (1 - \alpha^2 - \beta^2)^{1/2} \left\{\varphi\left(A_i{}^{Z}\right) \varphi\left(A_j{}^{Z+2}\right)\right\} + \alpha \left\{\varphi\left(A_i{}^{Z+!}\right) \varphi\left(A_j{}^{Z+1}\right)\right\} + \beta \left\{\varphi\left(A_i{}^{Z+2}\right) \varphi\left(A_j{}^{Z}\right)\right\}, \tag{8.2}$$

Among the mixed valence system for (Z, Z+2: ΔZ = 2), $Ba_{1-x}K_xBiO_3$ and $BaPb_{1-x}Bi_xO_3$ having perovskite structure have become of great interest since the discovery of superconductivity. In the case of $Ba_{1-x}K_xBiO_3$, the superconductivity appears on condition of $0.375 < x < 0.5$ and the maximum T_C is 34 K at x = 0.35 [28]. On the other hand, in the case of $BaPb_{1-x}Bi_xO_3$, the superconductivity appears on condition of $0.05 < x < 0.3$ and the maximum T_C is 11.7 K at x = 0.25 [29]. From the viewpoint of the dynamics of the mixed-valence system and the possibility of superconductivity for another mixed valence system with (Z, Z+2: ΔZ = 2), for example, $ATlX_3$ (A = Rb, Cs; X = F, Cl, Br) [30–32], and $Cs_2[Au^IX_2][Au^{III}X_4]$ [13–15, 33–49] having three-dimensional perovskite structure have extensively been investigated.

In the case of $Cs_2[Au^IX_2][Au^{III}X_4]$ (X = Cl, Br, I), the pressure-induced valence transition from the mixed-valence state of $Au^{I,\,III}$ to the single valence state of Au^{II} occurs, which is coupled with a structural phase transition. In the pressure region below the Au valence transition, these complexes show a metallic behavior under low temperatures, which would be attributed to a dynamic two-electron exchange between the Au^1 and Au^{III} states (i.e. highly mobile bipolarons). The metallic cubic phase of Au^{II} state appearing commonly for these complexes under high pressure and high temperature can be obtained as a metastable phase even at ambient pressure and room temperature [33, 36]. Moreover, $Cs_2[Au^IX_2][Au^{III}X_4]$ showing strong IVCT transition between the near infra-red and visible region has attracted much attention from the viewpoint of lead-free solar battery materials [50–52]. In addition to $Cs_2[Au^IX_2][Au^{III}X_4]$, $[NH_3(CH_2)_nNH_3]_2[(Au^II_2)(Au^{III}I_4)(I_3)_2]$ (n = 7, 8) and $(CH_3NH_3)_2[Au^II_2][Au^{III}I_4]$ having two-dimensional perovskite structure have been developed from the viewpoint of highly efficient lead-free solar battery materials [53, 54].

These interesting phenomena strongly depend on the kind of bridging halogen. For the sake of controlling the chemical bond and electronic state for the halogen-bridged Au mixed-valence system, we have tried to control the bridging halogens in the *ab*-plane and along the *c*-axis, individually. Recently, we have succeeded in synthesizing hetero-halogen bridged gold mixed-valence complexes having perovskite-type structure, $Cs_2[Au^IX_2][Au^{III}Y_4]$(X, Y = Cl, Br, and I; X ≠ Y), in an organic solvent by using a new method [55].

^{197}Au Mössbauer spectroscopy with 77.3 keV as the γ-ray source whose time scale is 1.91 ns is one of the most powerful methods to investigate the local crystal structure around Au site, the chemical bond between Au and the ligand, the Au valence state, and the dynamics of Au valence transition. In this chapter, we review the synthesis of gold mixed-valence complexes, $Cs_2[Au^IX_2][Au^{III}Y_4]$ (X, Y = Cl, Br, I), the crystal structure, the chemical bond, the Au valence state, and the Au valence transition from the mixed-valence state (Au^I, Au^{III}) to the single valence state (Au^{II}) from the viewpoint of ^{197}Au Mössbauer spectroscopy. Moreover, we describe the Au valence state of the layered perovskite-type gold mixed-valence complex system, $[NH_3(CH_2)_nNH_3]_2[(Au^II_2)(Au^{III}I_4)(I_3)_2]$ (n = 7, 8) by means of ^{197}Au Mössbauer spectroscopy.

The contents of this paper are as follows. In Section 8.2, experimental details are described. In Section 8.3, we describe the characteristic properties of the crystal structure of $Cs_2[Au^IX_2][Au^{III}Y_4]$ (X, Y = Cl, Br, I). In the case of $Cs_2[Au^II_2][Au^{III}Br_4]$, Raman spectroscopy proves the existence of $[Au^II_2]^-$ and $[Au^{III}Br_4]^-$. In Section 8.4, we discuss the nature of the chemical bond in Au-X for $[Au^IX_2]^-$ and $[Au^{III}X_4]^-$. In Section 8.5, we analyze the Mössbauer parameters of ^{197}Au in $(C_4H_9)_4N[Au^IX_2]$, $(C_4H_9)_4N[Au^{III}X_4]$, and $Cs_2[Au^IX_2][Au^{III}X_4]$ (X = Cl, Br, I). In Section 8.6, we analyze the charge transfer interaction in $Cs_2[Au^IX_2][Au^{III}X_4]$ (X = Cl, Br, I) by means of ^{197}Au Mössbauer spectroscopy. In Section 8.7, we analyze the ^{197}Au Mössbauer spectra of hetero-halogen-bridged gold mixed-valence complexes, $Cs_2[Au^IX_2][Au^{III}Y_4]$ (X, Y = Cl, Br, I), and discuss the charge transfer interaction between $[Au^IX_2]^-$ and $[Au^{III}Y_4]^-$. In Section 8.8, we describe the single-crystal ^{197}Au Mössbauer spectra of $Cs_2[Au^II_2][Au^{III}I_4]$, and discuss the sign of EFG for Au^I in $[Au^II_2]^-$ and Au^{III} in $[Au^{III}I_4]^-$. Moreover, we discuss the anisotropy of recoil-free fractions for $[Au^II_2]^-$ and $[Au^{III}I_4]^-$ in $Cs_2[Au^II_2][Au^{III}I_4]$. In Section 8.9, we describe the ^{197}Au Mössbauer spectra of $Cs_2[Au^IX_2][Au^{III}X_4]$ (X = Cl, I) under high pressures and the pressure-induced Au valence transition and the realization of Au^{II} state for $Cs_2[Au^II_2][Au^{III}I_4]$. In Section 8.10, we describe the ^{197}Au Mössbauer spectra of $[NH_3(CH_2)_nNH_3]_2[(Au^II_2)(Au^{III}I_4)(I_3)_2]$ (n = 7, 8) and discuss the charge transfer interaction between the Au^I and Au^{III} sites. In Section 8.11, we summarize the characteristic physical properties of $Cs_2[Au^IX_2][Au^{III}Y_4]$ (X, Y = Cl, Br, I) and $[NH_3(CH_2)_nNH_3]_2$ $[(Au^II_2)(Au^{III}I_4)(I_3)_2]$ (n = 7, 8) as a conclusion.

8.2 Experimental Procedure

8.2.1 Synthesis and Characterization

8.2.1.1 $Cs_2[Au^IX_2][Au^{III}Y_4]$ (X, Y = Cl, Br, I)

$(C_4H_9)_4N[Au^IX_2]$ (X = Cl, Br, I) and $(C_4H_9)_4N[Au^{III}X_4]$ (X = Cl, Br, I) were prepared according to the references [56–59]. $Cs_2[Au^IX_2][Au^{III}Y_4]$ (X, Y = Cl, Br, I) was prepared in the following way. Solutions of $(C_4H_9)_4N[Au^IX_2]$ and $(C_4H_9)_4N[Au^{III}Y_4]$ in 1,1,2-trichloroethane were stirred at −20 °C, individually. A solution of $C_6H_5SO_3Cs$ in methanol was stirred at −20 °C. To this, the solutions of $(C_4H_9)_4N[Au^IX_2]$ and $(C_4H_9)_4N[Au^{III}Y_4]$ in 1,1,2-trichloroethane were added and stirred at −20 °C for 1 hour. In this manner, $Cs_2[Au^IX_2][Au^{III}Y_4]$ (X, Y = Cl, Br, I) was obtained as black-colored precipitate [55, 60]. Single crystals were grown at −20 °C by a diffusion method using an H-type test tube. The single crystal X-ray diffraction analysis was carried out by using a RIGAKU RAXIS-RAPID Imaging Plate diffractometer equipped with graphite-monochromated Mo Kα radiation (λ = 0.071069 nm) at room temperature. Raman spectroscopic measurement was carried out by using an argon ion laser (514.5 nm) in a backward configuration in order to confirm the formation of hetero-halogen-bridged mixed-valence complex, $Cs_2[Au^II_2][Au^{III}Br_4]$.

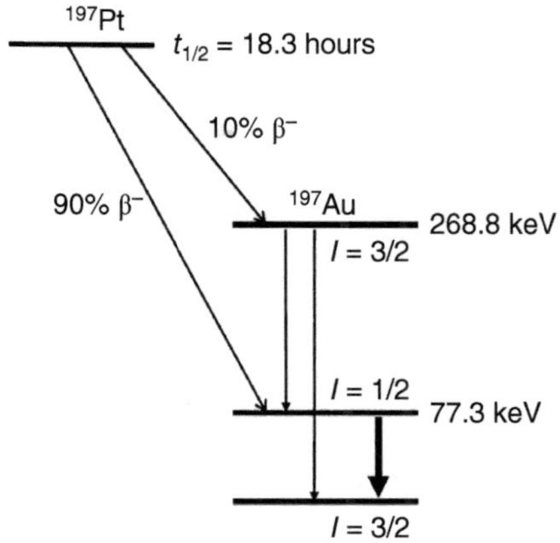

Figure 8.1 Nuclear decay scheme of ^{197}Au.

8.2.1.2 $[NH_3(CH_2)_nNH_3]_2[(Au^{I}I_2)(Au^{III}I_4)(I_3)_2]$ (n = 7, 8)

$[NH_3(CH_2)_nNH_3]_2[(Au^{I}I_2)(Au^{III}I_4)(I_3)_2]$ (n = 7, 8) were synthesized in a similar way as reported in reference [53]. Iodine was dripped into conc. HI. To this, the aqueous solution of $NH_2(CH_2)_nNH_2$ and $HAuCl_4$ was added at room temperature. Then a black precipitate was obtained.

8.2.2 ^{197}Au Mössbauer Spectroscopy

The γ-ray source of ^{197}Au Mössbauer spectroscopy, ^{197}Pt, is obtained by the neutron irradiation for ^{196}Pt-enriched material by the nuclear reaction of ^{196}Pt $[n, \gamma]$ ^{197}Pt, which is schematically shown in Figure 8.1. Since the half-life time ($t_{1/2}$) is 18.3 hours, a research reactor is necessary to perform ^{197}Au Mössbauer spectroscopy. We carried out ^{197}Au Mössbauer spectroscopy at the research reactor of the Institute for Integrated Radiation and Nuclear Science, Kyoto University. The γ-ray source (77.3 keV) was generated by the neutron irradiation to 98%-enriched ^{196}Pt metal foil. Since the recoil energy (77.3 keV) of γ-ray is relatively high, the γ-ray source and samples were cooled down to 16 K. The transmitted γ-ray was detected by a NaI(Tl) scintillation counter. The isomer shift (δ) of Au foil was referenced to 0 mm/s. The spectra were calibrated and referenced by using the six lines of a body-centered cubic Fe foil (α-Fe). The Mössbauer parameters of δ and quadrupole splitting (Δ) were determined by fitting the spectra using the program MossWin 3.0 [61].

The ^{197}Au Mössbauer spectroscopic measurements at 4.2 K and under high pressures up to 12.5 GPa were carried out at Marburg University, Germany. Powdered samples mixed with wax were placed in a pyrophyllite sample holder located between two B_4C anvils mounted in a CuBe clamp system, which was immersed in liquid He. The transmitted γ-ray was detected by a solid-state Ge detector. The pressure at the sample was determined by using the known critical dependence of the transition from the metallic phase to the superconducting phase of lead. A ^{197}Pt source with an activity of about 7×10^9 Bq was used. For this, a foil of ^{196}Pt (97% enriched) was exposed to a flux of neutrons at the research reactor of Hahn-Meitner Institute, Berlin, during 24 hours.

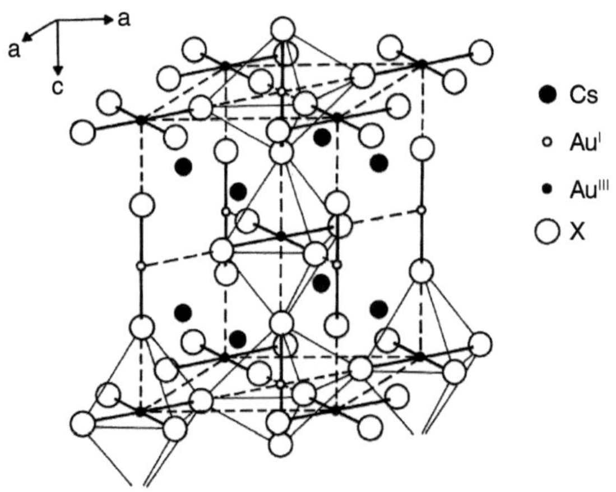

Figure 8.2 Crystal structure of $Cs_2[Au^IX_2][Au^{III}X_4]$ (X = Cl, Br, I). Source: Adapted from [36].

8.3 Crystal Structure of $Cs_2[Au^IX_2][Au^{III}Y_4]$ (X, Y = Cl, Br, I)

The crystal structure of $Cs_2[Au^IX_2][Au^{III}X_4]$ (X = Cl, Br, I) is a distorted perovskite-type structure which is shown in Figure 8.2. In the crystal, all the halogens are distorted from the midpoint between neighboring Au ions in the chains, and the linear $[Au^IX_2]^-$ and square-planar $[Au^{III}Y_4]^-$ complexes are stacked alternately. There are two relative orientations between $[Au^IX_2]^-$ and $[Au^{III}X_4]^-$, i.e., the $\cdots X–Au^I–X\cdots Au^{III}\cdots X–Au^I–X\cdots$ network lies along the *c*-axis, while the $\cdots X–Au^{III}–X\cdots Au^I\cdots X–Au^{III}–X\cdots$ network lies in the *ab* plane. Consequently, its structure consists of a three-dimensional metal-halogen framework forming elongated octahedra with Au^{III} and compressed octahedra with Au^I sharing their corners. All of $Cs_2[Au^IX_2][Au^{III}Y_4]$ (X, Y = Cl, Br, I) show similar powder X-ray diffraction (XRD) profiles, reflecting the isomorphous structure.

Figure 8.3 shows the powder X-ray diffraction patterns for $Cs_2[Au^IX_2][Au^{III}Br_4]$ (X = Cl, Br, I) as examples, in which each doublet (e.g. [(002), (110)], [(112), (200)]) is due to the difference between the $Au^I\cdots Au^{III}$ distances in the *ab*-plane and along the *c*-axis. The presence of weak superstructure reflections such as (103) and (301) indicates that the bridging halogen atoms are deviated from the symmetrical position between the Au atoms. Consequently, $Cs_2[Au^IX_2][Au^{III}Y_4]$ shows a tetragonal perovskite structure whose space group is *I*4/*mmm*. The lattice parameters of $Cs_2[Au^IX_2][Au^{III}Y_4]$ (X, Y = Cl, Br, I) are listed in Table 8.1.

In order to confirm the formation of the hetero-halogen bridged gold mixed-valence complex, $Cs_2[Au^II_2][Au^{III}Br_4]$, we investigated the Raman spectra for the single crystals of $Cs_2[Au^II_2][Au^{III}Br_4]$, $Cs_2[Au^IBr_2][Au^{III}Br_4]$, and $(C_4H_9)_4N[Au^II_2]$ [55], which is shown in Figure 8.4. In the case of $Cs_2[Au^IBr_2][Au^{III}Br_4]$, two A_{1g} modes are observed in the $c(a+b,\ a+b)c$ configuration, while one B_{1g} mode is observed in the $c(a+b,\ a-b)c$ configuration. The lower-lying two modes, i.e. B_{1g} (178 cm^{-1}) and A_{1g} (179 cm^{-1}) modes correspond to the stretching modes of the $[Au^{III}Br_4]^-$ molecule, while

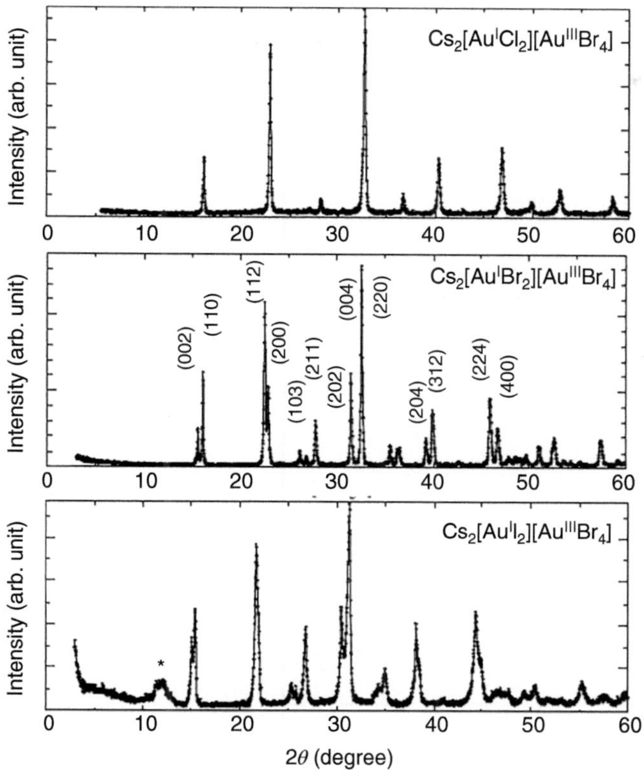

Figure 8.3 Powder X-ray diffraction patterns of $Cs_2[Au^IX_2][Au^{III}Br_4]$ (X = Cl, Br, I). The peak marked with (*) is the background reflection from the specimen holder. Source: Ref. [55]/with permission from Springer Nature.

Table 8.1 Lattice parameters of $Cs_2[Au^IX_2][Au^{III}Y_4]$ (X, Y = Cl, Br, I). $\sqrt{2} \times a/c$ shows the distortion degree from the cubic perovskite structure.

Complex	*a* (Å)	*c* (Å)	$\sqrt{2} \times a/c$
$Cs_2[Au^ICl_2][Au^{III}Cl_4]$	7.495	10.88	0.974
$Cs_2[Au^ICl_2][Au^{III}Br_4]$	7.741	10.936	1.001
$Cs_2[Au^ICl_2][Au^{III}I_4]$	7.918	11.164	1.003
$Cs_2[Au^IBr_2][Au^{III}Cl_4]$	7.539	11.028	0.967
$Cs_2[Au^IBr_2][Au^{III}Br_4]$	7.759	11.308	0.970
$Cs_2[Au^IBr_2][Au^{III}I_4]$	8.159	11.803	0.978
$Cs_2[Au^II_2][Au^{III}Cl_4]$	7.548	10.974	0.973
$Cs_2[Au^II_2][Au^{III}Br_4]$	7.849	11.407	0.973
$Cs_2[Au^II_2][Au^{III}I_4]$	8.284	12.092	0.969

Source: Ref. [55]/with permission from Springer Nature.

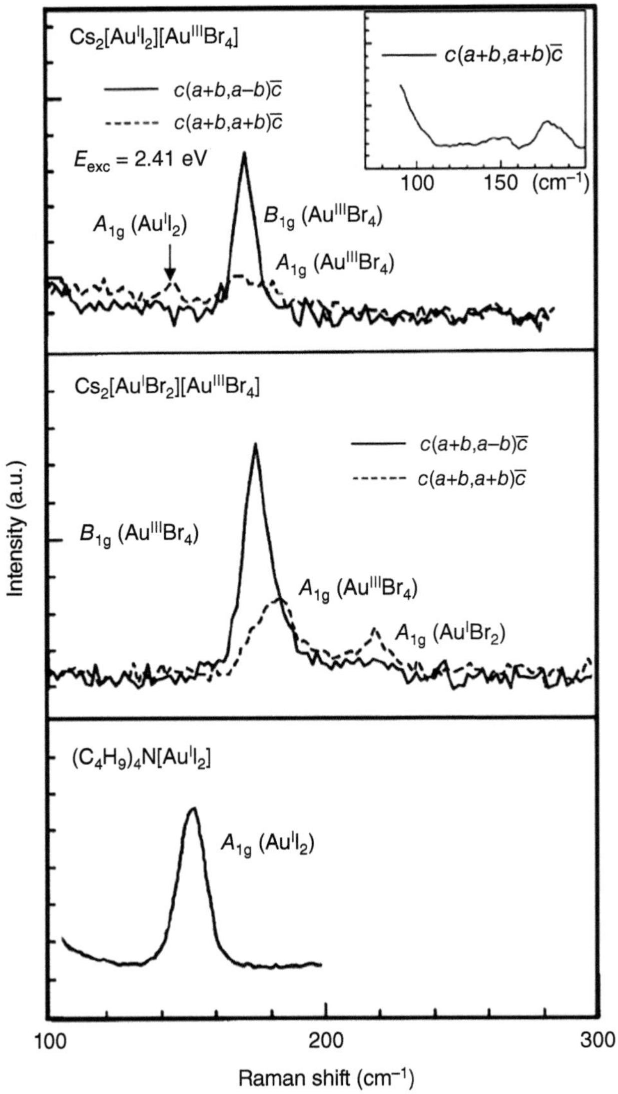

Figure 8.4 Raman spectra for $Cs_2[Au^{I}I_2][Au^{III}Br_4]$, $Cs_2[Au^{I}Br_2][Au^{III}Br_4]$, and $(C_4H_9)_4N[Au^{I}I_2]$ at 300 K. Excitation energy, E_{exc}, is 2.41 eV. Source: Ref. [55]/with permission from Springer Nature.

the higher-lying A_{1g} (220 cm^{-1}) mode corresponds to that of the $[Au^{I}Br_2]^-$ molecule. For $Cs_2[Au^{I}I_2][Au^{III}Br_4]$, the A_{1g} (220 cm^{-1}) mode corresponding to the $[Au^{I}Br_2]^-$ molecule disappears. Instead, the Raman spectrum corresponding to the $[Au^{I}I_2]^-$ molecule appears at about 150 cm^{-1}. In this manner, it is proved that the hetero-halogen bridged gold mixed-valence complex, $Cs_2[Au^{I}I_2][Au^{III}Br_4]$, is formed. In fact, the crystal structure of $Cs_2[Au^{I}I_2][Au^{III}Br_4]$ was determined by the single-crystal X-ray structural analysis [55]. In this crystal, all the halogens are distorted from the midpoint between neighboring Au atoms in the chains, and the linear $[Au^{I}I_2]^-$ and square-planar $[Au^{III}Br_4]^-$ complexes are stacked alternately. Consequently, its structure consists of a three-dimensional metal-halogen framework formed by elongated octahedra with Au^{III} and compressed octahedra with Au^I sharing their corners.

8.4 Chemical Bond of Au–X in $[Au^IX_2]^-$ and $[Au^{III}X_4]^-$

The relativistic effect, which is strongest at the position of Au in the periodic table of elements, enormously stabilizes the 6s level and destabilizes the 5d levels of Au ($[Xe]4f^{14}5d^{10}6s^1$) [62]. According to the molecular orbital calculation including the relativistic effect, the $5d_{z^2}6s$ hybridization rather than $6s6p_z$ hybridization is dominant for the chemical bonds in the linear $[Au^IX_2]^-$ complexes [63, 64]. However, many studies of ^{197}Au Mössbauer spectra of Au^I complexes have been interpreted in terms of $6s6p_z$ hybridization of Au^I [65–69], since the sign of EFG of Au^I in $KAu(CN)_2$ was determined to be negative [70].

When the ^{197}Au Mössbauer spectrum of Au complex is a simple doublet, the quadrupole splitting (Δ) is given by the following equation:

$$\Delta = \frac{1}{2}e^2qQ(1+\eta^2/3)^{1/2} = \frac{1}{2}eQV_{zz}(1+\eta^2/3)^{1/2} \tag{8.3}$$

where e is the charge on the proton, Q is the nuclear quadrupole moment (+54.7 fm^2 for the ground state of ^{197}Au) [71], $eq = V_{zz}$ is the principal component of the EFG, and η is the asymmetry parameter. In the cases of linear $[Au^IX_2]^-$, square-planar $[Au^{III}X_4]^-$, and tetragonal perovskite-type $Cs_2[Au^IX_2][Au^{III}X_4]$, η can be regarded as zero. The q contributed by 5d and 6p orbitals is described as the following equation [72]:

$$\begin{aligned} q &= \sum_i n_i \left\langle \varphi_i \left| \frac{3\cos^2\theta - 1}{r^3} \right| \varphi_i \right\rangle \\ &= \frac{2}{5}\{n(p_x) + n(p_y) - 2n(p_z)\}\langle r^{-3}\rangle_{6p} \\ &\quad + \frac{2}{7}\left\{2n(d_{xy}) + 2n\left(d_{x^2-y^2}\right) - n(d_{yz}) - n(d_{zx}) - 2n\left(d_{z^2}\right)\right\}\langle r^{-3}\rangle_{5d} \end{aligned} \tag{8.4}$$

where n_i is the electron population in the φ_i orbital. For 5d electrons, the $<r^{-3}>_{5d}$ for free Au atom was determined at 12.3 a_0^{-3} by W. J. Childs and L. S. Goodmann [73], while the $<r^{-3}>_{6p}$ for free Au atom was estimated at 16.48 a_0^{-3} by P. Machmer [74]. In connection with the orbital radius, the following should be noted. According to the SCF–MS–Xα (self-consistent field–multiple scattering–Xα) calculation of $[Au^IX_2]^-$ (X = Cl, Br, I) reported by G. A. Bownmaker et al. [64], there is a dramatic contraction in the $6p_z$ orbital of $[Au^IX_2]^-$ (X = Cl, Br, I) relative to that of the free Au atom. This contraction as well as a large amount of electron density in the $6p_z$ orbital of Au produces a large negative EFG in $[Au^IX_2]^-$. In the case of Au^{III} in $[Au^{III}X_4]^-$, the main contributions to EFG are given by the σ-bonding of $5d_{x^2-y^2}$ and $6p_{x,y}$ orbitals (positive), and by the hole of $5d_{x^2-y^2}$ in the 5d shell (negative). Since the $<r^{-3}>_{6p}$ is larger than $<r^{-3}>_{5d}$, a relatively small and positive EFG is expected for Au^{III} in $[Au^{III}X_4]^-$.

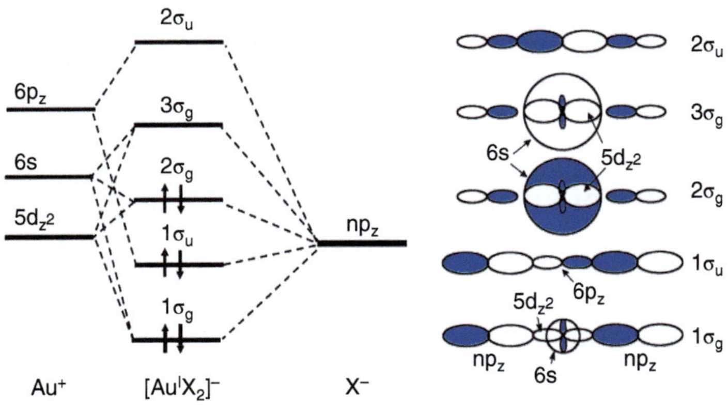

Figure 8.5 Schematic energy levels for linear coordinated complex, $[Au^IX_2]^-$.

In connection with the sign of EFG of Au^I in $[Au^IX_2]^-$ and Au^{III} in $[Au^{III}X_4]^-$, the determination of the sign of EFG is one of the most powerful methods to investigate the nature of chemical bonds of Au-X in $[Au^IX_2]^-$ and $[Au^{III}X_4]^-$. In $Cs_2[Au^IX_2][Au^{III}X_4]$, the concentration of negative charge in the *ab*-plane makes EFG positive and the ground doublet becomes $I_z = \pm 1/2$. On the other hand, the concentration of negative charge along the *c*-axis makes EFG negative and the ground doublet becomes $I_z = \pm 3/2$. In the case of linear $[Au^IX_2]^-$, the $5d_{z^2}$ hole due to $5d_{z^2}$26s hybridization produces a positive contribution to EFG, while the $6p_z$ electron produces a negative one. In the case of square-planar $[Au^{III}X_4]^-$, on the other hand, the hole in the $5d_{x^2-y^2}$ orbital produces a negative contribution to EFG, while the partially filled $6p_x$ and $6p_y$ orbitals produce a positive one. As will be described in Section 8.8, from the analysis of single crystal ^{197}Au Mössbauer spectroscopy, it is obvious that the contributions of $6p_z$ and $6p_{x,y}$ dominate the sign of EFG (negative) for Au^I in $[Au^II_2]^-$ and that (positive) for Au^{III} in $[Au^{III}I_4]^-$, respectively [40, 75]. However, this result does not exclude the existence of $5d_{z^2}$6s hybridization in $[Au^IX_2]^-$.

Figures 8.5 and 8.6 show the schematic energy diagrams of $[Au^IX_2]^-$ and $[Au^{III}X_4]^-$, respectively, where only the σ orbitals are picked up because the bonding of the halogeno complexes is predominantly σ in character.

In the case of the bonding in $[Au^IX_2]^-$, there are two main contributions to this. The first contribution is due to the hybridization of the p_z orbital of halogen, $5d_{z^2}$ and 6s orbitals of Au. In the absence of the Au-6s mixing, the overlap between the halogen-p_z and Au-$5d_{z^2}$ results in a bonding and anti-bonding pair of $1\sigma_g$ and $2\sigma_g$ with filled electrons, and there is no net bonding. The mixing with Au-6s, however, produces the stabilization of the $2\sigma_g$ orbital. The second contribution is due to the overlap of the p_z orbital of halogen and $6p_z$ orbital of Au, which produces the stabilization of the nonbonding $1\sigma_u$ orbital. In the case of $[Au^{III}X_4]^-$, as shown in Figure 8.5, the contributions of 6s and $6p_{x,y}$ orbitals of Au into the $1a_{1g}$ and $1e_u$ molecular orbitals stabilize the $1a_{1g}$ and $1e_u$ orbitals, respectively.

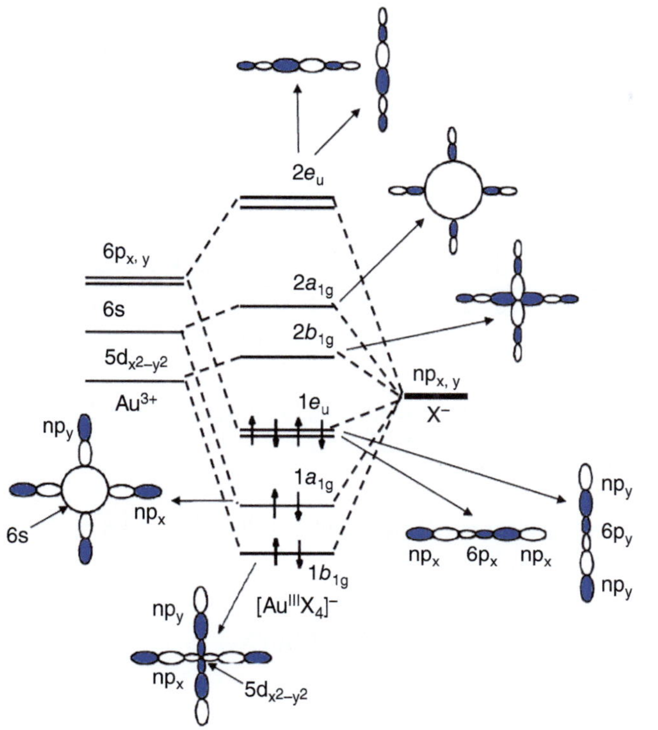

Figure 8.6 Schematic energy levels for square planar complex, $[Au^{III}X_4]^-$.

8.5 Mössbauer Parameters of ^{197}Au in $[Au^IX_2]^-$ and $[Au^{III}X_4]^-$

Figure 8.7 shows the isomer shifts and the quadrupole splittings of ^{197}Au in $(C_4H_9)_4N[Au^IX_2]$ (X = Cl, Br, I), $(C_4H_9)_4N[Au^{III}X_4]$ (X = Cl, Br, I) and $Cs_2[Au^IX_2][Au^{III}X_4]$ (X = Cl, Br, I). The isomer shift (δ) is given by the following equation:

$$\delta = C\,\Delta < R^2 > \Delta\rho(0), \tag{8.5}$$

where C is a constant containing nuclear parameters (>0), $\Delta<R^2>$ is the change in the square of the nuclear radius between the excited and ground states which has been estimated at $+8.8\times10^{-3}$ fm^2 for ^{197}Au [76], and $\Delta\rho(0)$ is the difference between Fermi contact densities of the source and the absorber. In the case of ^{197}Au, therefore, a positive shift in δ indicates an increase in the electron density at the nucleus, ρ(0). The contribution of valence electron to ρ(0) is well known to be affected in two ways: (i) the increase of 6s electron population in the valence shell directly increases ρ(0); (ii) the occupation of 6p or 5d orbitals decreases ρ(0) as a result of screening effect for 6s orbitals. As known for the highly covalent Au complexes, the Au orbitals due to $5d_{z^2}6s$ hybridization in $[Au^IX_2]^-$ having a larger s-electron character than those in $[Au^{III}X_4]^-$ yield larger δ in the term of (i).

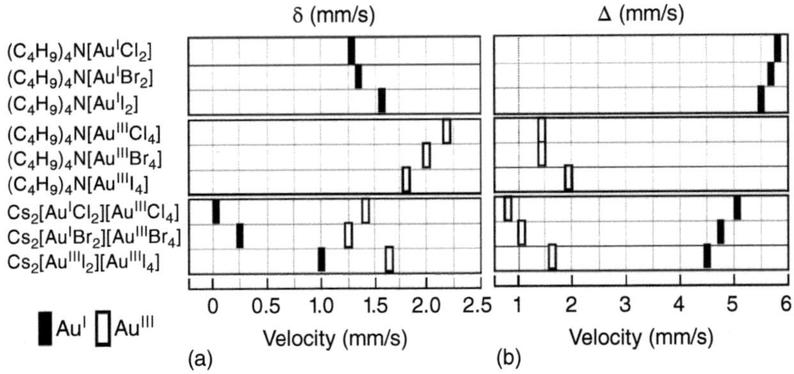

Figure 8.7 (a) Isomer shift (δ) and (b) quadrupole splitting (Δ) of $(C_4H_9)_4N[Au^IX_2]$ (X = Cl, Br, I), $(C_4H_9)_4N[Au^{III}X_4]$ (X = Cl, Br, I) and $Cs_2[Au^IX_2][Au^{III}X_4]$ (X = Cl, Br, I). Source: Ref. [55]/with permission from Springer Nature.

However, due to the screening term of (ii) ($Au^I(5d^{10})$, $Au^{III}(5d^8)$), the δ of Au^I in $[Au^IX_2]^-$ is smaller than that of Au^{III} in $[Au^{III}X_4]^-$.

8.5.1 Mössbauer Parameters of ^{197}Au in $(C_4H_9)_4N[Au^IX_2]$ and $(C_4H_9)_4N[Au^{III}X_4]$

8.5.1.1 Isomer Shift

Figure 8.7(a) shows the isomer shifts (δ) of ^{197}Au in $(C_4H_9)_4N[Au^IX_2]$ (X = Cl, Br, I), $(C_4H_9)_4N[Au^{III}X_4]$ (X = Cl, Br, I) and $Cs_2[Au^IX_2][Au^{III}X_4]$ (X = Cl, Br, I). In the case of $(C_4H_9)_4N[Au^IX_2]$ (X = Cl, Br, I), the covalent bond of $Au^I–X^-$ induces the 6s population and $5d_{z^2}$ hole in the $2\sigma_g$ orbital, and the $6p_z$ population in the $1\sigma_u$ orbital, which stabilizes the antibonding orbital of $2\sigma_g$ and the nonbonding orbital of $1\sigma_u$. The 6s population and $5d_{z^2}$ hole in the $2\sigma_g$ orbital, and the $6p_z$ population in the $1\sigma_u$ orbital increase and decrease the δ of Au^I, respectively. As shown in Figure 8.7(a), the δ of Au^I in $(C_4H_9)_4N[Au^IX_2]$ increases in the order of X = Cl < Br < I as follows, 1.27 → 1.34 → 1.57 mm/s, which is attributed to the increases of the 6s population and the $5d_{z^2}$-hole in the $2\sigma_g$ orbital in the order of X = Cl < Br < I.

In the case of $(C_4H_9)_4N[Au^{III}X_4]$ (X = Cl, Br, I), on the other hand, the covalent bond of $Au^{III}–X^-$ induces the 6s, $5d_{x^2-y^2}$, and $6p_{x,y}$ populations in the $1a_{1g}$, $1b_{1g}$, and $1e_g$ orbitals, respectively. The 6s population in the $1a_{1g}$ orbital increases δ, while the $5d_{x^2-y^2}$ and $6p_{x,y}$ populations decrease δ. As shown in Figure 8.7(a), the δ of Au^{III} in $(C_4H_9)_4N[Au^{III}X_4]$ decreases in the order of X = Cl < Br < I as follows, 2.20 → 1.99 → 1.80 mm/s, which is attributed to the increase of the $5d_{x^2-y^2}$ and $6p_{x,y}$ populations in the order of X = Cl < Br < I.

8.5.1.2 Quadrupole Splitting

Figure 8.7(b) shows the quadrupole splittings (Δ) of ^{197}Au in $(C_4H_9)_4N[Au^IX_2]$ (X = Cl, Br, I), $(C_4H_9)_4N[Au^{III}X_4]$ (X = Cl, Br, I), and $Cs_2[Au^IX_2][Au^{III}X_4]$ (X = Cl, Br, I). From the viewpoint of covalent bonding of $Au^I–X^-$ in $(C_4H_9)_4N[Au^IX_2]$, with increasing the σ-donor strength of the ligand in the order of X = Cl < Br < I, the

$6p_z$ population (negative contribution to the EFG) in Au^I increases and the quadrupole splitting is expected to increase in this order. However, the Δ of Au^I for $(C_4H_9)_4N[Au^IX_2]$ decreases in the order of X = Cl < Br < I as follows, 5.79 → 5.68 → 5.52 mm/s. This discrepancy can be explained as follows. According to the SCF-MS-Xα calculation [74], the $6p_z$ population in Au^I increases from $[Au^ICl_2]^-$ to $[Au^II_2]^-$. However, there is an expansion in the $6p_z$-orbital from $[Au^ICl_2]^-$ to $[Au^II_2]^-$, which results in a progressive decrease of Δ in this order. This phenomenon is closely related to the nephelauxetic series of ligands.

In the case of $(C_4H_9)_4N[Au^{III}X_4]$ (X = Cl, Br, I), the Δ of Au^{III} for X = Cl and Br are 1.46 and 1.47 mm/s, respectively. These almost same values are attributed to the competition between the expansion of radius and the increase of population for the $6p_{x,y}$ orbitals in the order of X = Cl < Br. The Δ increases from X = Br to I as follows, 1.47 → 1.92 mm/s, which is attributed to the overwhelming increase of $6p_{x,y}$ populations.

8.5.2 Mössbauer Parameters of ^{197}Au in $Cs_2[Au^IX_2][Au^{III}X_4]$ (X = Cl, Br, I)

8.5.2.1 Isomer Shift

As shown in Figure 8.7(a), in the case of $Cs_2[Au^IX_2][Au^{III}X_4]$ (X = Cl, Br, I), the δ of Au^I in $Cs_2[Au^IX_2][Au^{III}X_4]$ increases in the order of X = Cl < Br < I as follows, 0.05 → 0.25 → 1.01 mm/s. This tendency is quite similar to that of Au^I in $(C_4H_9)_4N[Au^IX_2]$ (X = Cl, Br, I). However, the δ values of Au^I in $Cs_2[Au^IX_2][Au^{III}X_4]$ are remarkably lower than those (1.27, 1.34, and 1.57 mm/s) of Au^I in $(C_4H_9)_4N[Au^IX_2]$ (X = Cl, Br, I). These significant decreases of the δ of Au^I in $Cs_2[Au^IX_2][Au^{III}X_4]$ compared with those in $(C_4H_9)_4N[Au^IX_2]$ are presumably attributed to the elongation of the Au^I–X distance owing to the formation of the $\cdots$X–Au^I–X$\cdots Au^{III} \cdots$X–Au^I–X$\cdots$ network lying along the *c*-axis. In fact, the Au^I–X distances in $Cs_2[Au^IX_2][Au^{III}X_4]$ are 2.281 Å [77], 2.403 Å [78], and 2.574 Å [79] for X = Cl, Br and I, respectively, while those in $(C_4H_9)_4N[Au^IX_2]$ are 2.257, 2.376, and 2.529 Å, respectively [80]. The increase of the Au^I–X distance in $Cs_2[Au^IX_2][Au^{III}X_4]$ causes the decrease of 6s population and $5d_{z^2}$-hole in the $2\sigma_g$ orbital, which is mainly responsible for the significant decrease of the δ of Au^I in $Cs_2[Au^IX_2][Au^{III}X_4]$. The δ of Au^I in $Cs_2[Au^IX_2][Au^{III}X_4]$ remarkably increases from X = Br to I, which is presumably caused by the significant increase of the charge transfer interaction between the Au^I ($5d_{x^2-y^2}$) and Au^{III} ($5d_{x^2-y^2}$) orbitals from X = Br to I.

Turning to the δ of Au^{III} in $Cs_2[Au^IX_2][Au^{III}X_4]$, it decreases from X = Cl to Br as follows, 1.40 → 1.26 mm/s in the same tendency as $(C_4H_9)_4N[Au^{III}X_4]$. However, the δ increases from X = Br to I as follows, 1.26 → 1.63 mm/s. It should be noted that the charge transfer interaction between the Au^I ($5d_{x^2-y^2}$) and Au^{III} ($5d_{x^2-y^2}$) orbitals induces the $5d_{x^2-y^2}$ population in Au^{III}, which decreases δ. The most plausible origin responsible for the significant increase of δ from X = Br to I is the effect of the increase of coordination number, 2 → 6, owing to the $\cdots$X–Au^I–X$\cdots Au^{III} \cdots$X–Au^I–X$\cdots$ network lying along the *c*-axis. The $5d_{z^2}$-6s mixing caused by the coordination of X^- along the *c*-axis induces the 6s population, which increases the δ of Au^{III} in $Cs_2[Au^IX_2][Au^{III}X_4]$. The $5d_{z^2}$-6s mixing is considered to be stronger in the order of X = Cl < Br < I.

8.5.2.2 Quadrupole Splitting

As shown in Figure 8.7(b), the Δ of Au^I in $Cs_2[Au^IX_2][Au^{III}X_4]$ decreases in the order of X = Cl < Br < I as follows, 5.06 → 4.73 → 4.48 mm/s, whose tendency is attributed to the increase of coordination number, 2 → 6, as well as the expansion of $6p_z$ orbital. The $6p_{x,y}$ populations caused by the coordination of X^- in the *ab*-plane decrease the Δ of Au^I. The Δ values of Au^I in $Cs_2[Au^IX_2][Au^{III}X_4]$ are lower than those of Au^I in $(C_4H_9)_4N[Au^IX_2]$ owing to the elongation of the Au^I–X distances due to the formation of the ···X–Au^I–X···Au^{III}···X–Au^I–X··· network lying along the *c*-axis.

On the other hand, the Δ of Au^{III} in $Cs_2[Au^IX_2][Au^{III}X_4]$ increases in the order of X = Cl < Br < I as follows, 0.76 → 1.10 → 1.64 mm/s. This tendency is attributed to the increase of $5d_{x^2-y^2}$ population in Au^{III} caused by the charge transfer interaction between the Au^I ($5d_{x^2-y^2}$) and Au^{III} ($5d_{x^2-y^2}$) orbitals, which increases Δ.

8.5.2.3 Analysis of ^{197}Au Mössbauer Parameters for $Cs_2[Au^IX_2][Au^{III}X_4]$

As discussed earlier, the Mössbauer parameters of δ and Δ for $Cs_2[Au^IX_2][Au^{III}X_4]$ (X = Cl, Br, I) depend on the following factors in addition to the factors responsible for the Mössbauer parameters of the mononuclear complexes, $(C_4H_9)_4N[Au^IX_2]$ and $(C_4H_9)_4N[Au^{III}X_4]$: (1) elongation of the Au–X distance; (2) increase of coordination number; (3) charge transfer interaction between Au^I and Au^{III}. The effects of these factors on the Mössbauer parameters of ^{197}Au in $Cs_2[Au^IX_2][Au^{III}X_4]$ are as follows:

(1) Elongation of the Au–X distance
The elongation of Au^I–X^- bond along the *c*-axis in $Cs_2[Au^IX_2][Au^{III}X_4]$ decreases the 6s and $6p_z$ populations and the $5d_{z^2}$ hole in Au^I. Among them, the decreases of the 6s and $6p_z$ populations decrease the δ and Δ of Au^I. The elongation of the Au^{III}–X^- bond in the *ab*-plane decreases the $5d_{x^2-y^2}$, 6s and $6p_{x,y}$ populations in Au^{III}. The decrease of the 6s population and the decreases of $5d_{x^2-y^2}$ and $6p_{x,y}$ populations decrease the δ and Δ of Au^{III}.

(2) Increase of coordination number
$Cs_2[Au^IX_2][Au^{III}X_4]$ consists of a three-dimensional metal-halogen framework formed by a compressed octahedron with Au^I and an elongated octahedron with Au^{III} and sharing their corners, where the Au^I site is surrounded by 2 + 4 halide ions (two shorter Au^I–X^- and four longer Au^I–X^- bonds) and the Au^{III} site is surrounded by 4 + 2 halide ions (four shorter Au^{III}–X^- and two longer Au^{III}–X^- bonds). In the case of Au^I site, the increase of coordination number, 2 → 6, induces the $6p_{x,y}$ populations through the σ-bonding in the *ab*-plane, which decreases the δ and Δ at the Au^I site. In the case of Au^{III} site, on the other hand, the increase of coordination number, 4 → 6, induces the 6s and $6p_z$ populations through the σ-bonding along the *c*-axis. The increase of 6s population and the increase of $6p_z$ one increases and decreases the δ and Δ, respectively.

(3) Increase of the charge transfer interaction between Au^I and Au^{III} in the *ab*-plane.

Table 8.2 Effects of three factors, ((1), (2), and (3)) responsible for the ^{197}Au Mössbauer parameters of $Cs_2[Au^IX_2][Au^{III}X_4]$ (X = Cl, Br, I). The up and down arrows represent the increase and decrease of the ^{197}Au Mössbauer parameters, respectively.

Au^I in $Cs_2[Au^IX_2][Au^{III}X_4]$	δ (mm/s)	Δ (mm/s)
(1) Increase of the Au^I-X^- distance	↓	↓
(2) Increase of the coordination number: 2 → 6	↓	↓
(3) Increase of the charge transfer interaction between Au^I ($5d_{x^2-y^2}$) and Au^{III} ($5d_{x^2-y^2}$) in the *ab*-plane.	↑	↑
Au^{III} in $Cs_2[Au^IX_2][Au^{III}X_4]$	**δ (mm/s)**	**Δ (mm/s)**
(1) Increase of the $Au^{III}-X^-$ distance	↓	↓
(2) Increase of the coordination number: 4 → 6	↑	↓
(3) Increase of the charge transfer interaction between Au^I ($5d_{x^2-y^2}$) and Au^{III} ($5d_{x^2-y^2}$) in the *ab*-plane.	↓	↑

The charge transfer interaction between Au^I and Au^{III} in the *ab*-plane induces the $5d_{x^2-y^2}$ hole at the Au^I site, which increases the δ and Δ at the Au^I site. On the other hand, this interaction induces the $5d_{x^2-y^2}$ population at the Au^{III} site, which decreases δ and increases Δ at the Au^{III} site.

The effects of these factors, (1), (2), and (3), on the δ and Δ at the Au^I and Au^{III} sites in $Cs_2[Au^IX_2][Au^{III}X_4]$ are summarized in Table 8.2.

8.6 Charge Transfer Interaction in $Cs_2[Au^IX_2][Au^{III}X_4]$ (X = Cl, Br, I)

In $Cs_2[Au^IX_2][Au^{III}X_4]$, there are two relative orientations between $[Au^IX_2]^-$ and $[Au^{III}X_4]^-$. In one orientation, the $\cdots$X–Au^I–X$\cdots$$Au^{III}$$\cdots$X–$Au^I$–X$\cdots$ network lies along the *c*-axis, while on the other, the $\cdots$X–Au^{III}–X$\cdots$$Au^I$$\cdots$X–$Au^{III}$–X$\cdots$ network lies in the *ab*-plane. According to the single-crystal reflectance spectra of $Cs_2[Au^IX_2][Au^{III}X_4]$, it is considered that the charge transfer interaction between Au^I and Au^{III} in $Cs_2[Au^IX_2][Au^{III}X_4]$ is two-dimensional [39, 81].

^{197}Au Mössbauer spectroscopy is one of the most powerful methods to elucidate the predominant charge transfer interaction in the ground state of $Cs_2[Au^IX_2][Au^{III}X_4]$. Figure 8.8 shows the two types of charge transfer interaction between Au^I and Au^{III} in $Cs_2[Au^IX_2][Au^{III}X_4]$. According to the molecular orbital calculation [40], the Au^I ($5d_{z^2}$) orbital of $[Au^IX_2]^-$ contains the 6s component because of $5d_{z^2}6s$ hybridization. However, the Au^I ($5d_{x^2-y^2}$) orbital of $[Au^IX_2]^-$ does not contain the 6s component because of the restriction of symmetry. Therefore, if the charge transfer interaction between Au^I ($5d_{z^2}$) and Au^{III} ($5d_{x^2-y^2}$) is predominant, the δ of Au^I becomes to be small because the s-electron population at the Au^I site decreases due to the charge transfer from Au^I to Au^{III}. On the other

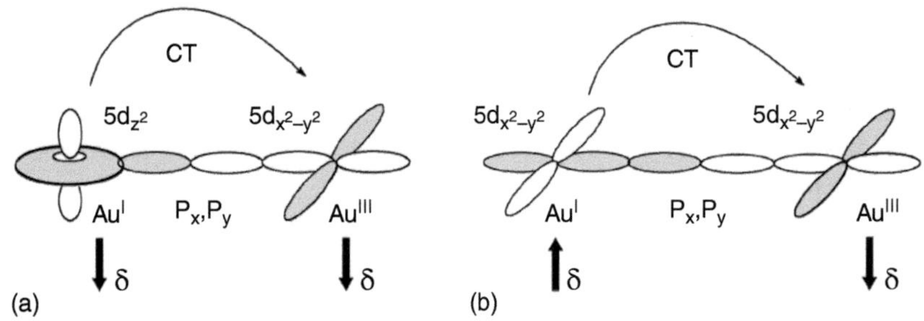

Figure 8.8 Two types of CT (charge transfer) interaction between Au^I and Au^{III} in $Cs_2[Au^IX_2][Au^{III}X_4]$. (a) CT interaction between the Au^I ($5d_{z^2}$) and Au^{III} ($5d_{x^2-y^2}$) orbitals. (b) CT interaction between the Au^I ($5d_{x^2-y^2}$) and Au^{III} ($5d_{x^2-y^2}$) orbitals. The δ denotes the isomer shift. The up and down arrows represent the increase and decrease of δ, respectively.

hand, if the charge transfer interaction between Au^I ($5d_{x^2-y^2}$) and Au^{III} ($5d_{x^2-y^2}$) is predominant, the δ of Au^I becomes to be large because of the decrease in the screening for the s-electron. Two types of charge transfer interaction between Au^I and Au^{III} in $Cs_2[Au^IX_2][Au^{III}X_4]$ are schematically shown in Figure 8.8.

8.7 ^{197}Au Mössbauer Spectra of $Cs_2[Au^IX_2][Au^{III}Y_4]$ (X, Y = Cl, Br, I)

As discussed earlier, the Mössbauer parameters, δ and Δ, of the Au^I and Au^{III} sites in $Cs_2[Au^IX_2][Au^{III}X_4]$ consist of various kinds of factors, which are quite difficult to analyze systematically. In order to elucidate the predominant factors responsible for the control of δ and Δ of the Au^I and Au^{III} sites and the charge transfer interaction for halogen-bridged gold mixed-valence complexes, we investigated the ^{197}Au Mössbauer spectra of $Cs_2[Au^IX_2][Au^{III}Y_4]$ (X, Y = Cl, Br, I). Figure 8.9 shows the ^{197}Au Mössbauer spectra of $Cs_2[Au^IX_2][Au^{III}Y_4]$ (X, Y = Cl, Br, I) at 16 K.

8.7.1 Isomer Shift of Au^I in $Cs_2[Au^IX_2][Au^{III}Y_4]$

Figure 8.10 shows the isomer shifts (δ) of Au^I and Au^{III} for $Cs_2[Au^IX_2][Au^{III}Y_4]$ at 16 K. In this section, we compare the δ of Au^I in $Cs_2[Au^IX_2][Au^{III}Y_4]$ with those in $(C_4H_9)_4N[Au^IX_2]$. As shown in Figure 8.7(a), the δ of Au^I in $(C_4H_9)_4N[Au^IX_2]$ increases in the order of X = Cl < Br < I owing to the increases of the 6s population and the $5d_{z^2}$-hole in the $2\sigma_g$ orbital in the order of X = Cl < Br < I. In the case of $Cs_2[Au^IX_2][Au^{III}Cl_4]$, the δ of Au^I decreases from X = Br to X = I, while that increases from X = Cl to Br. The decrease of δ of Au^I from X = Br to I is due to the effect of the increase of coordination number, 2 → 6, at the Au^I site, which induces the $6p_{x,y}$ populations through the σ-bonding in the *ab*-plane, resulting in the decreases of δ and Δ at the Au^I site. In the case of $Cs_2[Au^IX_2][Au^{III}Br_4]$, the δ of Au^I decreases from X = Cl to Br. However, it almost remains unchanged from X = Br to I, which is due to the competition between the effect of the increase of

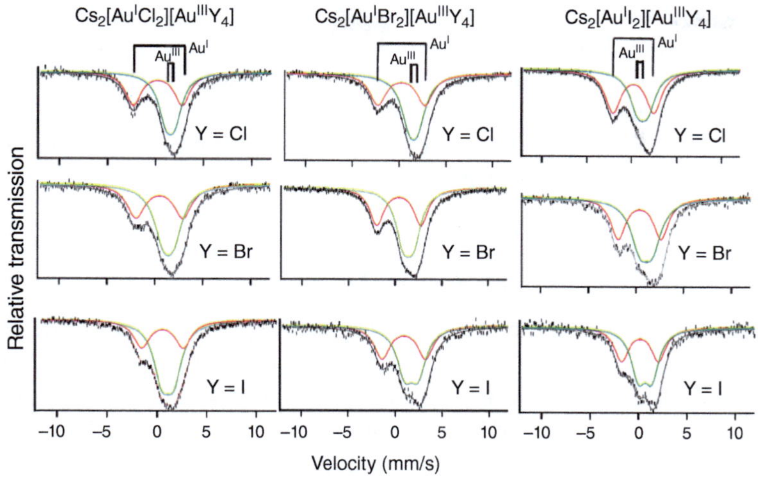

Figure 8.9 ^{197}Au Mössbauer spectra for $Cs_2[Au^IX_2][Au^{III}Y_4]$ (X, Y = Cl, Br, I) at 16 K. Source: Ref. [55]/with permission from Springer Nature.

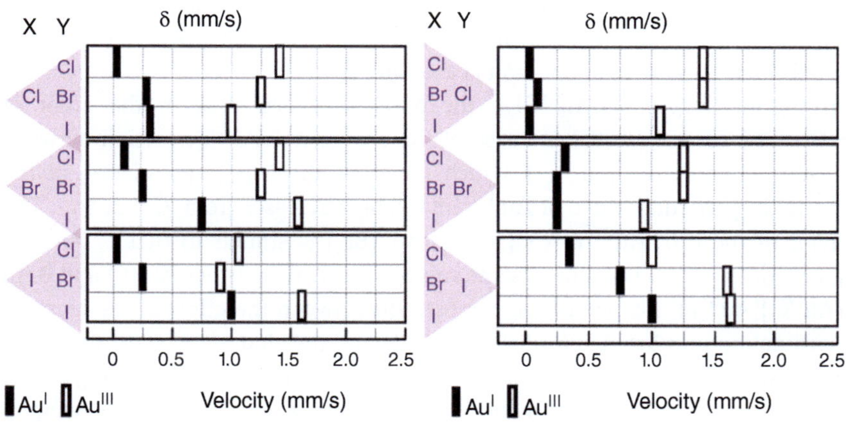

Figure 8.10 Isomer shift (δ) of $Cs_2[Au^IX_2][Au^{III}Y_4]$ (X, Y = Cl, Br, I) at 16 K. Source: Ref. [55]/with permission from Springer Nature.

coordination number from 2 to 6 at the Au^I site and the charge transfer interaction between Au^I ($5d_{x^2-y^2}$) and Au^{III} ($5d_{x^2-y^2}$). In the case of $Cs_2[Au^IX_2][Au^{III}I_4]$, the behavior of δ for Au^I is quite different from those of $Cs_2[Au^IX_2][Au^{III}Cl_4]$ and $Cs_2[Au^IX_2][Au^{III}Br_4]$. As shown in Figure 8.10, the δ of Au^I remarkably increases in the order of X = Cl < Br < I, which is due to the remarkable increase of the charge transfer interaction between Au^I ($5d_{x^2-y^2}$) and Au^{III} ($5d_{x^2-y^2}$).

Turning to $Cs_2[Au^ICl_2][Au^{III}Y_4]$, the δ of Au^I increases from Y = Cl to Br owing to the increase of the charge transfer interaction between Au^I ($5d_{x^2-y^2}$) and Au^{III} ($5d_{x^2-y^2}$). However, it almost remains unchanged from Y = Br to I, which is due to the competition between the increase of coordination number, 2 → 6, at the Au^I site and the charge transfer interaction between Au^I ($5d_{x^2-y^2}$) and Au^{III} ($5d_{x^2-y^2}$). In the case of $Cs_2[Au^IBr_2][Au^{III}Y_4]$, the δ of Au^I increases in the order of Y = Cl < Br < I owing to the increase of charge transfer interaction between Au^I ($5d_{x^2-y^2}$) and Au^{III} ($5d_{x^2-y^2}$). In particular, it is obvious that the charge transfer interaction between Au^I ($5d_{x^2-y^2}$) in $[Au^IBr_2]^-$ and Au^{III} ($5d_{x^2-y^2}$) in $[Au^{III}I_4]^-$ through the $5p_{x,y}$ orbitals of I^- is remarkably strong. In the case of $Cs_2[Au^II_2][Au^{III}Y_4]$, the behavior of δ of Au^I is quite similar to that in $Cs_2[Au^IBr_2][Au^{III}Y_4]$.

8.7.2 Isomer Shift of Au^{III} in $Cs_2[Au^IX_2][Au^{III}Y_4]$

In this section, we compare the δ of Au^{III} in $Cs_2[Au^IX_2][Au^{III}Y_4]$ with those in $(C_4H_9)_4N[Au^{III}X_4]$. As shown in Figure 8.7(a), the δ of Au^{III} in $(C_4H_9)_4N[Au^{III}X_4]$ decreases in the order of X = Cl < Br < I owing to the increase of the $5d_{x^2-y^2}$ and $6p_x$, $6p_y$ populations in the order of X = Cl < Br < I. The behavior of δ of Au^{III} in $Cs_2[Au^ICl_2][Au^{III}Y_4]$ is quite similar to that in $(C_4H_9)_4N[Au^{III}X_4]$. On the other hand, the behaviors of δ of Au^{III} in $Cs_2[Au^IBr_2][Au^{III}Y_4]$ and $Cs_2[Au^II_2][Au^{III}Y_4]$ are quite different from that of Au^{III} in $Cs_2[Au^ICl_2][Au^{III}Y_4]$. In the cases of $Cs_2[Au^IBr_2][Au^{III}Y_4]$ and $Cs_2[Au^II_2][Au^{III}Y_4]$, the δ of Au^{III} remarkably increases from Y = Br to I owing to the effect of the increase of coordination number, 4 → 6, at the Au^{III} site, where the $5d_{z^2}6s$ mixing caused by the coordination of X^- along the *c*-axis increases the 6s population at the Au^{III} site.

Turning to $Cs_2[Au^IX_2][Au^{III}Cl_4]$, the δ of Au^{III} decreases in the order of X = Cl < Br < I owing to the increase of the charge transfer interaction between Au^I ($5d_{x^2-y^2}$) and Au^{III} ($5d_{x^2-y^2}$), which increases the $5d_{x^2-y^2}$ population at the Au^{III} site. In particular, the δ of Au^{III} remarkably decreases from X = Br to I. The behavior of the δ of Au^{III} in $Cs_2[Au^IX_2][Au^{III}Br_4]$ is similar to that of Au^{III} in $Cs_2[Au^IX_2][Au^{III}Cl_4]$. The almost same value of δ of Au^{III} between X = Cl and Br in $Cs_2[Au^IX_2][Au^{III}Br_4]$ is due to the competition between the effect of the increase of coordination number, 4 → 6, at the Au^{III} site and the charge transfer interaction between Au^I ($5d_{x^2-y^2}$) and Au^{III} ($5d_{x^2-y^2}$). In the case of $Cs_2[Au^IX_2][Au^{III}I_4]$, on the other hand, the δ of Au^{III} remarkably increases from X = Cl to Br and I owing to the effect of the increase of coordination number, 4 → 6, at the Au^{III} site, where the $5d_{z^2}6s$ mixing caused by the coordination of X^- along the *c*-axis increases the 6s population at the Au^{III} site.

8.7.3 Quadrupole Splitting of Au^I in $Cs_2[Au^IX_2][Au^{III}Y_4]$

Figure 8.11 shows the Δ of Au^I and Au^{III} for $Cs_2[Au^IX_2][Au^{III}Y_4]$ at 16 K. In the cases of $Cs_2[Au^IX_2][Au^{III}Cl_4]$, $Cs_2[Au^IX_2][Au^{III}Br_4]$, and $Cs_2[Au^IX_2][Au^{III}I_4]$, the Δ of Au^I decreases in the order of Y = Cl < Br < I owing to the $6p_{x,y}$ population induced from the increase of coordination number, 2 → 6, as well as the expansion of $6p_z$ orbital. The almost same value of ΔE_Q of Au^I between X = Cl and Br for $Cs_2[Au^IX_2][Au^{III}Cl_4]$ is due to the competition between the effect of the increase of coordination number, 2 → 6, and the charge transfer interaction between Au^I ($5d_{x^2-y^2}$) and Au^{III} ($5d_{x^2-y^2}$).

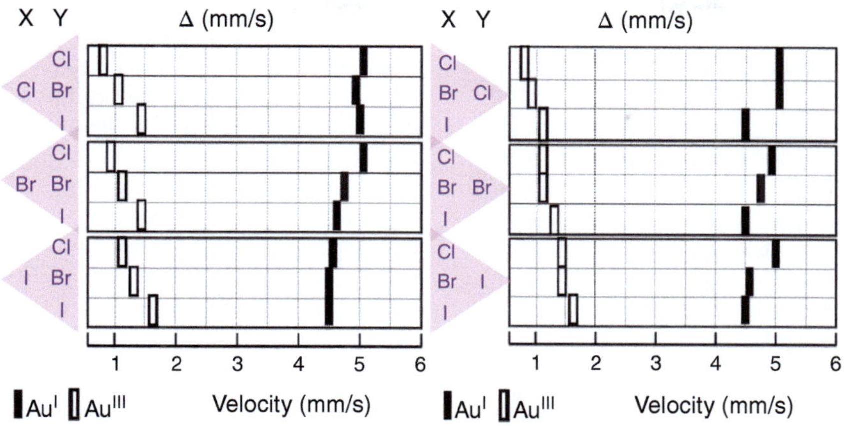

Figure 8.11 Quadrupole splitting (Δ) of $Cs_2[Au^{I}X_2][Au^{III}Y_4]$ (X, Y = Cl, Br, I) at 16 K. Source: Ref. [55]/with permission from Springer Nature.

In the cases of $Cs_2[Au^{I}Cl_2][Au^{III}Y_4]$, $Cs_2[Au^{I}Br_2][Au^{III}Y_4]$, and $Cs_2[Au^{I}I_2][Au^{III}Y_4]$, on the other hand, the Δ of Au^{I} is controlled by the competition between the increase of coordination number, $2 \rightarrow 6$, and the charge transfer interaction between Au^{I} ($5d_{x^2-y^2}$) and Au^{III} ($5d_{x^2-y^2}$).

8.7.4 Quadrupole Splitting of Au^{III} in $Cs_2[Au^{I}X_2][Au^{III}Y_4]$

In the cases of $Cs_2[Au^{I}Cl_2][Au^{III}Y_4]$, $Cs_2[Au^{I}Br_2][Au^{III}Y_4]$, and $Cs_2[Au^{I}I_2][Au^{III}Y_4]$, the Δ of Au^{III} increases in the order of X = Cl < Br < I owing to the increase of the charge transfer interaction between Au^{I} ($5d_{x^2-y^2}$) and Au^{III} ($5d_{x^2-y^2}$), which increases the $5d_{x^2-y^2}$ population at the Au^{III} site. This tendency is similar to that of Au^{III} in $Cs_2[Au^{I}X_2][Au^{III}Cl_4]$, $Cs_2[Au^{I}X_2][Au^{III}Br_4]$, and $Cs_2[Au^{I}X_2][Au^{III}I_4]$. The almost same values of Δ of Au^{I} between X = Cl and Br for $Cs_2[Au^{I}X_2][Au^{III}Br_4]$ and $Cs_2[Au^{I}X_2][Au^{III}I_4]$ are due to the competition between the $6p_z$ population caused by the coordination of X^- along the *c*-axis and the increase of the $5d_{x^2-y^2}$ population caused by the charge transfer interaction between Au^{I} ($5d_{x^2-y^2}$) and Au^{III} ($5d_{x^2-y^2}$).

8.8 Single Crystal ^{197}Au Mössbauer Spectra of $Cs_2[Au^{I}I_2][Au^{III}I_4]$

8.8.1 Comparison of ^{197}Au Mössbauer Spectra Between Single Crystal and Powder Crystal

Figure 8.12(a) and (b) show the ^{197}Au Mössbauer spectra of the powdered crystal and the single crystal of $Cs_2[Au^{I}I_2][Au^{III}I_4]$, respectively. As shown in Figure 8.12(a), the best fit is obtained with two doublets, the outer with lower intensity being assigned to the Au^{I} site and the inner to the Au^{III} site. In the case of powder crystals, the intensity ratio, $I\,(Au^{I})/I\,(Au^{III})$, is 0.72, which is due to the difference between the recoil-free

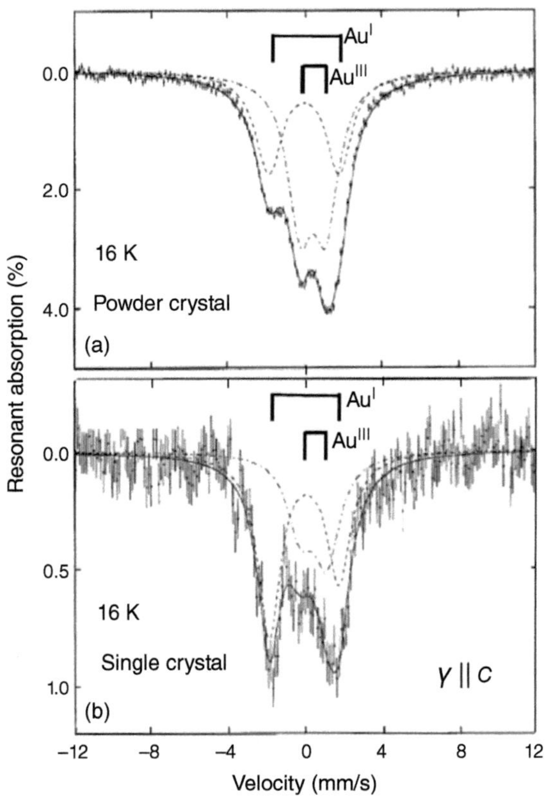

Figure 8.12 ^{197}Au Mössbauer spectra of $Cs_2[Au^{I}I_2][Au^{III}I_4]$ at 16 K [40]. (a) powder crystal, (b) single crystal. The direction of γ-ray is parallel to the *c*-axis.

fractions of Au^{I} and Au^{III}. Since the Au^{I} site is in twofold (linear) ligand coordination and the Au^{III} site is in fourfold (square planar) coordination, the environment of the Au^{III} site in $Cs_2[Au^{I}I_2][Au^{III}I_4]$ is considered to be more rigid than that of the Au^{I} site. However, as shown in Figure 8.12(b), $I\,(Au^{I})/I\,(Au^{III})$ of the single crystal is 1.56, which implies that the recoil-free fraction along the molecular axis of $[Au^{I}I_2]^-$ is larger than that perpendicular to the plane of $[Au^{III}I_4]^-$.

Now, we discuss the relationship between the anisotropic Debye temperature (Θ_D) and the anisotropy of recoil-free fraction for ^{197}Au. In the case of $T \ll \Theta_D$, the temperature dependence of the recoil-free fraction is expressed as follows [82]:

$$f = \exp\left\{\frac{-E_R}{k\Theta_D}\left(\frac{3}{2}+\frac{\pi^2T^2}{\Theta_D^2}\right)\right\}, \tag{8.6}$$

where E_R denotes the recoil energy and the value of ^{197}Au is 16.3×10^{-3} eV [83], and k is the Boltzmann constant. The anisotropic Debye temperatures, $\Theta_D{}^{\parallel}$ and $\Theta_D{}^{\perp}$, are derived from the following equation [82]:

$$\Theta_D^2 = \frac{3h^2T}{4\pi^2Mk < u^2 >}, \tag{8.7}$$

where h is Planck constant, M is the atomic weight of ^{197}Au, and $<u^2>$ is the mean-squared vibrational displacement. We have obtained the $<u^2>$ values of the Au^{I} and Au^{III} sites in $Cs_2[Au^{I}I_2][Au^{III}I_4]$ from the single-crystal X-ray diffraction

Table 8.3 Anisotropic mean-squared vibrational displacement at 293 K, and the estimated anisotropic Debye temperature at the Au^{I} and Au^{III} sites in $Cs_2[Au^{I}I_2][Au^{III}I_4]$. $\Theta_D^{\parallel}$ and $\Theta_D^{\perp}$ are the local Debye temperatures at the Au site whose directions are parallel and perpendicular to the *c*-axis, respectively.

	$<u_{\parallel}^2>$ (Å^2)	$<u_{\perp}^2>$ (Å^2)	$<u_{av.}^2>$ (Å^2)	$\Theta_D^{\parallel}$ (K)	$\Theta_D^{\perp}$ (K)	$\Theta_D^{av.}$ (K)
Au^{I}	0.0265	0.0350	0.0322	90.4	78.6	82.1
Au^{III}	0.0299	0.0269	0.0279	85.1	89.7	88.0

analysis at 293 K [79]. The anisotropic mean-squared vibrational displacement at 293 K and the estimated anisotropic Debye temperatures at the Au^{I} and Au^{III} sites in $Cs_2[Au^{I}I_2][Au^{III}I_4]$ are listed in Table 8.3. The anisotropic recoil-free fraction (f) and its relative ratio ($f(Au^{I})/f(Au^{III})$), and the relative intensity ratio ($I(Au^{I})/I(Au^{III})$) of ^{197}Au Mössbauer spectra for $Cs_2[Au^{I}I_2][Au^{III}I_4]$ at 16 K are listed in Table 8.4.

As shown in Tables 8.3 and 8.4, the average Debye temperature ($\Theta_D^{av.}$) at the Au^{I} site is lower than that at the Au^{III} site, which is responsible for the relative intensity ratio of recoil-free fractions, $f_{av.}(Au^{I})/f_{av.}(Au^{III}) < 1.0$, and that in the ^{197}Au Mössbauer spectra, $I(Au^{I})/I(Au^{III}) < 1.0$ for $Cs_2[Au^{I}I_2][Au^{III}I_4]$ at 16 K. However, when the incident γ-ray is parallel to the *c*-axis for the single crystal of $Cs_2[Au^{I}I_2][Au^{III}I_4]$, owing to the calculated result that the anisotropic Debye temperature ($\Theta_D^{\parallel}$) and anisotropic recoil-free fraction ($f_{\parallel}$) at the Au^{I} site is higher than that at the Au^{III} site, the intensity of the ^{197}Au Mössbauer spectrum at the Au^{I} site is stronger than that at the Au^{III} site, which is schematically shown in Figure 8.13.

Table 8.4 Anisotropic recoil-free fraction (f) and its relative ratio ($f(Au^{I})/f(Au^{III})$), and the relative intensity ratio ($I(Au^{I})/I(Au^{III})$) of ^{197}Au Mössbauer spectra for $Cs_2[Au^{I}I_2][Au^{III}I_4]$ at 16 K.

	$f_{\parallel}$	$f_{\perp}$	$f_{av.} = (f_{\parallel}+2f_{\perp})/2$	$\frac{f_{av.}(Au^{I})}{f_{av.}(Au^{III})}$	$\frac{I_{av.}(Au^{I})}{I_{av.}(Au^{III})}$	$\frac{f_{\parallel}(Au^{I})}{f_{\parallel}(Au^{III})}$	$\frac{I_{\parallel}(Au^{I})}{I_{\parallel}(Au^{III})}$
Au^{I}	0.0227	0.0101	0.0133	0.675	0.72	1.38	1.56
Au^{III}	0.0164	0.0218	0.0197				

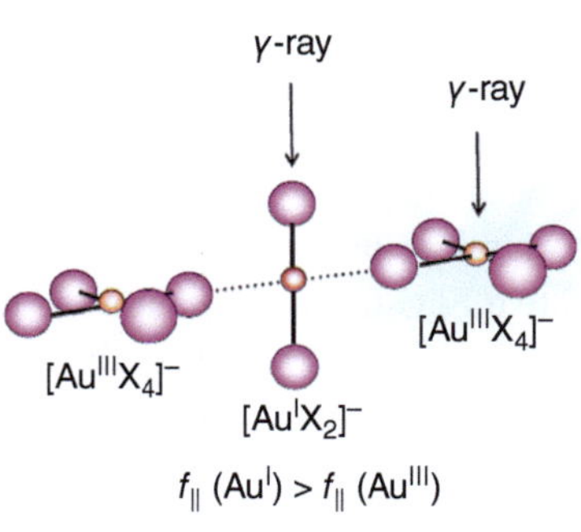

Figure 8.13 Schematic representation of the anisotropic recoil-free fraction ($f_{\parallel}$) at the $[Au^{I}X_2]^-$ and $[Au^{III}X_4]^-$ sites in $Cs_2[Au^{I}X_2][Au^{III}X_4]$ for the incident γ-ray parallel to the *c*-axis.

8.8.2 Sign of EFG for Au^I in $[Au^I I_2]^-$ and Au^{III} in $[Au^{III}X_4]^-$

Single-crystal ^{197}Au Mössbauer spectroscopy has been recognized as the most powerful method to determine the sign of EFG for the Au site and the nature of Au–X (X = ligand) chemical bond since the determination of negative EFG sign for Au^I in $KAu(CN)_2$ by means of single-crystal ^{197}Au Mössbauer spectroscopy [70]. In this section, we discuss the EFG for the Au^I and Au^{III} sites in $Cs_2[Au^I I_2][Au^{III}I_4]$ by means of single-crystal ^{197}Au Mössbauer spectroscopy.

In the case of $Cs_2[Au^I I_2][Au^{III}I_4]$, the quadrupole splitting for Au^1 and Au^{III} is given by $eqV_{zz}/2$, in which q is the nuclear quadrupole moment ($+0.59 \times 10^{-28}$ m^2 for the ground state of ^{197}Au) and V_{zz} is the principal component of EFG. Therefore, the concentration of negative charge in the xy plane makes EFG positive and the ground doublet becomes $I_z = \pm 1/2$. On the other hand, the concentration of negative charge along the z-axis makes EFG negative and the ground doublet becomes $I_z = \pm 3/2$. In the case of ^{197}Au, the 77.3 keV transition is the mixture of magnetic dipole transition (M1) and electric quadrupole transition (E2) with the mixing ratio E2/M1 (ξ) of -0.352 ± 0.005 [70]. Taking account of the M1–E2 mixing, the intensity ratio of $I(Au^I)/I(Au^{III})$ is expressed as follows [70]:

$$I(\pm 3/2 \rightarrow \pm 1/2)/I(\pm 1/2 \rightarrow \pm 1/2) = \{2(\sqrt{3} + \xi)^2 - 3(1 + 2\sqrt{3}\xi - \xi^2)\sin^2\theta\}/\{2(1 - \sqrt{3}\xi)^2 + 3(1 + 2\sqrt{3}\xi - \xi^2)\sin^2\theta\}. \tag{8.8}$$

In our experiment, the direction of the incident γ-ray is parallel to the c-axis ($\theta = 0$) of $Cs_2[Au^I I_2][Au^{III}I_4]$. Therefore, taking account of the M1–E2 mixing, the intensity ratio of $I(\pm 3/2 \rightarrow \pm 1/2)/I(\pm 1/2 \rightarrow \pm 1/2)$ for $\theta = 0$ should be 0.735. In the case of $Cs_2[Au^I I_2][Au^{III}I_4]$, the intensity ratio between the weaker and stronger branches is 0.69 and 0.71 for the Au^I and Au^{III} sites, respectively. These values are consistent with the theoretically predicted value of 0.735. Comparing the obtained spectral profile of Figure 8.12(b) with the theoretical transition probability ratio including the Ml–E2 mixing, we can determine the signs of EFG for Au^I and Au^{III} in $Cs_2[Au^I I_2][Au^{III}I_4]$ to be negative and positive, respectively. The negative sign of EFG for Au^I in $Cs_2[Au^I I_2][Au^{III}I_4]$ is consistent with that for Au^I in $KAu(CN)_2$ [70].

In general, in the case of linear coordinated $[Au^1X_2]^-$, the $5d_{z^2}$ hole due to 5d6s hybridization produces a positive contribution to EFG, while the $6p_z$ electron due to 6s6p hybridization produces a negative one. On the other hand, in the case of square-planar coordinated $[Au^{III}X_4]^-$, the electron hole in the $5d_{x^2-y^2}$ orbital produces a negative contribution to EFG, while the partly filled $5d_{x^2-y^2}6s6p^2$ hybridized orbital yields a positive one. Therefore, in the case of $Cs_2[Au^I I_2][Au^{III}I_4]$, the negative EFG of the Au^1 site in $[Au^I I_2]^-$ is attributed to the $6p_z$ electron population, while the positive EFG of the Au^{III} site in $[Au^{III}I_4]^-$ is attributed to the partly filled $5d_{x^2-y^2}6s6p^2$ hybridized orbital.

8.9 ^{197}Au Mössbauer Spectra of $Cs_2[Au^IX_2][Au^{III}X_4]$ (X = Cl, I) Under High Pressures

8.9.1 Phase Diagram of $Cs_2[Au^IX_2][Au^{III}X_4]$ (X = Cl, Br, I)

Investigations of the structural and electronic properties of $Cs_2[Au^ICl_2][Au^{III}Cl_4]$ under high pressures were started from 1974. In 1974, R. Keller, W. B. Holzapfel, et al. reported a strong continual decrease in the electrical resistivity by nearly nine orders of magnitude from ambient pressure to 12 GPa [84]. A gradual semiconductor-to-metal transition at room temperature and 6.5 GPa has been interpreted as evidence for the overlap of the filled $5d_{z^2}$ band of Au^I and the empty $5d_{x^2-y^2}$ one of Au^{III}. Being stimulated by this phenomenon, W. Denner, H. Schulz, et al. carried out the single-crystal X-ray diffraction measurement for $Cs_2[Au^ICl_2][Au^{III}Cl_4]$ under high pressures [85]. According to them, with increasing pressure, the Cl atom shifts gradually toward the midpoint of the Au atoms that is attained at 5.2 GPa. Therefore, they have proposed that the Au sites become indistinguishable and the valence state of Au is Au^{II} above 5.2 GPa. However, ^{197}Au Mössbauer spectra [86, 87] and Raman spectra [88] of $Cs_2[Au^ICl_2][Au^{III}Cl_4]$ under high pressures have shown that the Au^I and Au^{III} sites are clearly distinguishable even at 6.8 and 8.0 GPa, respectively.

In order to elucidate the Au valence states of $A_2[Au^IX_2][Au^{III}X_4]$ (A = Rb, Cs; X = Cl, Br, I) under high pressures, we carried out the powder X-ray diffraction measurement under high pressures and obtained the structural *P–T* phase diagram for $M_2[Au^IX_2][Au^{III}X_4]$ (M = Rb, Cs; X = Cl, Br, I) [36, 40, 41]. Figure 8.14 shows the phase diagram of $Cs_2[Au^IX_2][Au^{III}X_4]$ (X = Cl, Br, I) and $Rb_2[Au^II_2][Au^{III}I_4]$ under high pressures at room temperature, and Figure 8.15 shows the *P–T* phase diagram of $Cs_2[Au^IX_2][Au^{III}X_4]$ (X = Cl and I). In Figure 8.14, M, T, and C denote the monoclinic, tetragonal, and cubic phases, respectively. The phase marked by oblique lines shows the Au^{II} single valence state. The pressure-induced Au valence

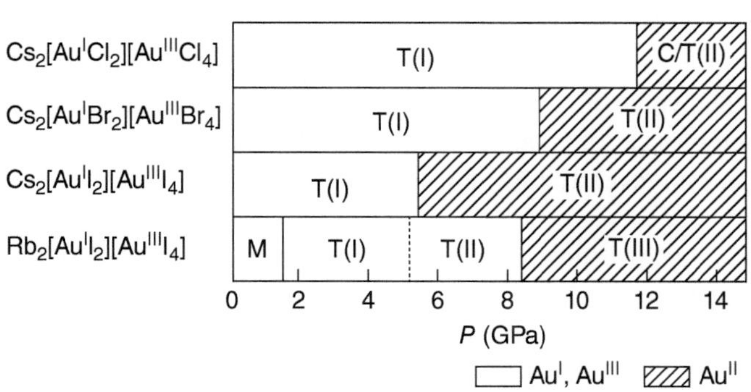

Figure 8.14 Phase diagram of $Cs_2[Au^IX_2][Au^{III}X_4]$ (X = Cl, Br, I) and $Rb_2[Au^II_2][Au^{III}I_4]$ under high pressures at room temperature [40]. M, T, and C represent the monoclinic, tetragonal, and cubic phases, respectively. In the second phase of $Cs_2[Au^ICl_2][Au^{III}Cl_4]$, the cubic phase appears by using He gas as a pressure medium, while the tetragonal (II) phase appears by using kerosene as a pressure medium.

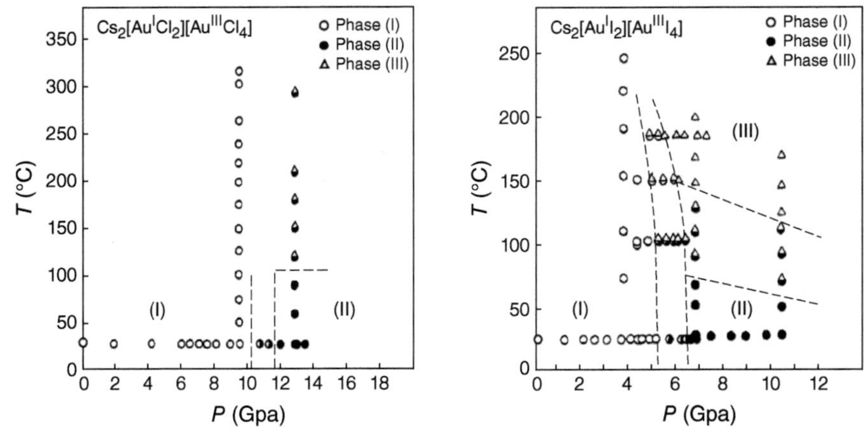

Figure 8.15 *P–T* phase diagram of $Cs_2[Au^IX_2][Au^{III}X_4]$ (X = Cl and I) [36]. Phase (I): tetragonal, phase (II): tetragonal, phase (III): cubic.

transition from the $Au^{I,\ III}$ mixed-valence state to the Au^{II} single valence state for $Cs_2[Au^IX_2][Au^{III}X_4]$ (X = Cl, Br, I) strongly depends on the kind of bridging halogen, and its critical pressure decreases 12.0 → 9.0 → 5.5 GPa in the order of Cl → Br → I.

8.9.2 Origin of Metallic Mixed-Valence State in $Cs_2[Au^ICl_2][Au^{III}Cl_4]$

For the case of $Cs_2[Au^ICl_2][Au^{III}Cl_4]$, we carried out the single-crystal X-ray diffraction analysis under high pressures up to 18.0 GPa by using He gas as a pressure medium and detected the superstructure reflections (*h l l*; *l* = odd) due to the shift of the Cl ions from the midpoint of the Au ions even just below 12.0 GPa. These superstructure reflections disappeared at the structural phase transition (*P* = 12.0 GPa). It is therefore concluded that the crystal structure changes from the tetragonal phase (*I4/mmm*) to the cubic phase (*Pm3m*) at 12.0 GPa [44]. As for the ^{197}Au Mössbauer spectroscopy, J. Stanek et al. observed the ^{197}Au Mössbauer spectra of $Cs_2[Au^ICl_2][Au^{III}Cl_4]$ under high pressures up to 10.2 GPa at 4.2 K [86, 87].

Figure 8.16 shows the ^{197}Au Mössbauer spectra of $Cs_2[Au^ICl_2][Au^{III}Cl_4]$ at 0.0 and 6.8 GPa [86]. As shown in Figure 8.16, the line profile of ^{197}Au Mössbauer spectra at 6.8 GPa resembles closely that at 0.0 GPa, and the Au^I and Au^{III} sites are clearly distinguishable. Moreover, the Au^I and Au^{III} site are distinguished even at 10.2 GPa [87].

On the other hand, we investigated the electrical resistivity of $Cs_2[Au^ICl_2]$ $[Au^{III}Cl_4]$ under high pressures and low temperatures cooled down to 5 K by using a standard four-probe method [40]. Figure 8.17 shows the temperature dependence of the electrical resistivity along the *c*-axis at *P* = 5.0–8.0 GPa [40]. At 5.0 GPa, a gradual semiconductor-to-metal transition occurs at around 250 K. With decreasing temperature, the resistivity decreases slowly from 250 to 100 K and then it decreases below 100 K. At 6.0 GPa, the metallic region becomes to be spread between 250 and 5 K, while the gradual semiconductor-to-metal transition at around 250 K still remains. Under the pressure region above 7.0 GPa, the broad peak due to the

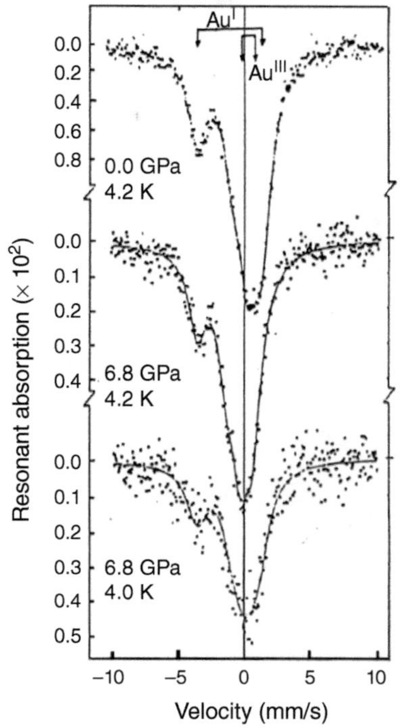

Figure 8.16 ^{197}Au Mössbauer spectra of $Cs_2[Au^ICl_2][Au^{III}Cl_4]$ at 0.0 GPa and 6.8 GPa. Source: Ref. [86]/with permission of Elsevier.

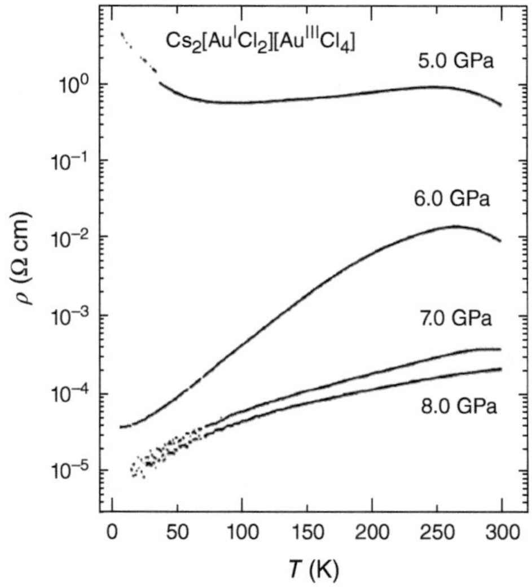

Figure 8.17 Temperature dependence of the electrical resistivity perpendicular to the *c*-axis of $Cs_2[Au^ICl_2][Au^{III}Cl_4]$ at several pressures. Source: Ref. [40]/with permission of Elsevier.

semiconductor-to-metal transition disappears and the resistivity shows the metallic behavior in the whole measuring temperature between 300 and 5 K.

In order to investigate the Au^I–Au^{III} mixed-valence state in the metallic region in $Cs_2[Au^ICl_2][Au^{III}Cl_4]$, we measured the pressure dependence of the Au–Cl atomic distances under high pressures up to 18.0 GPa by means of single-crystal X-ray diffraction analysis. Figure 8.18 shows the pressure dependence of the Au–Cl atomic distances in $Cs_2[Au^ICl_2][Au^{III}Cl_4]$ at 300 K [44]. Solid triangles, open squares, open

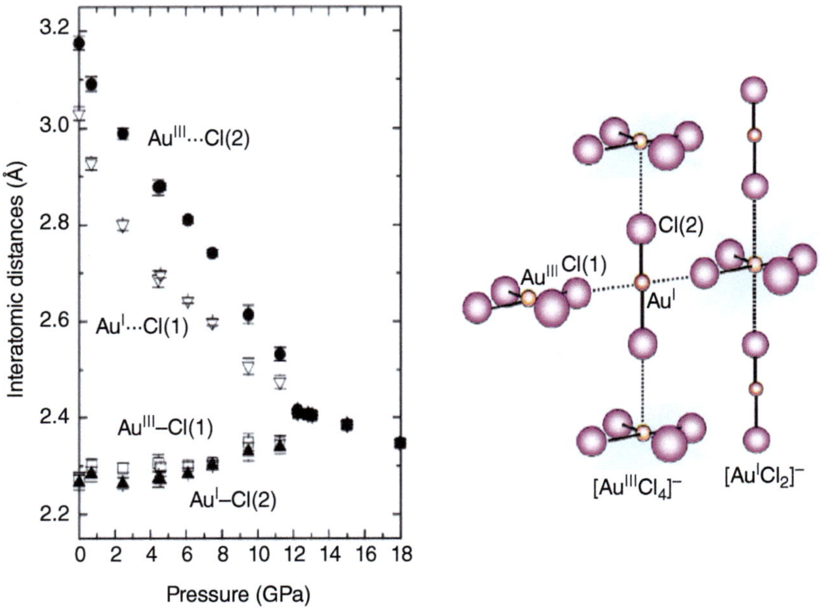

Figure 8.18 Pressure dependence of the Au–Cl atomic distances in $Cs_2[Au^{I}Cl_2][Au^{III}Cl_4]$ at 300 K. Source: Ref. [44]/with permission of Elsevier.

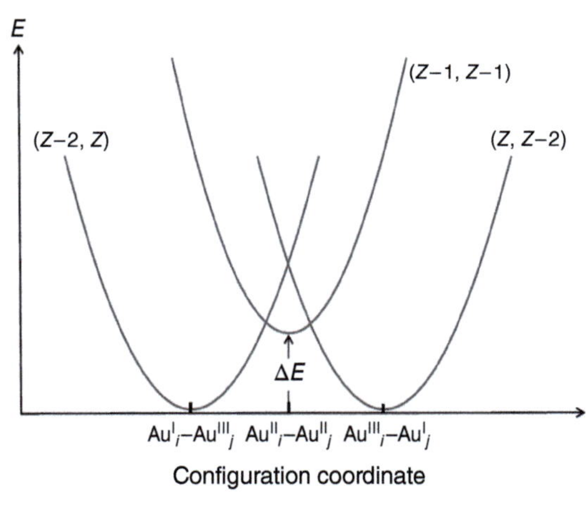

Figure 8.19 Potential energy surfaces of $(Z-2, Z)$, $(Z-1, Z-1)$, $(Z, Z-2)$ systems. Z, $Z-1$ and $Z-2$ denote the oxidation number of Au^{I}, Au^{II} and Au^{III}, respectively.

anti-triangles, and solid circles indicate the shorter apical distances, Au^{I}–Cl(2), the shorter planer distances, Au^{III}–Cl(1), the longer planer distances, $Au^{I}\cdots$Cl(1), and the longer apical distances, $Au^{III}\cdots$Cl(2), respectively. As shown in Figure 8.19, in the mixed-valence state (Phase (I)), the inter-molecular planer distance of $Au^{I}\cdots$Cl(1) and the inter-molecular apical distance of $Au^{III}\cdots$Cl(2) remarkably decrease with increasing pressure. On the other hand, the intra-molecular planer distance of Au^{III}–Cl(1) and the intra-molecular apical distance of Au^{I}–Cl(2) remain unchanged under the pressures between 0.0 GPa and ~5 GPa, then they expand by 0.06 Å with increasing pressure from 6.0 to 11.0 GPa.

In connection with the pressure dependence of intra-molecular vibration, X. J. Liu, Y. Moritomo, N. Kojima, et al. investigated the pressure dependence of Raman spectra for $Cs_2[Au^ICl_2][Au^{III}Cl_4]$ at 300 K [38, 39]. According to the Raman spectroscopy, the energy of B_{1g} mode attributed to the symmetrical stretching mode for $[Au^{III}Cl_4]^-$ decreases from 297 to 255 cm^{-1} with increasing pressure from 0.0 to 10.0 GPa. The pressure-induced softening of the stretching mode for $[Au^{III}Cl_4]^-$ is consistent with the expansion of the intra-molecular planer distance of Au^{III}—Cl(1) with increasing pressure. In contrast with the behavior of the B_{1g} mode for $[Au^{III}Cl_4]^-$, the A_{1g} mode attributed to the symmetrical stretching mode for $[Au^ICl_2]^-$ increases from 324 to 360 cm^{-1} with increasing pressure from 0.0 to 10.0 GPa.

From the analysis of the pressure dependence of inter-molecular distances between $[Au^ICl_2]^-$ and $[Au^{III}Cl_4]^-$ in $Cs_2[Au^ICl_2][Au^{III}Cl_4]$, it is considered that the isolated stretching modes in $[Au^ICl_2]^-$ and $[Au^{III}Cl_4]^-$ transform themselves into the cooperative elongated-compressed breathing mode, which induces the dynamical two-electron exchange (valence skipping) process and metallic behavior. The competition between one-electron exchange process and two-electron exchange process is schematically shown in Figure 8.19.

In Figure 8.19, it is considered that the one-electron exchange between ($Z-2$: $Au^I{}_i$, Z: $Au^{III}{}_j$) and ($Z-1$: $Au^{II}{}_i$, $Z-1$: $Au^{II}{}_j$) arises from a thermally activated electron hopping among Au sites, while the two-electron exchange (valence skipping) between ($Z-2$: $Au^I{}_i$, Z: $Au^{III}{}_j$) and (Z: $Au^{III}{}_i$, $Z-2$: $Au^I{}_j$) is induced by the tunneling process through the medium of the elongated-compressed breathing vibration mode of $AuCl_6$ octahedra. In the case of $Cs_2[Au^ICl_2][Au^{III}Cl_4]$ under high pressures (5–7 GPa), it is considered that the one-electron exchange between ($Z-2$: $Au^I{}_i$, Z: $Au^{III}{}_j$) and ($Z-1$: $Au^{II}{}_i$, $Z-1$: $Au^{II}{}_j$) is predominant above 250 K, while the two-electron exchange (valence skipping) between ($Z-2$: $Au^I{}_i$, Z: $Au^{III}{}_j$) and (Z: $Au^{III}{}_i$, $Z-2$: $Au^I{}_j$) is predominant below 250 K. Under the pressures above 7.0 GPa, the two-electron exchange (valence skipping) between ($Z-2$: $Au^I{}_i$, Z: $Au^{III}{}_j$) and (Z: $Au^{III}{}_i$, $Z-2$: $Au^I{}_j$) is predominant between 300 and 5 K. In order to support this model, the experimental proof of the existence of Au^{II} state under high pressures (5–7 GPa) and high temperatures by means of ^{197}Au Mössbauer spectroscopy is effective. However, unfortunately, this region of temperature is not available for the ^{197}Au Mössbauer spectroscopy because of high recoil energy (77.3 keV).

8.9.3 Au Valence Transition in $Cs_2[Au^II_2][Au^{III}I_4]$

As mentioned in Section 8.9.1, $Cs_2[Au^II_2][Au^{III}I_4]$ undergoes the structural first-order phase transition accompanied with the valence transition from the mixed-valence state of Au^I–Au^{III} to the single valence state of Au^{II} at about 5.5 GPa and 300 K [34–36]. As the pressure is increased from 0.0 to 5.0 GPa, the resistivity decreases by 10 orders of magnitude, while just above 5.0 GPa, both of the resistivity parallel and perpendicular to the c-axis increase rapidly by four orders of magnitude, then they remain unchanged between 6 and 8 GPa. The drastic resistivity jump around 5.0 GPa is attributed to the Mott insulator with Au^{II} ($5d^9$). Figure 8.20 shows

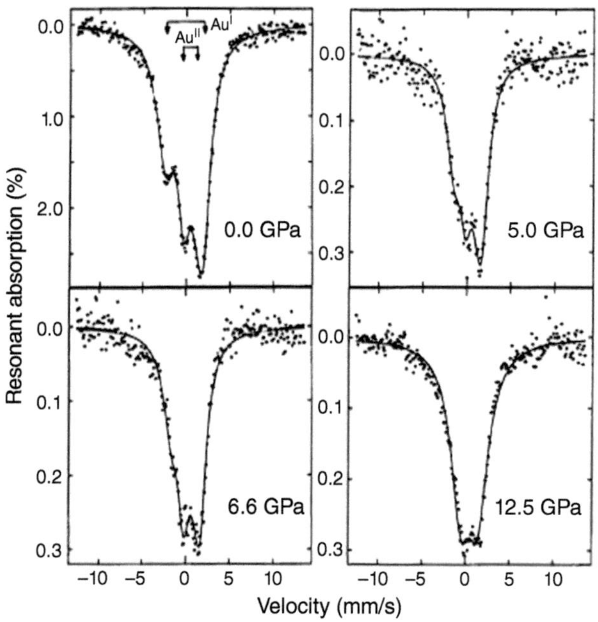

Figure 8.20 Pressure dependence of ^{197}Au Mössbauer spectra for $Cs_2[Au^I I_2][Au^{III} I_4]$ at 4.2 K. Source: Ref. [37]/with permission of Elsevier.

the ^{197}Au Mössbauer spectra of $Cs_2[Au^I I_2][Au^{III} I_4]$ under high pressures up to 12.5 GPa [37]. The valence states of Au^I and Au^{III} at ambient pressure are clearly distinguishable. As shown in Figure 8.20, with increasing pressure, a gradual increase in the overlap of the doublets is observed due to a substantial increase of the isomer shift of Au^I and a less pronounced decrease of that of Au^{III} with increasing pressure. Eventually, the ^{197}Au Mössbauer spectrum at 12.5 GPa shows only one electronic state for Au. Therefore, the Au valence state of phase (II) in $Cs_2[Au^I I_2][Au^{III} I_4]$ is concluded to be Au^{II}, which is the first direct observation of the Au^{II} state in relatively simple and inorganic compounds.

Figure 8.21 shows the pressure dependence of the correlation between the isomer shift and the quadrupole splitting for the ^{197}Au Mössbauer spectra of $Cs_2[Au^I I_2][Au^{III} I_4]$ at 4.2 K. The positions of the Au^I and Au^{III} sites of $Cs_2[Au^I I_2][Au^{III} I_4]$ in the δ–Δ correlation map are located in the grey-colored belts for linear-coordinated $[Au^I X_2]^-$ and square-planer-coordinated $[Au^{III} X_4]^-$ compounds, respectively. As shown in Figure 8.21, the Au^I position of $Cs_2[Au^I I_2][Au^{III} I_4]$ in the δ–Δ correlation map moves toward the Au^{III} area with increasing pressure, and eventually, the Au^I position edges into the Au^{III} area at 12.5 GPa. In contrast with the behavior of the Au^I position, the Au^{III} position remains almost unchanged with increasing pressure up to 6.5 GPa. It should be noted that the typical Au^{III} complexes are planarly coordinated by four ligands, while the single valence Au^{II} sites at 12.5 GPa for $Cs_2[Au^I I_2][Au^{III} I_4]$ are octahedrally coordinated by six ligands.

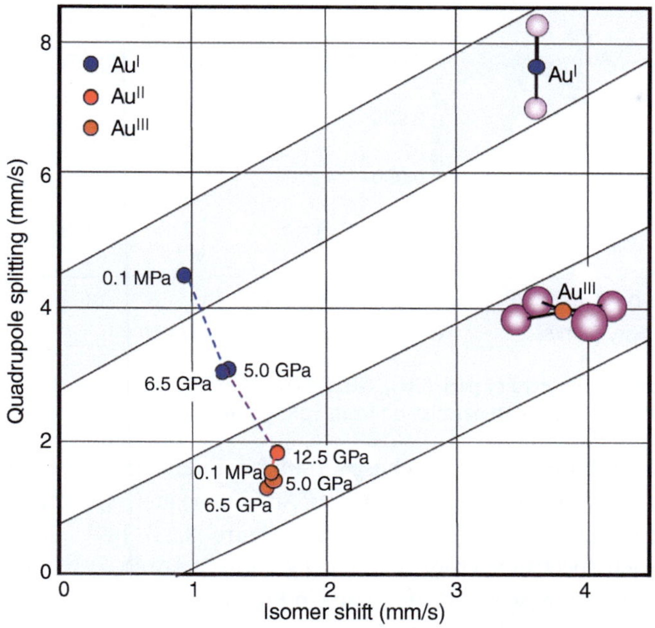

Figure 8.21 Pressure dependence of the correlation between the isomer shift and the quadrupole splitting for the ^{197}Au Mössbauer spectra of $Cs_2[Au^{I}I_2][Au^{III}I_4]$ at 4.2 K.

8.10 ^{197}Au Mössbauer Spectra of $[NH_3(CH_2)_nNH_3]_2[(Au^{I}I_2)(Au^{III}I_4)(I_3)_2]$ (n = 7, 8)

In 2003, M. Castro and A. M. Guloy synthesized layered perovskite-type gold mixed-valence complexes, $[NH_3(CH_2)_nNH_3]_2[(Au^{I}I_2)(Au^{III}I_4)(I_3)_2]$ (n = 7, 8), and determined the crystal structure [53]. Figure 8.22 shows the schematic structures of $Cs_2[Au^{I}I_2][Au^{III}I_4]$ and $[NH_3(CH_2)_nNH_3]_2[(Au^{I}I_2)(Au^{III}I_4)(I_3)_2]$ (n = 7, 8).

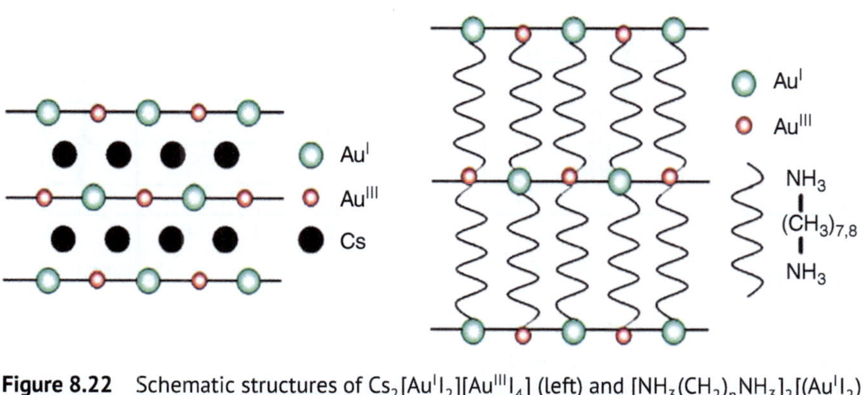

Figure 8.22 Schematic structures of $Cs_2[Au^{I}I_2][Au^{III}I_4]$ (left) and $[NH_3(CH_2)_nNH_3]_2[(Au^{I}I_2)(Au^{III}I_4)(I_3)_2]$ (n = 7, 8) (right). For simplicity, I_3^- is neglected, and $\cdots Au^{I}\cdots I{-}Au^{III}{-}I\cdots Au^{I}\cdots$ planes are guided by solid lines.

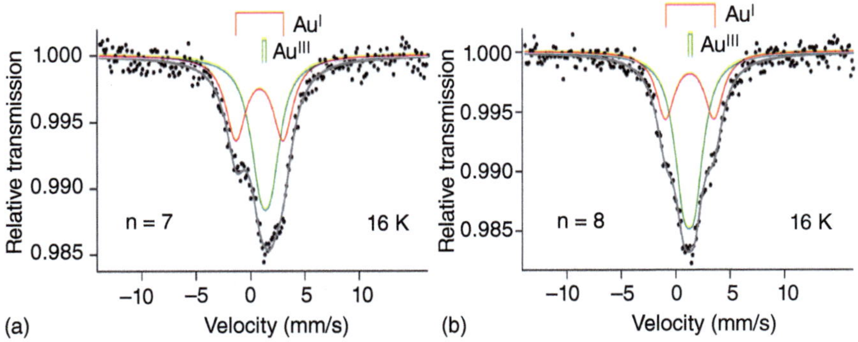

Figure 8.23 ^{197}Au Mössbauer spectra of $[NH_3(CH_2)_nNH_3]_2[(Au^{I}I_2)(Au^{III}I_4)(I_3)_2]$ (n = 7, 8). (a) n = 7, (b) n = 8. Source: Ref. [89]/with permission from Springer Nature.

In order to investigate the effect of charge transfer interaction between the Au^{I} and Au^{III} sites, we investigated the ^{197}Au Mössbauer spectra of $[NH_3(CH_2)_nNH_3]_2[(Au^{I}I_2)(Au^{III}I_4)(I_3)_2]$ (n = 7, 8) (Figure 8.23) [89]. There are two doublets in each spectrum. From the analogy with $Cs_2[Au^{I}X_2][Au^{III}Y_4]$ (X, Y = Cl, Br, I), we assigned these spectra as shown in Figure 8.23.

Figure 8.24 shows the isomer shift (a) and quadrupole splitting (b) of ^{197}Au for $Cs_2[Au^{I}I_2][Au^{III}I_4]$ and $[NH_3(CH_2)_nNH_3]_2[(Au^{I}I_2)(Au^{III}I_4)(I_3)_2]$ (n = 7, 8) at 16 K. Despite the center of isomer shift between the Au^{I} and Au^{III} sites in n = 7 is lower than that of $Cs_2[Au^{I}I_2][Au^{III}I_4]$, and the difference in the isomer shifts between the Au^{I} and Au^{III} sites of $Cs_2[Au^{I}I_2][Au^{III}I_4]$ is similar to that of $[NH_3(CH_2)_7NH_3]_2[(Au^{I}I_2)(Au^{III}I_4)(I_3)_2]$. In the case of $[NH_3(CH_2)_8NH_3]_2[(Au^{I}I_2)(Au^{III}I_4)(I_3)_2]$, however, the isomer shift of Au^{I} is quite similar to that of Au^{III}. In these complexes, there are two possible charge transfer interactions between Au^{I} and Au^{III} in the *ab* plane: (a) charge transfer interaction between Au^{I} ($5d_{z^2}$) and Au^{III} ($5d_{x^2-y^2}$), (b) charge transfer interaction between Au^{I} ($5d_{x^2-y^2}$) and Au^{III} ($5d_{x^2-y^2}$). In Au^{I}, the $5d_{z^2}$ orbital consists of a 6s component because of 5d6s hybridization, while $5d_{x^2-y^2}$ orbital does not consist of a 6s one from the symmetrical

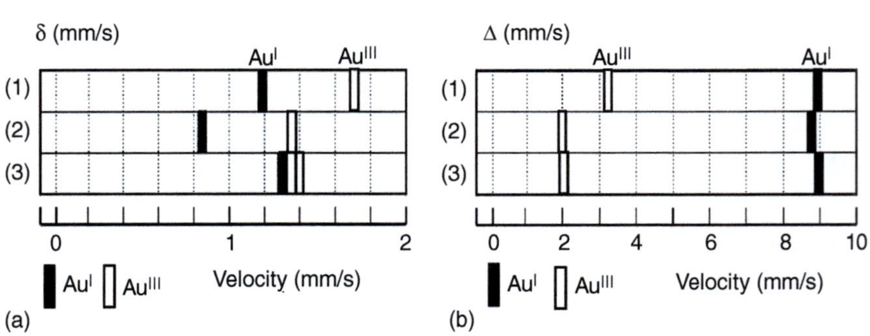

Figure 8.24 (a) Isomer shift (δ) and (b) quadrupole splitting (Δ) of ^{197}Au for (1) $Cs_2[Au^{I}I_2][Au^{III}I_4]$, (2) $[NH_3(CH_2)_7NH_3]_2[(Au^{I}I_2)(Au^{III}I_4)(I_3)_2]$, and (3) $[NH_3(CH_2)_8NH_3]_2[(Au^{I}I_2)(Au^{III}I_4)(I_3)_2]$ at 16 K. Source: Ref. [89]/with permission from Springer Nature.

restriction. Therefore, if the charge transfer interaction (a) is predominant, the isomer shift of Au^{I} becomes to be small because of the decrease of the 6s electron population at the Au^{I} site. On the other hand, if the charge transfer interaction (b) is predominant, the isomer shift of Au^{I} becomes to be large because of the decrease in the screening effect for the 6s orbital.

Comparing the ^{197}Au Mössbauer parameters of $[NH_3(CH_2)_7NH_3]_2[(Au^{I}I_2)(Au^{III}I_4)(I_3)_2]$ with those of $Cs_2[Au^{I}I_2][Au^{III}I_4]$, both δ of Au^{I} and Au^{III} in $[NH_3(CH_2)_7NH_3]_2[(Au^{I}I_2)(Au^{III}I_4)(I_3)_2]$ are smaller by 0.35 mm/s than those of Au^{I} and Au^{III} in $Cs_2[Au^{I}I_2][Au^{III}I_4]$, which implies that the charge transfer interaction (a) is stronger than that of $Cs_2[Au^{I}I_2][Au^{III}I_4]$. In the case of $[NH_3(CH_2)_8NH_3]_2[(Au^{I}I_2)(Au^{III}I_4)(I_3)_2]$, the δ of Au^{I} and Au^{III} are larger and smaller than those of Au^{I} and Au^{III} in $Cs_2[Au^{I}I_2][Au^{III}I_4]$. Therefore, the charge transfer interaction (b) is stronger than that in $Cs_2[Au^{I}I_2][Au^{III}I_4]$. Although n = 7 and 8 of $[NH_3(CH_2)_nNH_3]_2[(Au^{I}I_2)(Au^{III}I_4)(I_3)_2]$ have a small difference in geometric parameters, the effective charge transfer interaction between Au^{I} and Au^{III} is quite different from each other. In order to explain this difference, a comprehensive molecular orbital calculation including the tilting between the Au^{I} and Au^{III} octahedra is indispensable.

8.11 Conclusion

$Cs_2[Au^{I}X_2][Au^{III}X_4]$(X = Cl, Br, I) is well known for the perovskite-type gold mixed-valence system. In order to investigate the charge transfer interaction between Au^{I} and Au^{III} in this system, we have synthesized all of the hetero-halogen-bridged Au mixed-valence complexes having perovskite-type structure, $Cs_2[Au^{I}X_2][Au^{III}Y_4]$ (X, Y = Cl, Br, I), in an organic solvent. In this synthesizing process, we could independently control the coordinated halogens, X and Y. This hetero-halogen bridged gold mixed-valence system was confirmed by means of Raman spectroscopy for $Cs_2[Au^{I}I_2][Au^{III}Br_4]$.

In $Cs_2[Au^{I}X_2][Au^{III}Y_4]$, there are two relative orientations between $[Au^{I}X_2]^-$ and $[Au^{III}Y_4]^-$.

In one orientation, the $\cdots X–Au^{I}–X\cdots Au^{III}\cdots X–Au^{I}–X\cdots$ network lies along the *c*-axis, while on the other, the $\cdots X–Au^{III}–X\cdots Au^{I}\cdots X–Au^{III}–X\cdots$ network lies in the *ab*-plane. The charge transfer interaction between Au^{I} and Au^{III} in $Cs_2[Au^{I}X_2][Au^{III}Y_4]$ is considered to be two-dimensional because the $5d_{z^2}$ orbital (HOMO) of Au^{I} and the $5d_{x^2-y^2}$ orbital (LUMO) of Au^{III} are orthogonal to each other along the *c*-axis, which was confirmed by the single crystal reflectance spectra [81]. In the *ab*-plane, there are two types of charge transfer interaction between Au^{I} and Au^{III}. One is the charge transfer interaction between $Au^{I}(5d_{z^2})$ and Au^{III} $(5d_{x^2-y^2})$. The other is that between Au^{I} $(5d_{x^2-y^2})$ and Au^{III} $(5d_{x^2-y^2})$.

From the analysis of the isomer shift (δ) of Au^{I} for $Cs_2[Au^{I}Cl_2][Au^{III}Y_4]$, $Cs_2[Au^{I}Br_2][Au^{III}Y_4]$, and $Cs_2[Au^{I}I_2][Au^{III}Y_4]$, it is obvious that the charge transfer interaction between Au^{I} $(5d_{x^2-y^2})$ and Au^{III} $(5d_{x^2-y^2})$ is predominant. This charge transfer interaction increases in the order of Y = Cl < Br < I. Therefore, the δ of Au^{III} decreases in the order of Y = Cl < Br < I due to the screening effect on the 6s population. In the cases of $Cs_2[Au^IBr_2][Au^{III}Y_4]$ and $Cs_2[Au^II_2][Au^{III}Y_4]$, the δ of Au^I remarkably increases from Y = Br to I, which is attributed to the significant strong charge transfer interaction between Au^I ($5d_{x^2-y^2}$) and Au^{III} ($5d_{x^2-y^2}$) through the bridging halogen of I^- in the *ab*-plane. In connection with this, the following should be mentioned. $Cs_2[Au^IBr_2][Au^{III}Br_4]$ undergoes the pressure-induced valence transition from the mixed-valence state of Au^I and Au^{III} to the single valence state of Au^{II} at about 7.5 GPa [38, 40]. However, even at 8.1 GPa (above the valence transition), the bridging Br^- ions along the *c*-axis are still deviated from the symmetrical position between the Au ions, while the bridging Br^- ions in the *ab*-plane are located at the symmetrical position between the Au ions, which is attributed to the two-dimensional charge transfer interaction between Au^I and Au^{III} in $Cs_2[Au^IBr_2][Au^{III}Br_4]$ [40].

From the analysis of single crystal ^{197}Au Mössbauer spectra of $Cs_2[Au^II_2][Au^{III}I_4]$, the sign of EFG for Au^I in $[Au^II_2]^-$ and Au^{III} in $[Au^{III}X_4]^-$ were determined to be negative and positive, respectively. It is obvious that the contributions of $6p_z$ and $6p_{x,y}$ determine the sign of EFG for Au^I in $[Au^II_2]^-$ and Au^{III} in $[Au^{III}X_4]^-$, respectively.

$Cs_2[Au^IX_2][Au^{III}X_4]$(X = Cl, Br, I) undergo the pressure-induced Au valence transition from the mixed valence state (Au^I, Au^{III}) to the single valence state (Au^{II}) which is coupled with a structural phase transition. In the pressure region below the Au valence transition, these complexes show a metallic behavior due to the dynamical two-electron exchange (valence skipping) between the Au^I and Au^{III} sites, which was strongly supported by the electrical resistivity, ^{197}Au Mössbauer spectroscopy, Raman spectroscopy, and single-crystal X-ray diffraction analysis under high pressures for $Cs_2[Au^ICl_2][Au^{III}Cl_4]$. In the case of $Cs_2[Au^II_2][Au^{III}I_4]$, the single valence phase of Au^{II} was confirmed by the ^{197}Au Mössbauer spectroscopy at 12.5 GPa and 4.2 K.

Moreover, we investigated the valence state of the layered perovskite-type Au mixed-valence complex system, $[NH_3(CH_2)_nNH_3]_2[(Au^II_2)(Au^{III}I_4)(I_3)_2]$ (n = 7, 8). In the case of n = 8, the isomer shifts of Au^I and Au^{III} sites are close to each other. Therefore, the single valence state of Au^{II} might realize at ambient pressure for n = 8 with small perturbation such as an external field. Although n = 7 and 8 have a small difference in geometric parameter, the effective charge transfer interaction between Au^I and Au^{III} is quite different from each other. In order to explain this discrepancy, a comprehensive molecular orbital calculation including the tilting between the Au^I and Au^{III} octahedra is indispensable and is a future issue.

Acknowledgments

In this chapter, our contributed original research has been created in collaboration with Profs. Y. Moritomo (Tsukuba University), X. J. Liu (Nanjing University), N. Matsushita (Rikkyo University), H. Kitagawa (Kyoto University), J. Stanek (Jagiellonian University), S. S. Hafner (Marburg University), H. Takahashi (Nihon

University), N. Môri (University of Tokyo), O. Shimomura and T. Kikegawa (High Energy Accelerator Research Organization), Dr. K. Ikeda (Sumitomo Electric Industries, Ltd.), and Dr. H. Ahsbahs (Marburg University). We wish to thank all the collaborators. This work has partly been supported by Toyota Physical and Chemical Research Institute, and a Grant-in-Aid for Science Research from the Ministry of Education, Science, Sports, and Culture.

References

1 Robin, M.B. and Day, P. (1967). Mixed valence chemistry – a survey and classification. *Adv. Inorg. Chem.* 10: 248–422.

2 (a) Zener, C. (1951). Interaction between the d shells in the transition metals. *Phys. Rev.* 81: 440–444. (b) Zener, C. (1951). Interaction between the d-shells in the transition metals. II. Ferromagnetic compounds of manganese with perovskite structure. *Phys. Rev.* 82: 403–405.

3 Anderson, P.W. and Hasegawa, H. (1955). Considerations on double exchange. *Phys. Rev.* 100: 675–681.

4 de Gennes, P.G. (1960). Effects of double exchange in magnetic crystals. *Phys. Rev.* 118: 141–154.

5 Millis, A.J. (2000). Theory of CMR manganiters. In: *Colossal Magneto-Resistive Oxides* (ed. Y. Tokura), 53–86. Gordon and Breach Publishers, Chapter 2.

6 Larsson, S. (2008). Mixed valence and superconductivity. *Phil. Trans. R. Soc. A* 366: 47–54.

7 Sleight, A.W. (1988). Chemistry of high-temperature superconductors. *Science* 242: 1519–1527.

8 Cava, R.J., Batlogg, B., Krajewski, J.J. et al. (1988). Superconductivity near 30 K without copper: the $Ba_{0.6}K_{0.4}BiO_3$ perovskite. *Nature* 332: 814–816.

9 Matsushita, N., Kojima, N., Ban, T., and Tsujikawa, I. (1987). Photo-induced absorption band in one-dimensional halogen-bridged mixed-valence platinum complex: $[Pt(en)_2][PtI_2(en)_2](SO_4)_2{\cdot}6H_2O$ and $[Au_xPt_{1-x}I(en)_2]SO_4{\cdot}3H_2O$. *J. Phys. Soc. Jpn.* 56: 3808–3811.

10 Okamoto, H. and Yamashita, M. (1998). Photogeneration of solitons, polarons, and excitons in quasi-one-dimensional halogen-bridged transition metal compounds. *Bull. Chem. Soc. Jpn.* 71: 2023–2039.

11 Takaishi, S. and Yamashita, M. (2008). Solitons, polarons, and their dynamics in mixed-valence halogen-bridged MX-chains. *Phil. Trans. R. Soc. A* 366: 93–100. and related references therein.

12 Okamoto, H. and Matsuzaki, H. (2013). Structures and optical properties of Pt and Pd compounds with charge-density-waves. In: *Material Design and New Physical Properties in MX- and MMX-Chain Compounds* (ed. M. Yamashita and H. Okamoto), 9–30. Springer.

13 (a) Liu, X.J., Moritomo, Y., Ichida, M. et al. (2000). Photoinduced phase transition in a mixed-valence gold complex. *Phys. Rev. B* 61: 20–23. (b) Liu, X.J.,

Moritomo, Y., Ichida, M. et al. (2000). Critical nucleation in the phototransferred metastable phase of $Cs_2Au_2Br_6$. *J. Phys. Soc. Jpn.* 69: 1267–1270.

14 Liu, X.J., Moritomo, Y., Nakamura, A., and Kojima, N. (2001). Pressure- and photo-induced phase transition in mixed-valence gold complexes. *Phase Transit.* 74: 51–92.

15 Son, J.Y., Mizokawa, T., Quilty, J.W. et al. (2005). Photoinduced valence transition in gold complexes $Cs_2Au_2X_6$ (X = Cl and Br) probed by X-ray photoemission spectroscopy. *Phys. Rev. B* 72: 235105-1-4.

16 Herren, F., Fischer, P., Ludi, A., and Hälg, W. (1980). Neutron diffraction study of Prussian Blue, $Fe_4[Fe(CN)_6]_3 \cdot xH_2O$. Location of water molecules and long-range magnetic order. *Inorg. Chem.* 19: 956–959.

17 Varma, C.M. (1976). Mixed-valence compounds. *Rev. Mod. Phys.* 48: 219–238.

18 Day, P. (1980). Introduction to mixed-valence chemistry. In: *Mixed-Valence Compound – Theory and Applications in Chemistry, Physics, Geology, and Biology* (ed. D.B. Brown), 3–24. (D. Reidel Publishing Company.

19 Schatz, P.N. (1991). Vibronic coupling models of mixed valency: relation of the PKS and MO models for one- and two-electron systems. In: *Mixed Valency Systems: Applications in Chemistry, Physics and Biology* (ed. K. Prassides), 7–28. (Kluwer Academic.

20 Day, P., Hush, N.S., and Clark, R.J.H. (2008). Mixed valence: origins and developments. *Phil. Trans. R. Soc. A* 366: 5–14.

21 Anderson, P.W. (1975). Model for electronic structure of amorphous semiconductors. *Phys. Rev. Lett.* 34: 953–955.

22 Alexandrov, A. and Ranninger, J. (1981). Theory of bipolarons and bipolaronic bands. *Phys. Rev. B* 23: 1796–1801.

23 Emin, D. (1996). Pair breaking in semiclassical singlet small-bipolaron hopping. *Phys. Rev. B* 53: 1260–1268.

24 Bonča, J., Katrašnik, T., and Trugman, S.A. (2000). Mobile bipolaron. *Phys. Rev. Lett.* 84: 3153–3156.

25 Varma, C.M. (1988). Missing valence states, diamagnetic insulators, and superconductors. *Phys. Rev. Lett.* 61: 2713–2716.

26 Harrison, W.A. (2006). Valence-skipping compounds as positive-U electron systems. *Phys. Rev. B* 74: 245128-1-4.

27 Strand, H.U.R. (2014). Valence-skipping and negative-U in the d-band from repulsive local Coulomb interaction. *Phys. Rev. B* 90: 155108-1-6.

28 (a) Cava, R.J., Batlogg, B., Krajewski, J.J. et al. (1988). Superconductivity near 30 K without copper: the $Ba_{0.6}K_{0.4}BiO_3$ perovskite. *Nature* 332: 814–816. (b) Sleight, A.W. (2015). Bismuthates: $BaBiO_3$ and related superconducting phases. *Physica C* 514: 152–165.

29 Thanh, T.D., Koma, A., and Tanaka, S. (1980). Superconductivity in the $BaPb_{1-x}Bi_xO_3$ system. *Appl. Phys.* 22: 205–212.

30 Schoop, L.M., Müchler, L., Felser, C., and Cava, R.J. (2013). Lone pair effect, structural distortions, and potential for superconductivity in Tl perovskites. *Inorg. Chem.* 52: 5479–5483.

31 Retuerto, M., Emge, T., Hadermann, J. et al. (2013). Synthesis and properties of charge-ordered thallium halide perovskites, $CsTl^{+}{}_{0.5}Tl^{3+}{}_{0.5}X_3$ (X = F or Cl): theoretical precursors for superconductivity? *Chem. Mater.* 25: 4071–4079.

32 Hase, I., Yanagisawa, T., and Kawashima, K. (2017). One way to design a valence-skip compound. *Nanoscale Res. Lett.* 12: 127-1-5.

33 Kojima, N., Kitagawa, H., Ban, T. et al. (1990). Semiconductor-to-metal and metal-to-metal transitions in the three-dimensional mixed-valence compound $Cs_2Au_2I_6$ under high pressures. *Solid State Commun.* 73: 743–745.

34 Kitagawa, H., Sato, H., Kojima, N. et al. (1991). Metallization and phase transitions of the three-dimensional halogen-bridged mixed-valence complex $Cs_2Au_2I_6$ under high pressure. *Solid State Commun.* 78: 989–995.

35 Kitagawa, H., Kojima, N., Takahashi, H., and Mori, N. (1993). Electrical conductivity and its anisotropy of the mixed-valence compound $Cs_2Au_2I_6$ under high pressure and low temperature. *Synth. Met.* 56: 1726–1729.

36 Kojima, N., Hasegawa, M., Kitagawa, H. et al. (1994). P-T phase diagram and gold valence state of the perovskite-type mixed-valence compounds $Cs_2Au_2X_6$ (X = Cl, Br, and I) under high pressures. *J. Am. Chem. Soc.* 116: 11368–11374.

37 Hafner, S.S., Kojima, N., Stanek, J., and Zhang, L. (1994). Divalent gold in perovskite type $CsAuI_3$. *Phys. Lett. A* 192: 385–388.

38 Liu, X.J., Moritomo, Y., Nakamura, A., and Kojima, N. (1999). Pressure induced phase transition in mixed-valence gold complexes $Cs_2Au_2X_6$ (X = Cl and Br). *J. Chem. Phys.* 110: 9174–9178.

39 Liu, X.J., Matsuda, K., Moritomo, Y. et al. (1999). Electronic structure of the gold complexes $Cs_2Au_2X_6$ (X = I, Br, Cl). *Phys. Rev. B* 59: 7925–7930.

40 Kojima, N. (2000). Gold valence transition and phase diagram in the mixed-valence complexes, $M_2[Au^{I}X_2][Au^{III}X_4]$ (M = Rb, Cs; X = Cl, Br, and I). *Bull. Chem. Soc. Jpn.* 73: 1445–1460.

41 Kojima, N. and Matsushita, N. (2000). P-T Phase diagram and Au valence state of the perovskite-type Au mixed-valence complexes, $M_2[Au^{I}X_2][Au^{III}X_4]$ (M = K, Rb, Cs; X = Cl, Br, I). *Coord. Chem. Rev.* 198: 251–263.

42 Winkler, B., Pickard, C.J., Segall, M.D., and Milman, V. (2001). Density-functional study of charge disordering in $Cs_2Au(I)Au(III)Cl_6$ under pressure. *Phys. Rev. B* 63: 214103-1-4.

43 Sakata, M., Itsubo, T., Nishibori, E. et al. (2004). Charge density study under high pressure. *J. Phys. Chem. Solids* 65: 1973–1976.

44 Matsushita, N., Ahsbahs, H., Hafner, S.S., and Kojima, N. (2007). Single crystal X-ray diffraction study of a mixed-valence gold compound, $Cs_2Au^{I}Au^{III}Cl_6$ under high pressures up to 18 GPa: pressure-induced phase transition coupled with gold valence transition. *J. Solid State Chem.* 180: 1353–1364.

45 Kusmartseva, A.F., Yang, M., Arevalo-Lopez, A.M. et al. (2010). Pressure supression of charge order without metallisation in $Cs_2Au_2I_6$. *Chem. Commun.* 46: 6681–6683.

46 Wang, S., Kemper, A.F., Baldini, M. et al. (2014). Bandgap closure and reopening in $CsAuI_3$ at high pressure. *Phys. Rev. B* 89: 245109-1-7.

47 Trigo, M., Chen, J., Jiang, M.P. et al. (2012). Ultrafast pump-probe measurements of short small-polaron lifetimes in the mixed-valence perovskite $Cs_2Au_2I_6$ under high pressures. *Phys. Rev. B* 85: 081102-1-4.

48 Wang, S., Hirai, S., Shapiro, M.C. et al. (2013). Pressure-induced symmetry breaking in tetragonal $CsAuI_3$. *Phys. Rev. B* 87: 054104-1-6.

49 Naumov, P., Huangfu, S., Wu, X. et al. (2019). Large resistivity reduction in mixed-valent $CsAuBr_3$ under pressure. *Phys. Rev, B* 100: 155113-1-8.

50 Debbichi, L., Lee, S., Cho, H. et al. (2018). Mixed valence perovskite $Cs_2Au_2I_6$: a potential material for thin-film Pb-free photovoltaic cells with ultrahigh efficiency. *Adv. Mater.* 30: 1707001-1-6.

51 Giorgi, G., Yamashita, K., and Palummo, M. (2018). Two-dimensional optical excitations in the mixed-valence $Cs_2Au_2I_6$ fully inorganic double perovskite. *J. Mater. Chem. C* 6: 10197–10201.

52 Du, X., He, D., Mei, H. et al. (2020). Insights on electronic structures, elastic features and optical properties of mixed-valence double perovskites, $Cs_2Au_2X_6$ (X = F, Cl, Br, I). *Phys. Lett. A* 384: 126169-1-9.

53 Castro-Castro, L.M. and Guloy, A.M. (2003). Organic-based layered perovskites of mixed-valent gold (I)/gold (III) iodides. *Angew. Chem. Int. Ed.* 42: 2771–2774.

54 Ghosh, B., Febriansyah, B., Harikesh, P.C. et al. (2020). Direct band gap mixed-valence organic-inorganic gold perovskite as visible light absorbers. *Chem. Mater.* 32: 6318–6325.

55 Kojima, N., Ikeda, K., Kobayashi, Y., and Seto, M. (2013). Study on the electronic state of gold mixed-valence complexes, $Cs_2[Au(I)X_2][Au(III)Y_4]$ (X, Y = Cl, Br, I) by means of ^{197}Au Mössbauer spectroscopy. *Mössbauer Eff. Ref. Data J.* 36: 57–72.

56 Mason, W.R. and Gray, H.B. (1968). Electronic structures and spectra of square-planar gold (III) complexes. *Inorg. Chem.* 7: 55–58.

57 Ryan, J.L. (1969). Weak or unstable iodo complexes. II. Iodo complexes of titanium (IV), iron (III), and gold (III). *Inorg. Chem.* 8: 2058–2062.

58 Braunstein, P., Clark, R.J.H., Ramsay, W. et al. (1973). The preparation, properties, and vibrational spectra of complexes containing the $AuCl_2^-$, $AuBr_2^-$, and AuI_2^- ions. *J. Chem. Soc. Dalton Trans.* 1845–1848. https://doi.org/10.1039/DT9730001845.

59 Kitagawa, H., Kojima, N., and Nakajima, T. (1991). Studies of the mixed-valence states in three-dimensional halogen-bridged mixed-valence gold compounds, $Cs_2Au^IAu^{III}I_6$ (X = Cl, Br, and I). Part 2. X-ray photoelectron spectroscopic study of Au compounds. *J. Chem. Soc. Dalton Trans.* 3121–3125. https://doi.org/10.1002/chin.199211009.

60 Ikeda, K., Kojima, N., Ono, Y. et al. (2004). Study on chemical bond and electronic state of new gold mixed-valence complexes, $Cs_2[Au^IX_2][Au^{III}Y_4]$(X, Y = Cl, Br, I, etc.). *Hyperfine Interact.* 156–157: 311–314.

61 MossWin 3.0 (2011). Zoltan Klencsar. http://www.mosswinn.com/english/index.html (accessed 2011).

62 Pyykkö, P. and Desclaux, J.-P. (1979). Relativity and the periodic system of elements. *Acc. Chem. Res.* 12: 276–281.

63 Guenzburger, D. and Ellis, D.E. (1980). Electronic structure and Mössbauer hyperfine interactions of Au(I) compounds. *Phys. Rev. B* 22: 4203–4213.

64 Bowmaker, G.A., Boyd, P.D.W., and Sorrenson, R.J. (1985). An SCF-MS-Xα study of the bonding and nuclear quadrupole coupling in dihalogenocuprates(I), CuX_2^- (X = Cl, Br), dihalogenoaurates (I), AuX_2^- (X = Cl, Br, I), and bis(cyano)aurate(I), $Au(CN)_2^-$. *J. Chem. Soc. Faraday Trans. 2* 81: 1627–1641.

65 Braunstein, P., Schubert, U., and Burgard, M. (1984). Synthesis and spectroscopic studies of metal-metal bonded linear heterotrimetallic gold(I) complexes. Crystal structure of $[n\text{-}Bu_4N][Au[Cr(CO)_3\text{-}\eta\text{-}C_5H_5]_2]$. *Inorg. Chem.* 23: 4057–4064.

66 Jones, P.G., Maddock, A.G., Mays, M.J., and Muir, M.M. (1977). Structure and bonding in gold(I) compounds. Part 2. Mössbauer spectrum of linear gold(I) complexes. *J. Chem. Soc. Dalton Trans.* 1434–1439. https://doi.org/10.1039/DT9770001434.

67 Esquivel, D.M.S., Guenzburger, D., and Danon, J. (1979). Electronic configuration and nuclear quadrupole interaction in higher transition elements. *Phys. Rev. B* 19: 1357–1362.

68 Faltens, M.O. and Shirley, D.A. (1970). Mössbauer spectroscopy of gold compounds. *J. Chem. Phys.* 53: 4249–4262.

69 Bartunik, H.D., Potzel, W., and Mössbauer, R.L. (1970). Resonance spectroscopy of γ-radiation on Au (I) and Au (III) compounds. *Z. Phys.* 240: 1–16.

70 Prosser, H., Wagner, F.E., Wortman, G. et al. (1975). Mössbauer determination of the E2/M1 mixing ratio of the 77 keV transition in ^{197}Au and of the sign of the electric field gradient in $KAu(CN)_2$. *Hyperfine Interact.* 1: 25–32.

71 IUPAC (ed.) (2007). *Quantities, Units and Symbols in Physical Chemistry, Third Edition*. (RSC Publishing.

72 Sham, T.K., Watson, R.E., and Perlman, M.L. (1980). Electronic configuration and electric field gradient in Au complexes. *Phys. Rev. B* 21: 1457–1461.

73 Childs, W.J. and Goodman, L.S. (1966). Hyperfine structure of the 9161 cm^{-1} $^2D_{5/2}$ state of ^{197}Au and the nuclear electric-quadrupole moment. *Phys. Rev.* 141: 176–141.

74 Machmer, P. (1966). Über die Chemische Bindung in Gold(I)-Jodid. *Z. Naturforsch.* 21b: 1025–1027.

75 Kojima, N., Seto, M., and Maeda, Y. (1996). Single crystal ^{197}Au Mössbauer spectroscopy of gold mixed-valence compound $Cs_2Au_2I_6$. *Societa Italiana di Fisica* 50: 43–46.

76 Parish, R.V. (1984). Gold-197 Mössbauer spectroscopy in the characterization of gold compounds. In: *Mössbauer Spectroscopy Applied to Inorganic Chemistry*. Volume 1 (ed. G.J. Long), 577–617. Plenum Press.

77 Eijndhoven, J.C.M.T.-v. and Verchoor, G.C. (1974). Redetermination of the crystal structure of $Cs_2AuAuCl_6$. *Mater. Res. Bull.* 9: 1667–1670.

78 Matsushita, N., Fukuhara, F., and Kojima, N. (2005). A three-dimensional bromo-bridged mixed-valence gold(I, III) compound, $Cs_2Au^IAu^{III}Br_6$. *Acta Cryst. E* 61: i123–i125.

79 Matsushita, N., Kitagawa, H., and Kojima, N. (1997). Three-dimensional iodo-bridged mixed-valence gold(I,III) compound, $Cs_2Au^IAu^{III}I_6$. *Acta Cryst. C* 53: 663–666.

80 Braunstein, P., Müller, A., and Bögge, H. (1986). Crystal structure of the tetra-*n*-butylammonium salts of the $AuCl_2^-$, $AuBr_2^-$, and AuI_2^- ions. *Inorg. Chem.* 25: 2104–2106.

81 Kojima, N. and Kitagawa, H. (1994). Optical investigation of the inter-valence charge-transfer interaction in the three-dimensional Au mixed-valence compounds $Cs_2Au_2X_6$ (X = Cl, Br and I). *J. Chem. Soc. Dalton Trans.* 327–331. https://doi.org/10.1039/DT9940000327.

82 Greenwood, N.N. and Gibb, T.C. (ed.) (1971). The Mössbauer effect. In: *Mössbauer Spectroscopy*, 11. Chapman & Hall, Chapter 1.

83 Greenwood, N.N. and Gibb, T.C. (ed.) (1971). Table of nuclear data for Mössbauer transitions. In: *Mössbauer Spectroscopy*, 610. Chapman & Hall, Appendix 1.

84 Keller, R., Fenner, J., and Holzapfel, W.B. (1974). The resistivity of mixed valence compound $Cs_2Au_2C_{16}$ at high pressure and low temperatures. *Mater. Res. Bull.* 9: 1363–1370.

85 Denner, W., Schulz, H., and D'Amour, H. (1979). The influence of high hydrostatic pressure on the crystal structure of cesium gold chloride ($Cs_2Au^IAu^{III}Cl_6$) in the pressure range up to 52×10^8 Pa. *Acta Cryst. A* 35: 360–365.

86 Stanek, J., Hafner, S.S., and Schulz, H. (1980). Electronic state of Au in $Cs_2Au_2Cl_6$ at high pressure. *Phys. Lett. A* 76: 333–334.

87 Stanek, J. (1982). High pressure Mössbauer study of ^{197}Au in $Cs_2Au_2Cl_6$, $Cs_2AgAuCl_6$, and Au I. *J. Chem. Phys.* 76: 2315–2320.

88 Tanino, H., Syassen, K., Wang, Z. et al. (1990). The mixed valence compound $Cs_2Au(I)Au(III)Cl_6$ under pressure: Raman modes and optical reflectivity. *High Pressure Res.* 3: 183–185.

89 Ikeda, K., Kojima, N., Kobayashi, Y., and Seto, M. (2005). Study on charge transfer interaction and valence state of layered perovskite-type mixed valence complex $[NH_3(CH_2)nNH_3]_2$ $[(AuI_2)(AuI_4)(I_3)_2]$ (n =7, 8) by means of ^{197}Au Mössbauer spectroscopy. *Hyperfine Interact.* 166: 403–407.

9

Temperature- and Photo-Induced Spin-Crossover in Molecule-Based Magnets

Hiroko Tokoro[1,2], Kenta Imoto[1], and Shin-ichi Ohkoshi[1]

[1] *The University of Tokyo, School of Science, Department of Chemistry, 7-3-1 Hongo, Bunkyo-ku, Tokyo 113-0033, Japan*
[2] *University of Tsukuba, Department of Materials Science, Faculty of Pure and Applied Sciences, 1-1-1 Tennodai, Tsukuba, Ibaraki 305-8577, Japan*

9.1 Introduction

Bistable functional materials have potential in sensor and switching materials due to their physical properties such as switching [1–14]. As practical examples, magnetic switching should contribute to magnetic storage, while color switching should enhance optical storage [13–56]. Additionally, bistable functional materials are interesting from fundamental viewpoints in materials chemistry and theoretical physics (i.e. non-equilibrium statistical physics, etc.) [57–60]. Many studies have revealed that cyano-bridged metal complexes [26–56] are magnetic materials that exhibit switching properties in response to external stimuli. For example, photo-magnetic, ferroelectric-ferromagnetic, and humidity-sensitive materials have been reported [33–56]. Cyano-bridged metal complexes have superior characteristics. Such characteristics not only permit the development of long-range magnetic ordering by allowing arbitrary combinations of metal ions and organic ligands but also can control the network structure of metal-cyano frameworks. In addition, cyano-bridged metal complexes are tolerant to structural changes due to switching because the phonon mode of the cyano group allows for structural flexibility [61–63].

Observations of the magnetic states before and after switching are essential for developing magnetic materials with tunable functions. Various technologies have been used to observe magnetic states. For example, a magnetometer can detect magnetic moments. Among the available technologies, Mössbauer spectroscopy is a superior technique to observe fine structures in a magnetic state of a material [64–66].

In this paper, we focus on the unique switching function of our synthesized cyanide-metal magnetic complexes. We present the results of Mössbauer spectroscopy and a switching phenomenon by a light-induced spin crossover, which is a LIESST effect (light-induced excited spin-state trapping effect). First, we introduce

Mössbauer Spectroscopy: Applications in Chemistry and Materials Science,
First Edition. Edited by Yann Garcia, Junhu Wang, and Tao Zhang.

a temperature-induced spin-crossover phenomenon in hexacyanido complex $CsFeCr(CN)_6{\cdot}1.3H_2O$ [18] followed by a photomagnetic effect due to light-induced spin crossover in octacyanido complex $Fe_2[Nb(CN)_8](4\text{-pyridinealdoxime})_8{\cdot}2H_2O$ [48]. Finally, we explain a photomagnetic effect by a reversible light-induced spin crossover in $Fe_2[Nb(CN)_8](4\text{-bromopyridine})_8{\cdot}2H_2O$ [50] octacyanido complexes and optical control of the second harmonic generation (SHG).

9.2 Spin-Crossover Phenomena in Cesium Iron Hexacyanidochromate Prussian Blue Analog

Many studies in solid-state chemistry research thermal phase transition phenomena induced by spin-crossover because they are responsive to external stimuli such as light, humidity, and so on. Herein, we focus on the cesium iron hexacyanidochromate [18], which is a Prussian blue analog, exhibiting a spin-crossover phenomenon.

Our research began by synthesizing the target compound. A mixed aqueous solution of $K_3[Cr^{III}(CN)_6]$ and Cs^ICl was added to a mixed aqueous solution of $Fe^{II}Cl_2$ and Cs^ICl to yield a precipitate with a composition of $CsFeCr(CN)_6{\cdot}1.3H_2O$ (Figure 9.1a).

$CsFeCr(CN)_6{\cdot}1.3H_2O$ shows a temperature-induced spin crossover phenomenon. A plot of the product of the molar magnetic susceptibility and the temperature ($\chi_M T$) as a function of temperature reveals two phases (Figure 9.1b). The high-temperature (HT) phase has a $\chi_M T$ value of 6.11 K cm^3 mol^{-1} at 280 K. The $\chi_M T$ value drops around 210 K as the temperature decreases (i.e. the low-temperature [LT] phase appears below 200 K). Upon warming the LT phase, the $\chi_M T$ value increases to around 230 K until the HT phase is restored at 250 K. The transition temperature ($T_{1/2}$) occurs when half of a material is in the HT phase and the other half is in

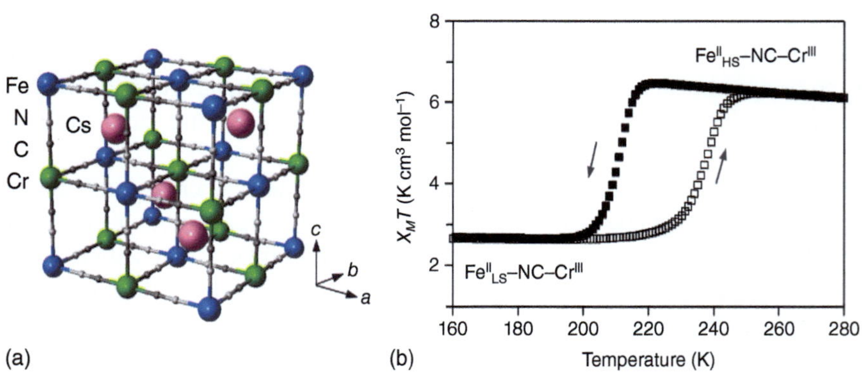

Figure 9.1 (a) Schematic crystal structure of $CsFeCr(CN)_6$ Prussian blue analog. (b) The $\chi_M T - T$ plot for $CsFeCr(CN)_6{\cdot}1.3H_2O$ measured in an external magnetic field of 5000 Oe where closed and open squares denote cooling and heating, respectively. Source: Panels have been modified from Ref. [18]. Copyright 2005 American Chemical Society.

the LT phase. $CsFeCr(CN)_6{\cdot}1.3H_2O$ has a HT → LT ($T_{1/2\downarrow}$) of 211 K and a LT → HT ($T_{1/2\uparrow}$) of 238 K. This thermal phase transition can be observed several times.

Next, we investigated the temperature dependence of the infrared (IR) spectra. The results of IR spectra show that the HT phase has an electronic state of $Cs^I\{Fe^{II}{}_{HS}[Cr^{III}(CN)_6]\}_{0.94}\{Fe^{II}{}_{LS}[Cr^{III}(NC)_6]\}_{0.06}{\cdot}1.3H_2O$, where 94% of Fe^{II} is in the high spin (HS) state ($S=2$) and 6% is the low spin (LS) state ($S=0$). On the other hand, the LT phase has an electronic state of $Cs^I\{(Fe^{II}{}_{HS})_{0.12}(Fe^{II}{}_{LS})_{0.88}[Cr^{III}(CN)_6]\}_{0.94}\{Fe^{II}{}_{LS}[Cr^{III}(NC)_6]\}_{0.06}{\cdot}1.3H_2O$ where 11% of Fe^{II} is the HS state and 89% of Fe^{II} is the LS state. Then we used ^{57}Fe Mössbauer spectroscopy (Figure 9.2, Table 9.1) to investigate the Fe^{II} spin-crossover. The velocity scale and isomer shifts were calibrated using a metallic α-Fe foil absorber at room temperature. The HT and LT phases display distinct spectra. The HT phase has a doublet peak ($\delta = 1.07(1)\,mm\,s^{-1}$; $\Delta E_Q = 0.74(1)\,mm\,s^{-1}$). In contrast, the LT phase shows a singlet peak ($\delta = 0.438(2)\,mm\,s^{-1}$). These spectra confirm that spin-crossover occurs on the Fe^{II} sites because the peak in the HT phase is attributed to $Fe^{II}{}_{HS}$ while that in the LT phase is attributed to $Fe^{II}{}_{LS}$.

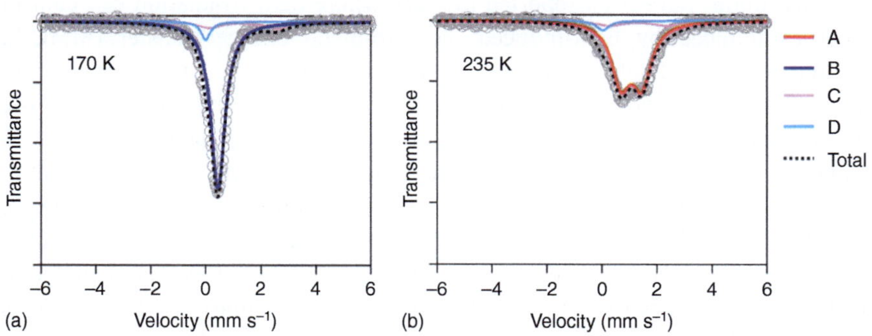

Figure 9.2 ^{57}Fe Mössbauer spectra of (a) LT phase (170 K) and (b) HT phase (235 K) in the cooling process. Doublet peak (A) in the HT phase is assigned to $Fe^{II}{}_{HS}$ at 235 K. This peak disappears at 170 K and is replaced by a singlet peak (B) assigned to $Fe^{II}{}_{LS}$, indicating an occurrence of spin-crossover on the Fe^{II} sites. A small doublet peak (C) and a singlet peak (D) observed at both temperatures are assigned to the remaining $Fe^{II}{}_{HS}$ in the LT phase and the $Fe^{II}{}_{LS}$ of cyano flip, respectively. Source: Panels have been modified from Ref. [18]. Copyright 2005 American Chemical Society.

Table 9.1 ^{57}Fe Mössbauer parameters of LT phase and HT phase of $CsFeCr(CN)_6{\cdot}1.3H_2O$.

	LT phase (170 K)			HT phase (235 K)			
	Area	**Δ**	**ΔE_Q**	**Area**	**δ**	**ΔE_Q**	**Assignment**
A	—	—	—	85%	1.07(1)	0.74(1)	$Fe^{II}{}_{HS}$–NC
B	84%	0.438(2)	0	—	—	—	$Fe^{II}{}_{LS}$–NC
C	10%	1.36(3)	2.43(6)	10%	1.32	2.12	$Fe^{II}{}_{HS}$–NC
D	6%	−0.01(2)	0	5%	0.02(7)	0	$Fe^{II}{}_{LS}$–CN

To verify whether a structural phase transition accompanies this spin-crossover, we measured the temperature dependence of the X-ray powder diffraction (XRD) patterns. Indeed, a structural phase transition occurs during spin-crossover. The cubic $F\bar{4}3m$ structure with a lattice constant of 10.708(1) Å is transformed into the cubic $F\bar{4}3m$ structure with a lattice constant of 10.330(1) Å.

In summary, this study demonstrates that $CsFeCr(CN)_6{\cdot}1.3H_2O$ displays a spin-crossover behavior. Prussian blue analogs with a spin-crossover phenomenon should lead to new functionalities. For example, the next section demonstrates an octacyanidometallate-based system with photoinduced magnetization due to the light-induced excited spin-state trapping (LIESST) effect.

9.3 Light-Induced Spin-Crossover Magnet in Iron Octacyanidoniobate Bimetal Assembly

An active area of research in materials science is the optical control of ferromagnetism. Several works on photoinduced magnetization of cyanido-bridged metal assemblies have been reported [33–52], where light irradiation can control properties induced by light-induced metal-to-metal charge transfer (MMCT). Examples of controllable properties include magnetization, Curie temperature (T_C), and the magnetic coercive field (H_c). Spin-crossover materials exhibit the LIESST effect [21–25]. However, usually, the LIESST effect does not exhibit a spontaneous magnetization. To achieve photoinduced spontaneous magnetization, an appropriate spin-crossover material with a three-dimensional (3D) magnetic network, which exhibits photo-induced HS states and a strong exchange coupling, must be employed. Here we discuss our studies on an octacyanidoniobate-based metal–organic framework, $Fe_2[Nb(CN)_8]$(4-pyridinealdoxime)$_8{\cdot}2H_2O$ [48]. This framework exhibits not only spin-crossover but also photoinduced spontaneous magnetization due to the distinct magnetic ordering of the Fe(II) centers, through the –NC–Nb^{IV}($S = 1/2$)–CN– bridges.

Adding a mixed aqueous solution of $FeCl_2{\cdot}4H_2O$ and 4-pyridinealdoxime to an aqueous solution of $K_4[Nb(CN)_8]{\cdot}2H_2O$ gave $Fe_2[Nb(CN)_8]$(4-pyridinealdoxime)$_8{\cdot}$ $2H_2O$ as a 3D Fe–Nb bimetallic assembly. The assembly has a tetragonal structure ($a = 20.2001(4)$ Å and $c = 14.9565(5)$ Å) (Figure 9.3a).

Similar to $CsFeCr(CN)_6{\cdot}1.3H_2O$, $Fe_2[Nb(CN)_8]$(4-pyridinealdoxime)$_8{\cdot}2H_2O$ displays a temperature-induced spin-crossover effect. Figure 9.3b plots $\chi_M T$ as a function of temperature. The HT phase has a $\chi_M T$ value of 7.15 K cm^3 mol^{-1} at 290 K, whereas the LT phase has a value of 1.72 K cm^3 mol^{-1} at 50 K. $T_{1/2}$ is 130 K.

Next, we measured the ultraviolet-visible (UV-vis) absorption spectra. As the temperature decreases, bands appear at 480 and 650 nm. These bands are assigned to the $^1A_1 \rightarrow {}^1T_2$ and $^1A_1 \rightarrow {}^1T_1$ optical transitions on the $Fe^{II}{}_{LS}$ ($S = 0$) site, respectively. The spin-crossover from $Fe^{II}{}_{HS}$ to $Fe^{II}{}_{LS}$ is responsible for the thermal conversion from the HT phase to the LT phase. We also measured the ^{57}Fe Mössbauer spectra to verify Fe^{II} spin-crossover (Figure 9.4). The velocity scale and isomer shifts were calibrated using a metallic α-Fe foil absorber at room temperature. The HT

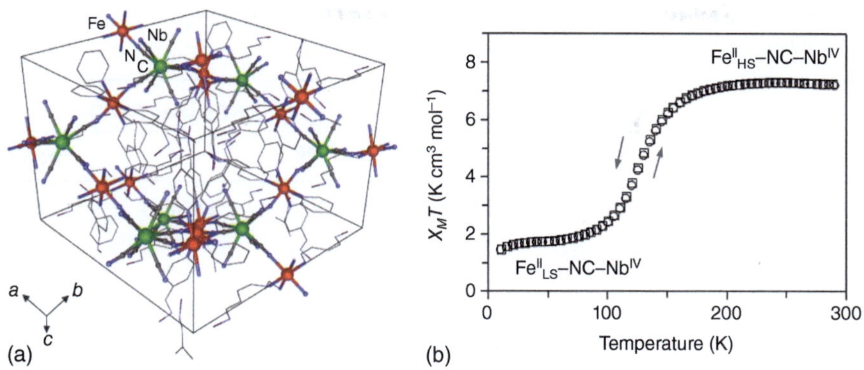

Figure 9.3 (a) Crystal structure of $Fe_2[Nb(CN)_8](4\text{-pyridinealdoxime})_8\cdot 2H_2O$ viewed along the *c*-axis. Non-coordinated water molecules are omitted for clarity. (b) The $\chi_M T-T$ plot for $Fe_2[Nb(CN)_8](4\text{-pyridinealdoxime})_8\cdot 2H_2O$ measured in an external magnetic field of 5000 Oe where circles and squares denote cooling and heating, respectively. Source: Panels have been modified from Ref. [48]. Copyright 2011 Springer Nature.

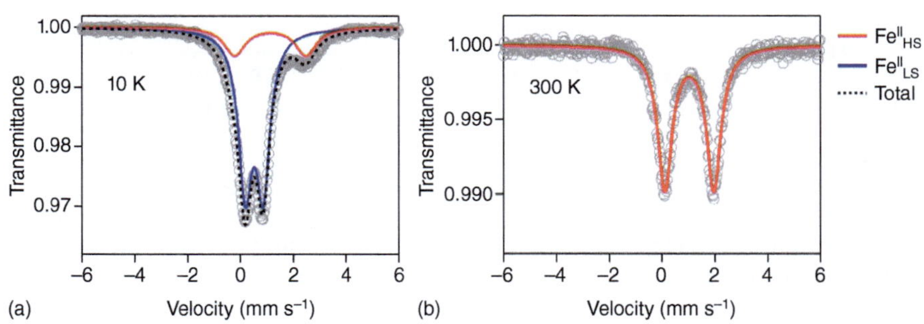

Figure 9.4 ^{57}Fe Mössbauer spectra of $Fe_2[Nb(CN)_8](4\text{-pyridinealdoxime})_8\cdot 2H_2O$ at (a) 10 K and (b) 300 K. HT phase has a doublet due to the Fe^{II}_{HS} ($\delta = 1.034(2)$ mm s^{-1}; $\Delta E_Q = 1.852(4)$ mm s^{-1}), whereas the LT phase has peaks due to Fe^{II}_{LS} (78%, $\delta = 0.524(2)$ mm s^{-1}; $\Delta E_Q = 0.681(4)$ mm s^{-1}) and the remaining Fe^{II}_{HS} (22%; $\delta = 1.12(5)$ mm s^{-1}; $\Delta E_Q = 2.72(3)$ mm s^{-1}). Source: Panels have been modified from Ref. [48]. Copyright 2011 Springer Nature.

phase has a doublet due to Fe^{II}_{HS} ($\delta = 1.034(2)$ mm s^{-1}; $\Delta E_Q = 1.852(4)$ mm s^{-1}). The LT phase contains mainly a doublet due to Fe^{II}_{LS} ($\delta = 0.524(2)$ mm s^{-1}; $\Delta E_Q = 0.681(4)$ mm s^{-1}) (Table 9.2). Hence, the electronic states of the HT and LT phases are attributed to $(Fe^{II}_{HS})_2[Nb^{IV}(CN)_8](4\text{-pyridinealdoxime})_8\cdot 2H_2O$ and $(Fe^{II}_{HS})_{0.44}(Fe^{II}_{LS})_{1.56}[Nb^{IV}(CN)_8](4\text{-pyridinealdoxime})_8\cdot 2H_2O$, respectively.

After confirming spin-crossover, we focused our attention on the photomagnetic effect. Irradiating 473-nm laser light (17 mW cm^{-2}) at 2 K excites the LT phase absorption band due to the $^1A_1 \rightarrow {}^1T_2$ transition of Fe^{II}_{LS} at 480 nm, inducing a large spontaneous magnetization. Figure 9.5a plots the magnetization vs. temperature curve, while Figure 9.5b plots the magnetization vs. the external magnetic field. The photo-induced state of $Fe_2[Nb(CN)_8](4\text{-pyridinealdoxime})_8\cdot 2H_2O$ has a T_C of 20 K. A magnetic hysteresis loop with H_c of 240 Oe appears at 2 K. The measured

Table 9.2 ^{57}Fe Mössbauer parameters of HT phase and LT phase of $Fe_2[Nb(CN)_8]$(4-pyridinealdoxime)$_8$·$2H_2O$.

	LT phase (10 K)			HT phase (300 K)		
	Area	δ	ΔE_Q	Area	δ	ΔE_Q
$Fe^{II}{}_{HS}$	22%	1.12(5)	2.72(3)	100%	1.034(2)	1.852(4)
$Fe^{II}{}_{LS}$	78%	0.524(2)	0.681(4)	—	—	—

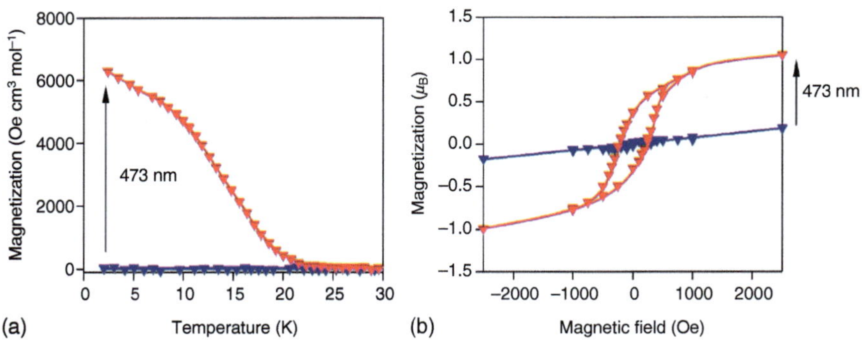

Figure 9.5 Photoinduced magnetization caused by a light-induced spin-crossover in $Fe_2[Nb(CN)_8]$(4-pyridinealdoxime)$_8$·$2H_2O$. (a) Magnetization vs. temperature curves at 100 Oe. (b) Magnetic hysteresis curves at 2 K. Blue and red triangles denote before and after irradiation with 473-nm light. Source: Panels have been modified from Ref. [48]. Copyright 2011 Springer Nature.

saturation magnetization (M_s) at 7 T (7.4 μ_B) agrees with the theoretical M_s value of 7.7 μ_B predicted from the ferrimagnetic coupling between Nb^{IV} ($S = 1/2$) and the photoproduced $Fe^{II}{}_{HS}$ ($S = 2$) using g-values of $g_{Nb}{}^{IV} = 1.99$ and $g_{Fe}{}^{II}{}_{HS} = 2.17$. Additionally, light irradiation affects the Mössbauer spectra as irradiation increases $Fe^{II}{}_{HS}$ but decreases $Fe^{II}{}_{LS}$ (Figure 9.6). Consequently, the observed magnetization originates from a photoinduced spin-crossover from $Fe^{II}{}_{LS}$ to $Fe^{II}{}_{HS}$ (i.e. the LIESST effect).

Figure 9.7 depicts the proposed mechanism for the observed photomagnetism in $Fe_2[Nb(CN)_8]$(4-pyridinealdoxime)$_8$·$2H_2O$. Prior to irradiation, a CN ligand connects paramagnetic Nb^{IV} and diamagnetic $Fe^{II}{}_{LS}$ of the LT phase in an alternating fashion. Consequently, paramagnetism is exhibited. Irradiating with 473-nm light causes a LIESST effect on the Fe^{II} site. In the photoinduced metastable HT phase, antiferromagnetic interactions occur due to magnetic spins on the photoproduced $Fe^{II}{}_{HS}$ ($S = 2$) and neighboring Nb^{IV} ($S = 1/2$), which are induced by a strong superexchange interaction through the CN ligand. This leads to spontaneous magnetization.

In $CsFe[Cr(CN)_6]$·$1.3H_2O$ [18] and $Fe_2[Nb(CN)_8]$·(3-pyCH$_2$OH)$_8$·$4.6H_2O$ [29], spin-crossover and ferromagnetism (or ferrimagnetism) coexist, but these assemblies do not exhibit light-induced spin-crossover ferromagnetism. $CsFe[Cr(CN)_6]$·

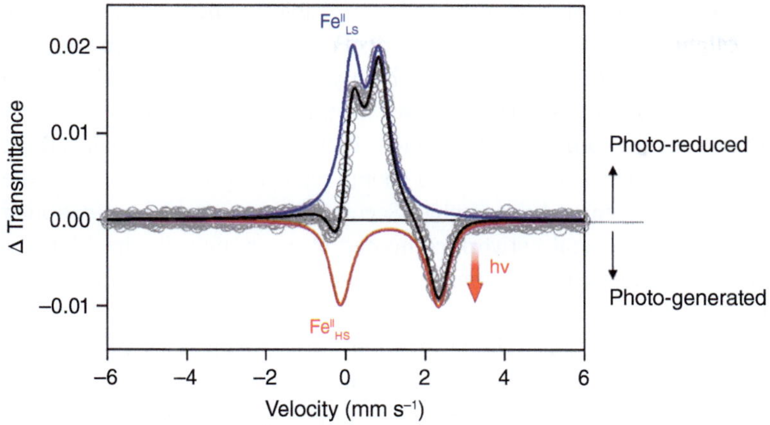

Figure 9.6 Light-irradiation effect on the ^{57}Fe Mössbauer spectra at 10 K (gray circles) of $Fe_2[Nb(CN)_8]$(4-pyridinealdoxime)$_8 \cdot 2H_2O$. Black line is the fit composed of Fe^{II}_{HS} (red line) and Fe^{II}_{LS} (blue line). Irradiation induces a transition from Fe^{II}_{LS} to Fe^{II}_{HS}, suggesting that the observed magnetization is due to a photoinduced spin-crossover from Fe^{II}_{LS} to Fe^{II}_{HS}. Source: Panels have been modified from Ref. [48]. Copyright 2011 Springer Nature.

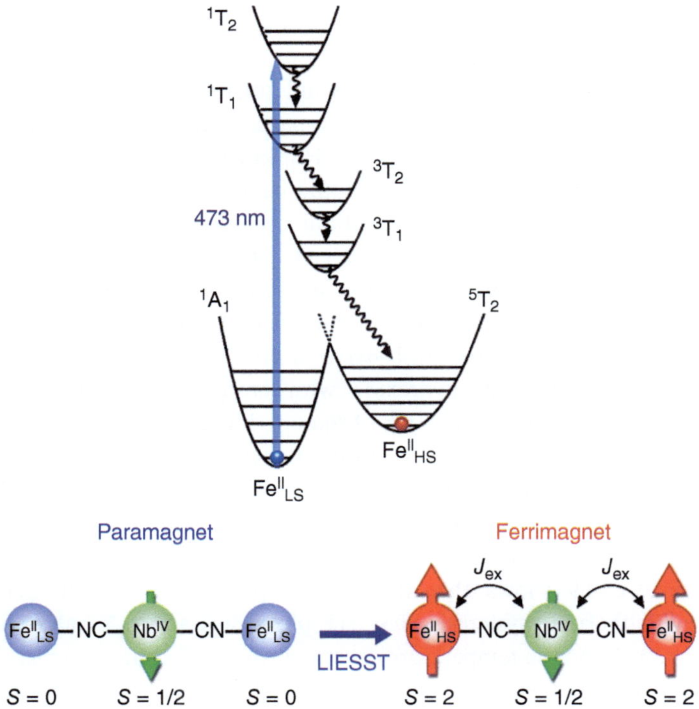

Figure 9.7 Proposed mechanism of light-induced spin-crossover ferromagnetism in $Fe_2[Nb(CN)_8]$(4-pyridinealdoxime)$_8 \cdot 2H_2O$. (Upper) Schematic of the light-induced spin-crossover of Fe^{II}, which is the light-induced excited spin-state trapping (LIESST) effect. (Lower) Schematic of the ferrimagnetic ordering between Nb^{IV} ($S = 1/2$) and Fe^{II}_{HS} ($S = 2$) due to a light-induced spin-crossover. Source: Panels have been modified from Ref. [48]. Copyright 2011 Springer Nature.

$1.3H_2O$ shows an intense MMCT absorption in the visible and near-infrared region. Consequently, light irradiation does not excite its $Fe^{II}{}_{LS}$ d–d transition. In contrast, $Fe_2[Nb(CN)_8]\cdot(3\text{-pyCH}_2OH)_8\cdot 4.6H_2O$ has a highly symmetric and stable crystal structure (cubic, $Ia\bar{3}d$) with an alternating arrangement of $\cdots$–HS–LS–$\cdots$. This arrangement inhibits the formation of the photo-induced HS domain. Hence, there are three factors to realize LIESST-induced spontaneous magnetization. First, the d–d transition of $Fe^{II}{}_{LS}$ must be excitable. Second, a strong exchange coupling must link a three-dimensional network composed of magnetic spin sites. Finally, the crystal structure must have a symmetry amenable to induce distortion in the photo-induced HS domain.

In summary, $Fe_2[Nb(CN)_8](\text{4-pyridinealdoxime})_8\cdot 2H_2O$ realizes a spin-crossover magnet with a 3D metal–organic framework, which can be controlled via light irradiation. The change in the magnetic properties is attributed to two factors: the LIESST effect on the Fe(II) centers in the metal–organic framework and the strong superexchange interactions of the magnetic centers in the 3D network.

9.4 Chiral Photomagnetism and Light-Controllable Second Harmonic Light in Iron Octacyanidoniobate Bimetal Assembly

Many studies aim to regulate the physical properties and functionalities of materials via external stimuli such as light [21–25, 33–52]. In particular, the LIESST effect can control photoswitching from the LS state to the HS state [21–25]. Additionally, photoinduced magnetization in a 3D network should be achieved by connecting spin crossover sites. Here we describe a 3D chiral cyanido-bridged bimetallic assembly, (+)-$Fe_2[Nb(CN)_8](\text{4-bromopyridine})_8\cdot 2H_2O$ [50]. It not only shows spin-crossover phenomenon and photoreversible spin-crossover magnetism but also photoreversible switching of the magnetization-induced SHG (MSHG) effect.

Adding a mixed aqueous solution of $FeCl_2\cdot 4H_2O$, 4-bromopyridine hydrochloride, ascorbic acid, and potassium hydroxide with an aqueous solution of $K_4[Nb(CN)_8]\cdot 2H_2O$ yielded (+)-$Fe_2[Nb(CN)_8](\text{4-bromopyridine})_8\cdot 2H_2O$ as a single crystal. Figure 9.8 shows that $Fe_2[Nb(CN)_8](\text{4-bromopyridine})_8\cdot 2H_2O$ is a 3D cyanido-bridged Fe–Nb bimetallic assembly with a chiral structure in the $I4_122$ space group ($a = 20.4607(9)$ Å and $c = 14.0173(6)$ Å).

$Fe_2[Nb(CN)_8](\text{4-bromopyridine})_8\cdot 2H_2O$ displays a temperature-induced spin-crossover effect. Figure 9.9 shows that the $\chi_M T$–T plot displays a thermal phase transition with thermal hysteresis. The HT phase has an $\chi_M T$ value of $7.30\,K\,cm^3\,mol^{-1}$ at 300 K. As the temperature decreases, the $\chi_M T$ value decreases. The LT phase has a value of $0.98\,K\,cm^3\,mol^{-1}$ at 10 K. This phase transition can be observed repeatedly. It has an HT → LT ($T_{1/2\downarrow}$) of 112 K and an LT → HT ($T_{1/2\uparrow}$) of 124 K. Next we measured the UV-vis absorption spectra. Bands appear at 430 and 560 nm as the temperature decreases. These are attributed to ${}^1A_1 \rightarrow {}^1T_2$ and ${}^1A_1 \rightarrow {}^1T_1$ transitions on the $Fe^{II}{}_{LS}$ site, respectively. These results indicate that spin-crossover from $Fe^{II}{}_{HS}$ ($S = 2$) to $Fe^{II}{}_{LS}$ ($S = 0$) is responsible for the HT to

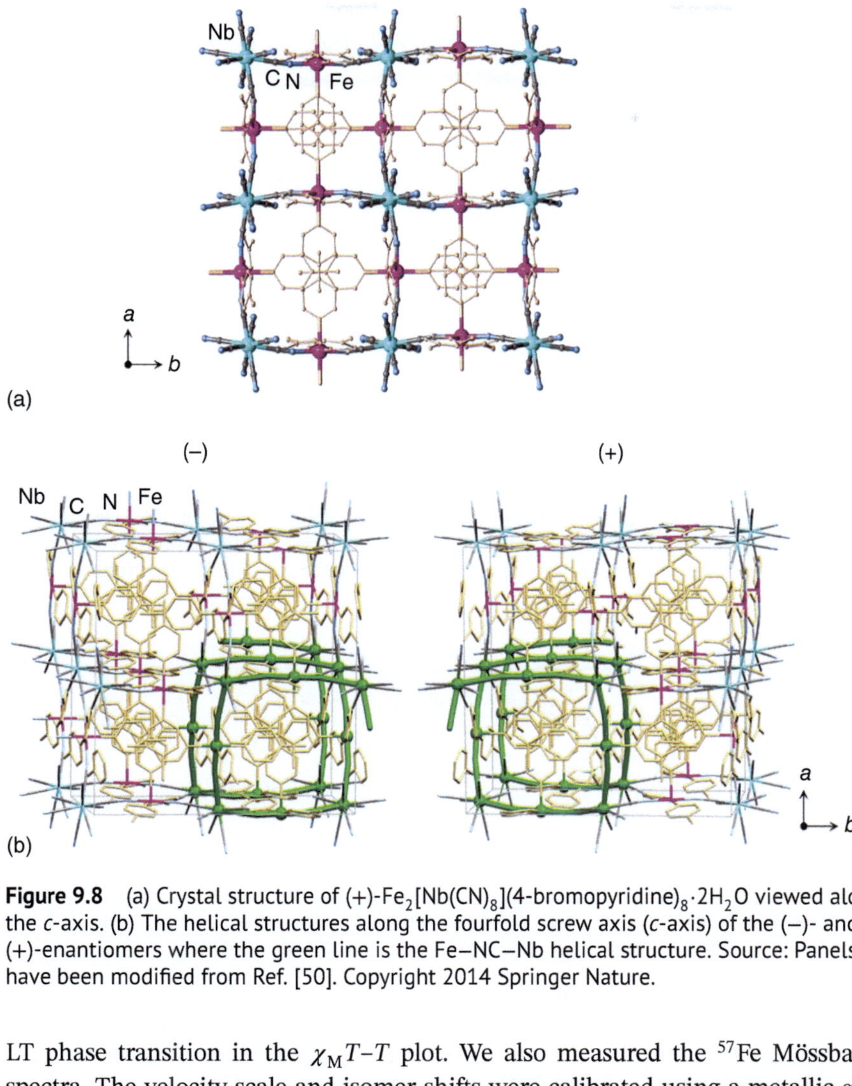

Figure 9.8 (a) Crystal structure of (+)-$Fe_2[Nb(CN)_8](4\text{-bromopyridine})_8 \cdot 2H_2O$ viewed along the *c*-axis. (b) The helical structures along the fourfold screw axis (*c*-axis) of the (−)- and (+)-enantiomers where the green line is the Fe—NC—Nb helical structure. Source: Panels have been modified from Ref. [50]. Copyright 2014 Springer Nature.

LT phase transition in the $\chi_M T$–T plot. We also measured the ^{57}Fe Mössbauer spectra. The velocity scale and isomer shifts were calibrated using a metallic α-Fe foil absorber at room temperature. Figure 9.10 shows the spectra for the HT and LT phases. The HT phase contains a doublet peak due to $Fe^{II}{}_{HS}$. The LT phase is attributed to the doublet peak due to $Fe^{II}{}_{LS}$ (Figure 9.10, Table 9.3). Hence, the HT and LT phases are assigned to $(Fe^{II}{}_{HS})_2[Nb^{IV}(CN)_8](4\text{-bromopyridine})_8 \cdot 2H_2O$ and $(Fe^{II}{}_{LS})_{1.74}(Fe^{II}{}_{HS})_{0.26}$ $[Nb^{IV}(CN)_8](4\text{-bromopyridine})_8 \cdot 2H_2O$, respectively, indicating that spin-crossover occurs on the Fe^{II} sites. Next, we measured the XRD pattern of the LT phase. From the Rietveld analysis, the LT phase has an orthorhombic crystal structure in the chiral *F*222 space group ($a' = 13.746(1)$ Å, $b' = 26.253(3)$ Å, $c' = 30.133(3)$ Å, $Z = 8$). This is a distorted crystal structure of the HT phase.

We then focused our attention to the spin-crossover induced SHG in $Fe_2[Nb(CN)_8](4\text{-bromopyridine})_8 \cdot 2H_2O$. We used a 1064-nm femtosecond laser as the incident light to measure SHG in a single crystal. Second harmonic (SH) light

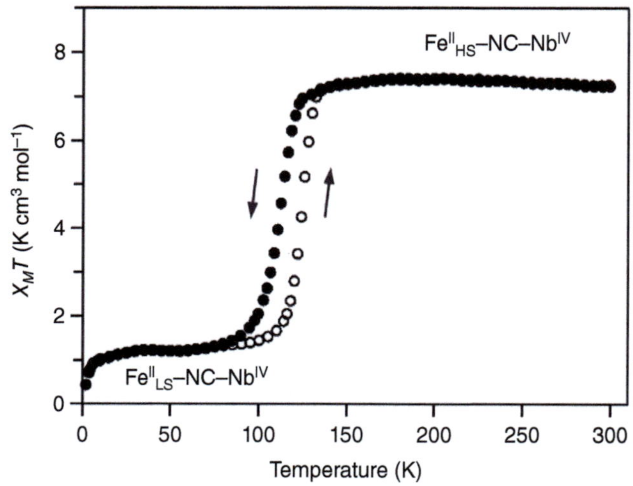

Figure 9.9 The $\chi_M T-T$ plots measured in an external magnetic field of 5000 Oe for $Fe_2[Nb(CN)_8](4\text{-bromopyridine})_8 \cdot 2H_2O$ where open and closed circles denote cooling and heating, respectively. Source: Panels have been modified from Ref. [50]. Copyright 2014 Springer Nature.

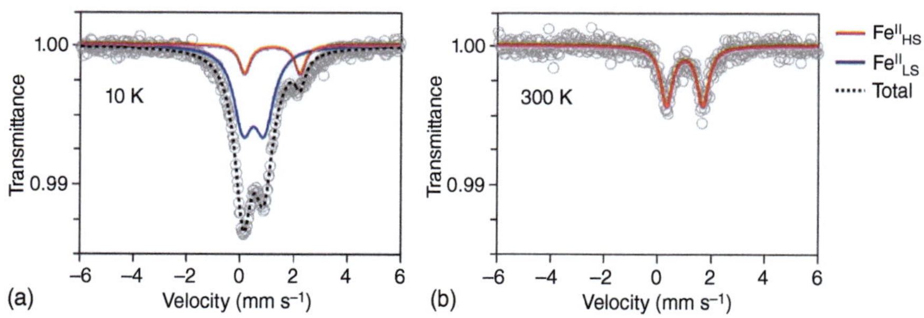

Figure 9.10 ^{57}Fe Mössbauer spectra of (a) the LT phase and (b) HT phase of (±)-$Fe_2[Nb(CN)_8](4\text{-bromopyridine})_8 \cdot 2H_2O$. HT phase has a doublet peak due to the Fe^{II}_{HS} ($\delta = 1.00(3)$ mm s^{-1}; $\Delta E_Q = 1.35(5)$ mm s^{-1}), whereas the LT phase has peaks due to Fe^{II}_{LS} (87%; $\delta = 0.49(1)$ mm s^{-1}; $\Delta E_Q = 0.77(1)$ mm s^{-1}) and the remaining Fe^{II}_{HS} (13%; $\delta = 1.18(2)$ mm s^{-1}; $\Delta E_Q = 2.09(3)$ mm s^{-1}). Source: Panels have been modified from Ref. [50]. Copyright 2014 Springer Nature.

Table 9.3 ^{57}Fe Mössbauer parameters of LT phase and HT phase of (±)-$Fe_2[Nb(CN)_8]$ $(4\text{-bromopyridine})_8 \cdot 2H_2O$.

	LT phase (10 K)			**HT phase (300 K)**		
	Area	**δ**	**ΔE_Q**	**Area**	**δ**	**ΔE_Q**
Fe^{II}_{HS}	13%	1.18(2)	2.09(3)	100%	1.00(3)	1.35(5)
Fe^{II}_{LS}	87%	0.49(1)	0.77(1)	—	—	—

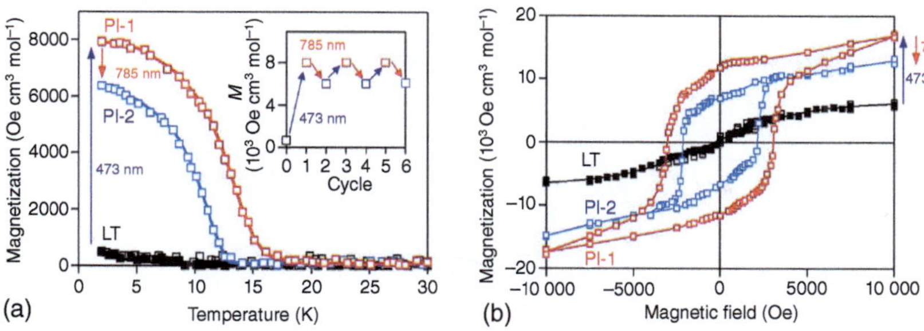

Figure 9.11 Photoinduced magnetization in $Fe_2[Nb(CN)_8](4\text{-bromopyridine})_8{\cdot}2H_2O$. (a) Magnetization vs. temperature curves at 100 Oe. (b) Magnetic hysteresis curves at 2 K. Black, red, and blue squares denote the measurements before irradiation, after irradiation with 473-nm light, and after irradiation with 785-nm light, respectively. Source: Panels have been modified from Ref. [50]. Copyright 2014 Springer Nature.

($\lambda = 532$ nm) is detected as *s*-polarized output SH light (*s*-out, where the analyzer angle of maximum SH intensity (θ_{max}) is 0°) with *p*-polarized fundamental input light (*p*-in). The SH intensity remains fairly constant between 290 and 140 K but abruptly increases at $T_{1/2}$. Thermal hysteresis of the SH intensity agrees well with the thermal hysteresis loop of the spin-crossover behavior in the $\chi_M T$–T plot.

We also investigated the photomagnetic effect. $Fe_2[Nb(CN)_8](4\text{-bromopyridine})_8{\cdot}2H_2O$ was irradiated with 473-nm laser light (17 mW cm^{-2}) at 2 K to excite the LT phase absorption band due to the $^1A_1 \rightarrow {}^1T_2$ transition of $Fe^{II}{}_{LS}$ at 430 nm. A large spontaneous magnetization is observed. Hereafter, this photoinduced phase is called PI-1 because $Fe_2[Nb(CN)_8](4\text{-bromopyridine})_8{\cdot}2H_2O$ has two photoinduced phases, as discussed later. Figure 9.11a plots the magnetization *M* vs. *T* curve, while Figure 9.11b shows the *M* vs. external magnetic field H_0. PI-1 phase has a T_C value of 15 K. A magnetic hysteresis loop with a H_c of 3 kOe appears at 2 K. The measured M_s at 5 T (7.6 μ_B) agrees with the theoretical M_s value (7.8 μ_B) predicted from the ferrimagnetic coupling between Nb^{IV} ($S = 1/2$) and the photoproduced $Fe^{II}{}_{HS}$ ($S = 2$) with g-factors of $g_{Nb}{}^{IV} = 2.0$ and $g_{Fe}{}^{II}{}_{HS} = 2.2$. Inset of Figure 9.11a shows reversible photoswitching between PI-1 and PI-2 by alternately irradiating with 473 and 785 nm light. Next, we measured the UV-vis spectra. Light irradiation reduces the absorption bands of the $^1A_1 \rightarrow {}^1T_2$ and $^1A_1 \rightarrow {}^1T_1$ transitions on $Fe^{II}{}_{LS}$. Additionally, light irradiation affects the Mössbauer spectra as irradiation increases $Fe^{II}{}_{HS}$ but decreases $Fe^{II}{}_{LS}$ (Figure 9.12). Consequently, the observed magnetization originates from a photoinduced spin-crossover from $Fe^{II}{}_{LS}$ to $Fe^{II}{}_{HS}$ (i.e. the LIESST effect).

We studied the XRD pattern. The PI-1 phase has a tetragonal crystal structure, which is the same space group as that of the HT phase. Thus, the PI-1 phase was irradiated with 785-nm laser light (20 mW cm^{-2}), which corresponds to the excitation of the d–d transition ($^5T_2 \rightarrow {}^5E$) of $Fe^{II}{}_{HS}$. PI-1 phase disappears upon irradiating with 785-nm light. The T_C value is reduced from 15 to 12 K. This is the second photoinduced phase, which is referred to as PI-2. After irradiating with

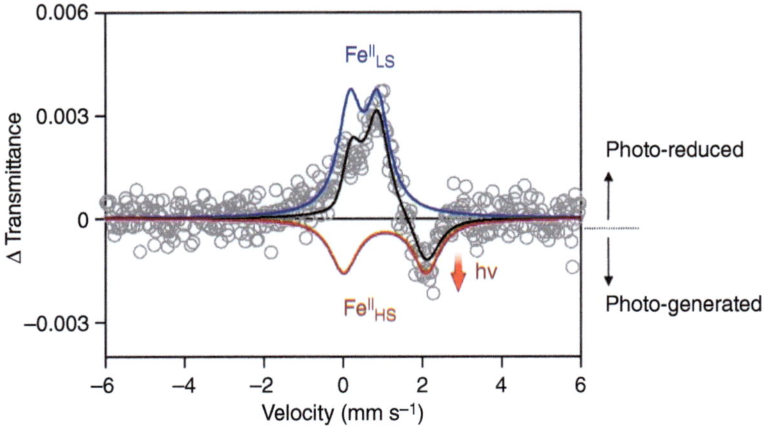

Figure 9.12 Light-irradiation effect on the ^{57}Fe Mössbauer spectra at 10 K (gray circles) of $Fe_2[Nb(CN)_8](4\text{-bromopyridine})_8\cdot 2H_2O$. Black lines are the fit composed of $Fe^{II}{}_{HS}$ (red line) and $Fe^{II}{}_{LS}$ (blue line). Irradiation induces a transition from $Fe^{II}{}_{LS}$ to $Fe^{II}{}_{HS}$, suggesting that the observed magnetization is due to a photoinduced spin-crossover from $Fe^{II}{}_{LS}$ to $Fe^{II}{}_{HS}$. Source: Panels have been modified from Ref. [50]. Copyright 2014 Springer Nature.

785-nm light, 38% of $Fe^{II}{}_{HS}$ in the UV-vis spectrum returns to the original $Fe^{II}{}_{LS}$. PI-2 has a formula of $(Fe^{II}{}_{LS})_{0.66}(Fe^{II}{}_{HS})_{1.34}[Nb^{IV}(CN)_8](4\text{-bromopyridine})_8\cdot 2H_2O$. These results are consistent with the calculated T_C value of the PI-2 phase based on the molecular field model. Although the XRD pattern shows that the PI-2 phase has an orthorhombic crystal structure in the *F*222 space group, it can be regarded as a tetragonal crystal (i.e. a pseudo-tetragonal structure).

Figure 9.13 depicts the proposed mechanism for the observed photoreversible magnetization. Similar to $Fe_2[Nb(CN)_8](4\text{-pyridinealdoxime})_8\cdot 2H_2O$, prior to irradiation a CN ligand connects paramagnetic Nb^{IV} and diamagnetic $Fe^{II}{}_{LS}$ of the LT phase in an alternating fashion. Irradiating with 473-nm light induces the PI-1 phase due to the LIESST effect. The magnetic spins on the photoproduced $Fe^{II}{}_{HS}$ ($S = 2$) and neighboring Nb^{IV} ($S = 1/2$) interact antiferromagnetically via a superexchange interaction through the CN ligand. This results in bulk magnetization due to ferrimagnetism. On the other hand, the magnetization decreases upon 785-nm light irradiation due to the reverse-LIESST effect from $Fe^{II}{}_{HS}$ to $Fe^{II}{}_{LS}$ caused by the photoexcitation of $^5T_2 \rightarrow {}^5E$ on $Fe^{II}{}_{HS}$, producing the PI-2 phase.

Finally, we investigated the optical switching effect on MSHG. First, the SHG for the LT phase of the paramagnetic state was measured prior to irradiation. Figure 9.14 plots the SH intensity as a function of the analyzer rotation angle (θ). Before irradiation, θ_{max} is 0° at $+H_0$ (+2.3 kOe along the sample z-axis). In the PI-1 phase, θ_{max} at $+H_0$ is $+88 \pm 3°$, whereas θ_{max} is $-86 \pm 4°$ at $-H_0$. The θ_{max} values in the PI-2 phase are restored to $+3 + 1°$ and $-3 + 1°$ at $+H_0$ and $-H_0$, respectively. Consequently, LIESST and reverse-LIESST effects control the polarization plane of the output SH light in $Fe_2[Nb(CN)_8](4\text{-bromopyridine})_8\cdot 2H_2O$.

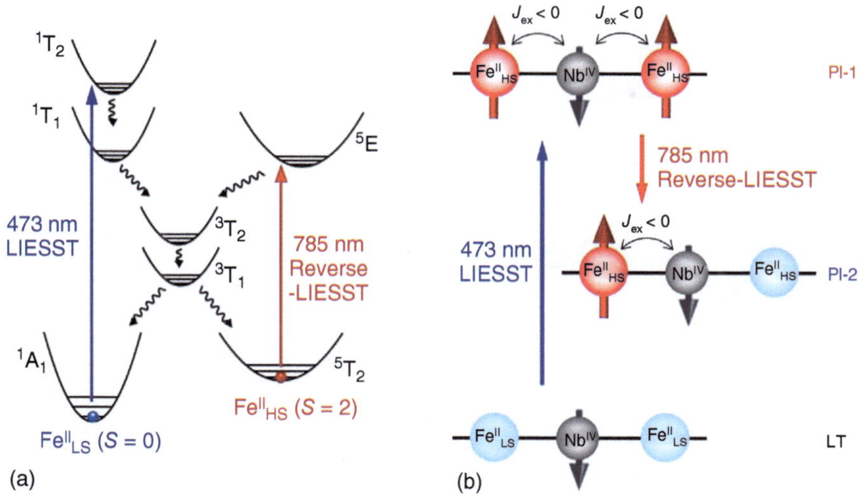

Figure 9.13 Proposed mechanism of light-reversible spin-crossover of Fe^{II} in (±)-$Fe_2[Nb(CN)_8]$(4-bromopyridine)$_8$·$2H_2O$. (a) Schematic of LIESST and reverse-LIESST effects. (b) Schematic of the mechanisms for the photoinduced magnetization between the paramagnetic LT phase and the ferrimagnetic PI-1 phase as well as for the photoreversible magnetization between the PI-1 and PI-2 phases. In the PI-1 phase, ferrimagnetic ordering occurs between Nb^{IV} ($S = 1/2$) and $Fe^{II}{}_{HS}$ ($S = 2$) due to LIESST. Decreased magnetization in the PI-2 phase is due to the $Fe^{II}{}_{LS}$ ($S = 0$) regenerated by reverse-LIESST. Source: Panels have been modified from Ref. [50]. Copyright 2014 Springer Nature.

Here a mechanism for the observed perpendicular switching of the polarization plane of SH light is proposed. Prior to irradiation, the sample is in the LT phase, which is a paramagnetic state. The χ_{yzx}^{cry} term is responsible for the SH polarization (s-out). In contrast, the magnetic space group is $I4_122$ under an external magnetic field H_0 ($H_0//z$) for the tetragonal PI-1 phase under the magnetized state ($T_C = 15$ K). The θ dependence of the SH intensity $I^{SH}(\theta)$ can be described as $I^{SH}(\theta) \propto \{\chi^{cry}\cos\theta + \chi^{mag}(M)\sin\theta\}^2$, where $\chi^{cry} = \chi_{yzx}^{cry}$ and $\chi^{mag}(M) = \left(2\chi_{xzx}^{mag}(M) + \chi_{zxx}^{mag}(M) + \chi_{zzz}^{mag}(M)\right)/2\sqrt{2}$, and the polarization planes for χ^{cry} and $\chi^{mag}(M)$ terms are s-out ($\theta_{max} = 0°$) and p-out ($\theta_{max} = 90°$), respectively. The crystal and magnetic terms are orthogonal. Consequently, the ratio of χ^{cry} and $\chi^{mag}(M)$ at the magnetized state below T_C determines the polarization plane of the output SH light. In the PI-1 phase, the MSHG rotation angle is enormous (almost 90°). Because $\chi_{PI-1}^{mag}(M)$ only contributes to the SH polarization, p-out SH light is radiated. On the other hand, in the PI-2 phase, the magnetic space group is $F222$, which is the same as that of the LT phase. Consequently, χ_{yzx}^{cry} becomes the dominant SHG susceptibility tensor element.

In summary, $Fe_2[Nb(CN)_8]$(4-bromopyridine)$_8$·$2H_2O$, a cyanido-bridged metal–organic framework, demonstrates a chiral optical-switching magnet. This chiral 3D spin-crossover assembly exhibits a spin-crossover-induced SHG, light-reversible spin-crossover long-range magnetic ordering, and photoswitching of MSHG.

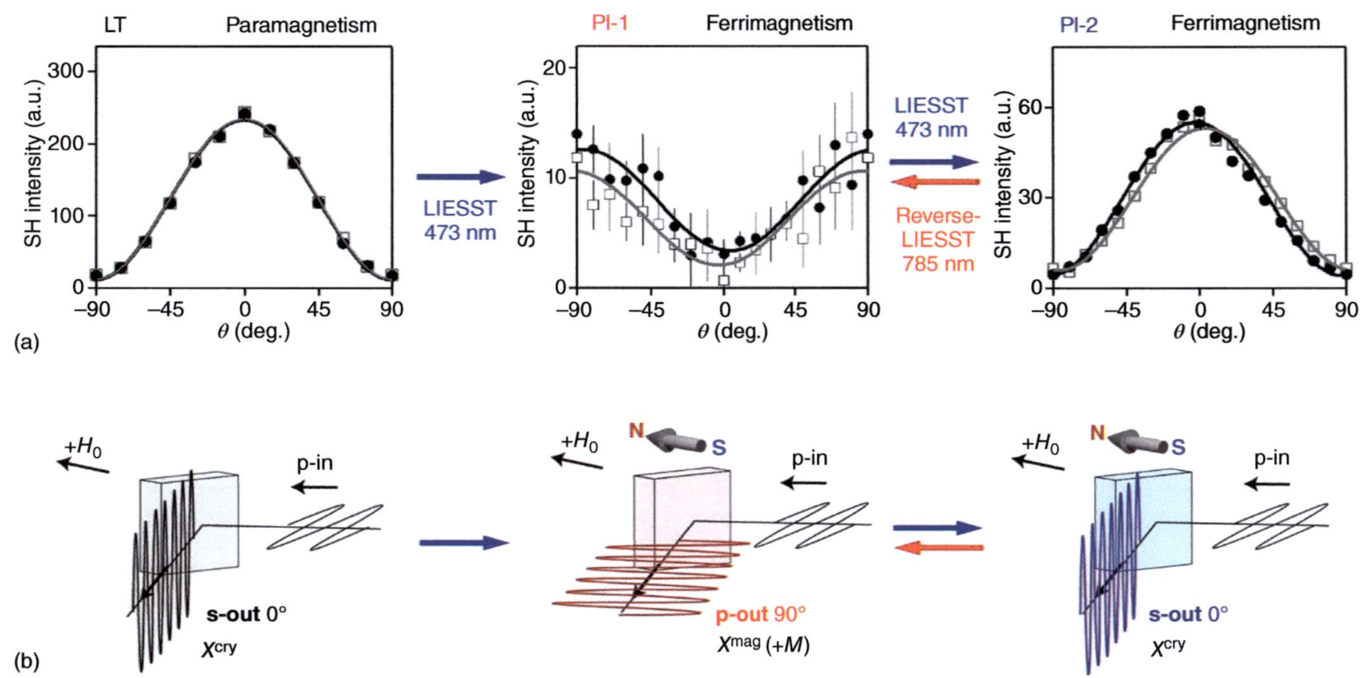

Figure 9.14 Optical switching of MSHG in (±)-$Fe_2[Nb(CN)_8]$(4-bromopyridine)$_8 \cdot 2H_2O$. (a) (left) SH intensity versus θ plot of the LT phase at $+H_0$ (closed circles) and $-H_0$ (open squares) at 2 K. (middle) SH intensity versus θ plot of the PI-1 phase, which is produced by 473-nm light irradiation (LIESST), at $+H_0$ (closed circles) and $-H_0$ (open squares) at 2 K. (right) SH intensity versus θ plot of the PI-2 phase, which is produced by 785-nm irradiation (Reverse-LIESST) at $+H_0$ (closed circles) and $-H_0$ (open squares) at 2 K. (b) Schematic illustrations of SH polarization plane of (left) $\theta_{max} = 0°$ (s-out) on the LT phase, (middle) $\theta_{max} = +88°$ (near p-out) on the PI-1 phase, and (right) $\theta_{max} = +3°$ (near s-out) on the PI-2 phase. Gray, red, and blue objects show the crystals of LT, PI-1, and PI-2 phases, respectively, and gray arrows represent the magnetic pole direction of the magnetized sample. SH polarization planes for χ^{cry} and $\chi^{mag}(M)$ terms are s-out ($\theta_{max} = 0°$) and p-out ($\theta_{max} = 90°$), respectively. Source: Panels have been modified from Ref. [50]. Copyright 2014 Springer Nature.

9.5 Conclusion and Perspective

Spin-crossover phenomena in cyano-bridged metal complexes are introduced. $CsFeCr(CN)_6 \cdot 1.3H_2O$ is the first Prussian blue analog to show a spin-crossover phenomenon. We also present $Fe_2[Nb(CN)_8]$(4-pyridinealdoxime), which is the first photomagnetic material showing spontaneous magnetization generated by a light-induced spin-crossover mechanism, and $Fe_2[Nb(CN)_8]$(4-bromopyridine), which is the first chiral photomagnetic material, exhibiting optical switching effect on SHG and MSHG. Currently, only these two compounds have been reported as a light-induced spin-crossover magnet. In these studies on functional magnetic complexes, Mössbauer spectroscopy plays a crucial role as it can identify the spin states before and after switching.

Many studies actively investigate stimuli-responsive materials such as changing the material properties by sensitively responding to external stimuli. In the future, these studies should realize materials with various characteristics. It is an extremely important work to develop stimuli-responsive materials with in-demand characteristics by utilizing academic knowledge and experience. It is a challenge both academically and practically for the future development of sensor and switching materials.

References

1 Gütlich, P. and Goodwin, H.A. (ed.) (2004), *Top. Curr. Chem.* **2004**). *Spin Crossover in Transition Metal Compounds I, II, III*, 233–235. Berlin: Springer.
2 Gütlich, P., Hauser, A., and Spiering, H. (1994). Thermal and optical switching of iron(II) complexes. *Angew. Chem. Int. Ed. Engl.* 33: 2024–2054.
3 Kahn, O. and Martinez, C. (1998). Spin-transition polymers: from molecular materials toward memory devices. *Science* 279: 44–48.
4 Verdaguer, M., Bleuzen, A., Marvaud, V. et al. (1999). Molecules to build solids: high TC molecule-based magnets by design and recent revival of cyano complexes chemistry. *Coord. Chem. Rev.* 190–192: 1023–1047.
5 Gütlich, P., Garcia, Y., and Goodwin, H.A. (2000). Spin crossover phenomena in Fe(II) complexes. *Chem. Soc. Rev.* 29: 419–427.
6 Gütlich, P., Garcia, Y., and Woike, T. (2001). Photoswitchable coordination compounds. *Coord. Chem. Rev.* 219: 839–879.
7 Ohkoshi, S. and Hashimoto, K. (2001). Photo-magnetic and magneto-optical effects of functionalized metal polycyanides. *J. Photochem. Photobiol. C* 2: 71–88.
8 Real, J.A., Gaspar, A.B., Niel, V., and Muñoz, M. (2003). Communication between iron(II) building blocks in cooperative spin transition phenomena. *Coord. Chem. Rev.* 236: 121–141.
9 Bousseksou, A., Molnar, G., Salmon, L., and Nicolazzi, W. (2011). Molecular spin crossover phenomenon: recent achievements and prospects. *Chem. Soc. Rev.* 40: 3313–3335.
10 Tokoro, H. and Ohkoshi, S. (2011). Novel magnetic functionalities of Prussian blue analogs. *Dalton Trans.* 40: 6825–6833.

11 Halcrow, M.A. (2011). Structure:function relathionships in molecular spin-crossover complexes. *Chem. Soc. Rev.* 40: 4119–4142.

12 Ohkoshi, S. and Tokoro, H. (2012). Photomagnetism in cyano-bridged bimetal assemblies. *Acc. Chem. Res.* 45: 1749–1758.

13 Collet, E., Lemée-Cailleau, M.-H., Cointe, M.B.-L. et al. (2003). Laser-induced ferroelectric structural order in a organic charge-transfer crystal. *Science* 300: 612–615.

14 Ohkoshi, S., Tsunobuchi, Y., Matsuda, T. et al. (2010). Synthesis of a metal oxide with a room-temperature photoreversible phase transition. *Nat. Chem.* 2: 539–545.

15 Létard, J.F., Guionneau, P., Codjovi, E. et al. (1997). Wide thermal hysteresis for the mononuclear spin-crossover compound cis-Bis(thiocyanato)bis [N-(2'-pyridylmethylene)-4-(phenylethynyl)anilino]iron(II). *J. Am. Chem. Soc.* 119: 10861–10862.

16 Boukheddaden, K., Shteto, I., Hoo, B., and Varret, F. (2000). Dynamical model for spin-crossover solids. I. Relaxation effects in the mean-field approach. *Phys. Rev. B* 62: 14796–14805.

17 Niel, V., Thompson, A.L., Muñoz, M.C. et al. (2003). Crystalline-state reaction with allosteric effect in spin-crossover, interpenetrated networks with magnetic and optical bistability. *Angew. Chem. Int. Ed.* 42: 3760–3763.

18 Kosaka, W., Nomura, K., Hashimoto, K., and Ohkoshi, S. (2005). Observation of an Fe(II) spin-crossover in a cesium iron hexacyanochromate. *J. Am. Chem. Soc.* 127: 8590–8591.

19 Ould-Moussa, N., Trzop, E., Mouri, S. et al. (2007). Wavelength selective light-induced magnetic effects in the binuclear spin crossover compound $\{[Fe(bt)(NCS)_2]_2(bpym)\}$. *Phys. Rev. B* 75: 054101.

20 van der Veen, R.M., Kwon, O.-H., Tissot, A. et al. (2013). Single-nanoparticle phase transitions visualized by four-dimensional electron microscopy. *Nat. Chem.* 5: 395–402.

21 Decurtins, S., Gütlich, P., Kohler, C. et al. (1984). Light-induced excited spin state trapping in a transition-metal complex: the hexa-1-propyltetrazole-iron (II) tetrafluoroborate spin-crossover system. *Chem. Phys. Lett.* 105: 1–4.

22 Létard, J.F., Guionneau, P., Rabardel, L. et al. (1998). Structural, magnetic, and photomagnetic studies of a mononuclear iron(II) derivative exhibiting an exceptionally abrupt spin transition. *Inorg. Chem.* 37: 4432–4441.

23 Garcia, Y., Kahn, O., Rabardel, L. et al. (1999). Two-step spin conversion for the three-dimensional compound Tris(4,4′-bis-1,2,4-triazole)iron(II) diperchlorate. *Inorg. Chem.* 38: 4663–4670.

24 Renz, F., Oshio, H., Ksenofontov, V. et al. (2000). Strong field iron(II) complex converted by light into a long-lived high-spin state. *Angew. Chem. Int. Ed.* 39: 3699–3700.

25 Breuning, E., Ruben, M., Lehn, J.-M. et al. (2000). Spin crossover in a supramolecular Fe_4II [2×2] grid triggered by temperature, pressure, and light. *Angew. Chem. Int. Ed.* 39: 2504–2507.

26 Ferlay, S., Mallah, T., Ouahès, R. et al. (1995). A room-temperature organometallic magnet based on Prussian blue. *Nature* 378: 701–703.

27 Holmes, S.M. and Girolami, G.S. (1999). Sol–gel synthesis of $KV^{II}[Cr^{III}(CN)_6]\cdot 2H_2O$: a crystalline molecule-based magnet with a magnetic ordering temperature above 100 °C. *J. Am. Chem. Soc.* 121: 5593–5594.

28 Hatlevik, Ø., Buschmann, W.E., Zhang, J. et al. (1999). Enhancement of the magnetic ordering temperature and air stability of a mixed valent vanadium hexacyanochromate(III) magnet to 99 °C (372 K). *Adv. Mater.* 11: 914–918.

29 Arai, M., Kosaka, W., Matsuda, T., and Ohkoshi, S. (2008). Observation of an iron(II) spin-crossover in an iron octacyanoniobate-based magnet. *Angew. Chem. Int. Ed.* 47: 6885–6887.

30 Margadonna, S., Prassides, K., and Fitch, A.N. (2004). Large lattice responses in a mixed-valence Prussian blue analogue owing to electronic and spin transitions induced by X-ray irradiation. *Angew. Chem. Int. Ed.* 43: 6316–6319.

31 Kaye, S.S. and Long, J.R. (2005). Hydrogen storage in the dehydrated Prussian blue analogues $M_3[Co(CN)_6]_2$ (M = Mn, Fe, Co, Ni, Cu, Zn). *J. Am. Chem. Soc.* 127: 6506–6507.

32 Coronado, E., Giménez-López, M.C., Levchenko, G. et al. (2005). Pressure-tuning of magnetism and linkage isomerism in iron(II) hexacyanochromate. *J. Am. Chem. Soc.* 127: 4580–4581.

33 Ohkoshi, S., Yorozu, S., Sato, O. et al. (1997). Photoinduced magnetic pole inversion in a ferro–ferrimagnet: $(Fe^{II}{}_{0.40}Mn^{II}{}_{0.60})_{1.5}Cr^{III}(CN)_6$. *Appl. Phys. Lett.* 70: 1040–1042.

34 Bleuzen, A., Lomenech, C., Escax, V. et al. (2000). Photoinduced ferrimagnetic systems in Prussian blue analogues $C^I{}_xCo_4[Fe(CN)_6]_y$ (C^I = alkali cation). 1. Conditions to observe the phenomenon. *J. Am. Chem. Soc.* 122: 6648–6652.

35 Ohkoshi, S. and Hashimoto, K. (1999). Design of a novel magnet exhibiting photoinduced magnetic pole inversion based on molecular field theory. *J. Am. Chem. Soc.* 121: 10591–10597.

36 Pejakovic, D.A., Manson, J.L., Miller, J.S., and Epstein, A.J. (2000). Photoinduced magnetism, dynamics, and cluster glass behavior of a molecule-based magnet. *Phys. Rev. Lett.* 85: 1994–1997.

37 Rombaut, G., Verelst, M., Golhen, S. et al. (2001). Structural and photomagnetic studies of two compounds in the system $Cu^{2+}/Mo(CN)_8{}^{4-}$: from trinuclear molecule to infinite network. *Inorg. Chem.* 40: 1151–1159.

38 Tokoro, H., Ohkoshi, S., and Hashimoto, K. (2003). One-shot-pulse-induced demagnetization in rubidium manganese hexacyanoferrate. *Appl. Phys. Lett.* 82: 1245–1247.

39 Sato, O., Hayami, S., Einaga, Y., and Gu, Z.-Z. (2003). Control of the magnetic and optical properties in molecular compounds by electrochemical, photochemical and chemical methods. *Bull. Chem. Soc. Jpn.* 76: 443–470.

40 Moore, J.G., Lochner, E.J., Ramsey, C. et al. (2003). Transparent, superparamagnetic $KCo[Fe^{III}(CN)_6]$–silica nanocomposites with tunable photomagnetism. *Angew. Chem. Int. Ed.* 42: 2741–2743.

41 Arimoto, Y., Ohkoshi, S., Zhong, Z.J. et al. (2003). Photoinduced magnetization in a two-dimensional cobalt octacyanotungstate. *J. Am. Chem. Soc.* 125: 9240–9241.

42 Herrera, J.M., Marvaud, V., Verdaguer, M. et al. (2004). Reversible photoinduced magnetic properties in the heptanuclear complex $[Mo^{IV}(CN)_2(CN\text{–}CuL)_6]^{8+}$: a photomagnetic high-spin molecule. *Angew. Chem. Int. Ed.* 43: 5468–5471.

43 Ohkoshi, S., Ikeda, S., Hozumi, T. et al. (2006). Photoinduced magnetization with a high Curie temperature and a large coercive field in a cyano-bridged cobalt–tungstate bimetallic assembly. *J. Am. Chem. Soc.* 128: 5320–5321.

44 Ohkoshi, S., Tokoro, H., Hozumi, T. et al. (2006). Photoinduced magnetization in copper octacyanomolybdate. *J. Am. Chem. Soc.* 128: 270–277.

45 Bleuzen, A., Marvaud, V., Mathoniere, C. et al. (2009). Photomagnetism in clusters and extended molecule-based magnets. *Inorg. Chem.* 48: 3453–3466.

46 Pajerowski, D.M., Andrus, M.J., Gardner, J.E. et al. (2010). Persistent photoinduced magnetism in heterostructures of Prussian blue analogues. *J. Am. Chem. Soc.* 132: 4058–4059.

47 Avendano, C., Hilfiger, M.G., Prosvirin, A. et al. (2010). Temperature and light induced bistability in a $Co_3[Os(CN)_6]_2$ $6H_2O$ Prussian blue analog. *J. Am. Chem. Soc.* 132: 13123–13125.

48 Ohkoshi, S., Imoto, K., Tsunobuchi, Y. et al. (2011). Light-induced spin-crossover magnet. *Nat. Chem.* 3: 564–569.

49 Koumousi, E.S., Jeon, I.-R., Gao, Q. et al. (2014). Metal-to-metal electron transfer in Co/Fe Prussian blue molecular analogues: the ultimate miniaturization. *J. Am. Chem. Soc.* 136: 15461–15464.

50 Ohkoshi, S., Takano, S., Imoto, K. et al. (2014). 90-degree optical switching of output second-harmonic light in chiral photomagnet. *Nat. Photon.* 8: 65–71.

51 Bertoni, R., Lorenc, M., Cailleau, H. et al. (2016). Elastically driven cooperative response of a molecular material impacted by a laser pulse. *Nat. Mater.* 15: 606–610.

52 Felts, A.C., Slimani, A., Cain, J.M. et al. (2018). Control of the speed of a light-induced spin transition through mesoscale core–shell architecture. *J. Am. Chem. Soc.* 140: 5814–5824.

53 Ohkoshi, S., Tokoro, H., Matsuda, T. et al. (2007). Coexistence of ferroelectricity and ferromagnetism in a rubidium manganese hexacyanoferrate. *Angew. Chem. Int. Ed.* 46: 3238–3241.

54 Ohkoshi, S., Arai, K., Sato, Y., and Hashimoto, K. (2004). Humidity-induced magnetization and magnetic pole inversion in a cyano-bridged metal assembly. *Nat. Mater.* 3: 857–861.

55 Milon, J., Daniel, M.-C., Kaiba, A. et al. (2007). Nanoporous magnets of chiral and racemic $[\{Mn(HL)\}_2Mn\{Mo(CN)_7\}_2]$ with switchable ordering temperatures (T_C = 85 K ↔ 106 K) driven by H_2O sorption (L = N,N-dimethylalaninol). *J. Am. Chem. Soc.* 129: 13872–13878.

56 Pinkowicz, D., Podgajny, R., Gaweł, B. et al. (2011). Double switching of a magnetic coordination framework through intraskeletal molecular rearrangement. *Angew. Chem. Int. Ed.* 50: 3973–3977.

57 Azzolina, G., Bertoni, R., Ecolivet, C. et al. (2020). Landau theory for non-symmetry-breaking electronic instability coupled to symmetry-breaking order parameter applied to Prussian blue analog. *Phys. Rev. B* 102: 134104.

58 Nishino, M., Enachescu, C., and Miyashita, S. (2019). Multistep spin-crossover transitions induced by the interplay between short- and long-range interactions with frustration on a triangular lattice. *Phys. Rev. B.* 100: 134414.

59 Enachescu, C., Nishino, M., Miyashita, S. et al. (2012). Monte Carlo Metropolis study of cluster evolution in spin-crossover solids within the framework of a mechanoelastic model. *Phys. Rev. B* 86: 054114.

60 Nishino, M., Enachescu, C., Miyashita, S. et al. (2011). Macroscopic nucleation phenomena in continuum media with long-range interactions. *Sci. Rep.* 1: 162.

61 Tokoro, H., Nakagawa, K., Imoto, K. et al. (2012). Zero thermal expansion fluid and oriented film based on a bistable metal-cyanide polymer. *Chem. Mater.* 24: 1324–1330.

62 Ohkoshi, S., Yoshikiyo, M., Namai, A. et al. (2017). Cesium ion detection by terahertz light. *Sci. Rep.* 7: 8088.

63 Tokoro, H., Namai, A., Yoshikiyo, M. et al. (2018). Theoretical prediction of a charge-transfer phase transition. *Sci. Rep.* 8: 63.

64 Gütlich, P., Bill, E., and Trautwein, A.X. (2011). *Mössbauer Spectroscopy and Transition Metal Chemistry: Fundamentals and Applications*. Berlin: Springer.

65 Sharma, V.K., Klingelhöfer, G., and Nishida, T. (2013). *Mössbauer Spectroscopy: Applications in Chemistry, Biology, and Nanotechnology*. New York: Wiley.

66 Long, G.J. (1984). *Mössbauer Spectroscopy Applied to Inorganic Chemistry*. New York: Plenum Press.

10

Developing a Methodology to Obtain New Photoswitchable Fe(II) Spin Crossover Complexes

Varun Kumar and Yann Garcia

Université catholique de Louvain, Institute of Condensed Matter and Nanosciences, Molecular Chemistry, Materials and Catalysis (IMCN/MOST), Place Louis Pasteur 1, 1348 Louvain-la-Neuve, Belgium

List of Abbreviations

SCO	spin crossover
LIESST	Light-Induced Excited Spin-State Trapping
LD-LISC	Ligand Driven Light Induced Spin Change
GD-LISC	Guest Driven Light Induced Spin Change
HS	high-spin
LS	low-spin
ptz	1-propyl-tetrazole
hyetrz	4-2′-hydroxyethyl-1,2,4-triazole
stpy	styrylpyridine
bpz	dihydrobis(1-pyrazolyl)borate
o-btphen	open form-5,6-bis(2,5-dimethyl-3-thienyl)-1,10-phenanthroline
o-L	open form-1,2-bis(2′-methyl-5′-(pyrid-4″-yl)thien-3′-yl) perfluorocyclopentene
psca	*para*-sulfocinnamic acid
dsta	4,4′-disulfonate-truxillic acid
DMPP	3,5-dimethyl-1-(2′-pyridyl) pyrazole
JMAK	Johnson–Mehl–Avrami–Kolmogorov
ps	phenyl sulfonate
tos	*p*-toluenesulfonate
nitps	*m*-nitrophenylsulfonate
NH_2trz	4-amino-1,2,4-triazole

10.1 Introduction and Context

The pursuit of new functional materials is a current need for the development of new technologies and smart devices. Stimuli-responsive materials can offer the

Mössbauer Spectroscopy: Applications in Chemistry and Materials Science, First Edition. Edited by Yann Garcia, Junhu Wang, and Tao Zhang.

possibilities to combine various functionalities in a single compound. Spin crossover (SCO) compounds belong to one such class of advanced materials [1, 2], which exhibit a remarkable change in magnetic and optical properties, and are thus lauded as potential candidates for various applications including data storage, display devices, and sensing applications [3–17]. The most common approach to induce SCO has been to use temperature as an external stimulus, but this is not very practical when imagining its use in commercial devices. Alternatively, using light irradiation as an external trigger to induce a spin state change would provide the possibility for faster switching as well as the possibility to control it remotely [18]. Till date, literature has reported two types of light-induced spin transition in iron coordination compounds depending on the molecular component targeted by light irradiation. Inducing spin transition through selective light irradiation of the metal center was called LIESST (Light-Induced Excited Spin-State Trapping), whereas selective light irradiation of ligands was termed LD-LISC (Ligand-Driven Light-Induced Spin Change). The LIESST effect, first observed in the solid state by Decurtins et al. on an Fe(II) complex, $[Fe(ptz)_6](BF_4)_2$ (ptz = 1-propyl-tetrazole) [19] and investigated later in detail by Hauser, Gütlich et al. pose a major limitation regarding its practical use given that the effect needs to be initiated at very low temperature (≤20 K) [20], irrespective of the thermal relaxation temperature. In this respect, the use of **^{57}Fe Mössbauer spectroscopy** is very useful since it allows to combine in a single experiment, cryogenic temperature as well as light irradiation, i.e. using a laser source. We present here the study of $[Fe(hyetrz)_3]$(4-chlorophenylsulfonate)$_2$ (hyetrz = 4-2′-hydroxyethyl-1,2,4-triazole), which allowed recording the LIESST effect in an Fe(II) 1D 1,2,4-triazole chain compound, by Mössbauer spectroscopy, for the first time [21]. This material reveals at 20 K, two quadrupole doublets: a major one with a small quadrupole splitting, corresponding to LS Fe(II) ions (91%), and a second one, with a larger quadrupole splitting, corresponding to HS Fe(II) ions (9%) (Figure 10.1). Upon green-light irradiation ($\lambda = 514.5$ nm), the population of the HS doublet grows (61%) at the expense of the LS one (39%), thus evidencing a LS to HS transition (i.e. LIESST effect). Upon warming to 70 K, a thermal relaxation from HS to LS occurred, and no dramatic change was observed in the Mössbauer spectrum compared to the one before irradiation at 20 K, showing that there is no photochemical decomposition of the sample. Thus, this example nicely demonstrates how useful can be Mössbauer spectroscopy to precisely detect a spin state change, after LIESST in an iron complex [22].

An alternative was presented by the LD-LISC effect [23], first observed in 1996 by Boillot et al. [24], which uses photo-responsive ligands to induce indirect spin transition in Fe(II) ions, although it has no restriction to be observed in other metal complexes. The principle behind the LD-LISC effect is to alter the ligand field strength around the metal center through photoisomerization of the ligand by light irradiation, which, in turn, results in a change in the spin state of the metal ions [25]. This methodology has been yet applied to *cis-trans* isomerization in stilbenoïd molecules [23–25], ring open-closure in diarylethene derivatives [28–30], and [2 + 2] cycloaddition [31, 32], as photochemical reactions. The first example of an LD-LISC effect was observed in a mononuclear neutral complex, $[Fe(stpy)_4(NCBPh_3)]$

Figure 10.1 ^{57}Fe Mössbauer spectra of $[Fe(hyetrz)_3](4\text{-chlorophenylsulfonate})_2$ at 20 K (top), at 20 K after λ = 514 nm (middle), and 70 K after thermal relaxation (bottom). Source: Ref. [21]/with permission from Springer Nature.

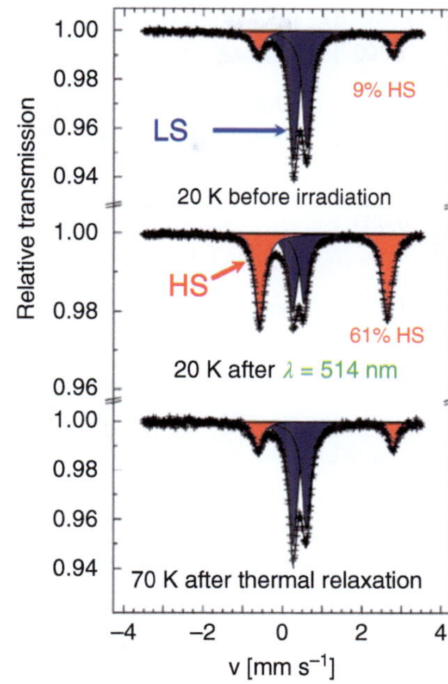

(stpy = styrylpyridine), dispersed in cellulose acetate film at 140 K [24]. It was later observed at room temperature and extended to Fe(III) SCO complexes [26]. The LD-LISC effect saw major improvements in recent years [27]. In an interesting study, Khusniyarov et al. were able to observe a partial LD-LISC effect in a solid state at room temperature in a diarylethene-based Fe(II) mononuclear complex viz. $[Fe(bpz)_2(o\text{-btphen})]$ (bpz = dihydrobis(1-pyrazolyl)borate; and *o*-btphen = open form-5,6-bis(2,5-dimethyl-3-thienyl)-1,10-phenanthroline) [28], which was similarly reported by Oshio [29]. However, no magnetic response could be observed. This result contrasts with the one observed by Garcia et al. who reported room temperature photomagnetism on the coordination polymer $[Fe(o\text{-L})_2(NCS)_2]\cdot 2CH_3OH$ with *o*-L = open form-1,2-bis(2′-methyl-5′-(pyrid-4″-yl)thien-3′-yl)perfluorocyclopentene [30]. Such phenomenon was observed by SQUID magnetometry as well as by ^{57}Fe Mössbauer spectroscopy at room temperature, on an enriched sample of $[^{57}Fe_{0.1}{}^{56}Fe_{0.9}(L)_2(NCS)_2]\cdot 2CH_3OH$, before and after irradiation. The spectrum reveals a HS quadrupole doublet with $\delta^{HS} = 0.99(1)$ mm.s^{-1} and $\Delta E_Q{}^{HS} = 2.73(1)$ mm.s^{-1} with a population of 70% (Figure 10.2). Such isomer shift matches one of the mononuclear coordination Fe(II) compounds, $[Fe(py)(NCS)_2]$ with a similar FeN_4S_2 core [33], with $\delta^{HS} = 1.00(4)$ mm.s^{-1}. A singlet of minor intensity (11%) is observed for LS Fe(II) ions with $\delta^{LS} = 0.086$ mm.s^{-1}. A central signal (19%), originating for HS Fe(III) ions, is also observed with $\delta = 0.62(1)$ mm.s^{-1} and $\Delta E_Q = 0.40$ mm.s^{-1}. The sample was irradiated at $\lambda = 365$ nm using a Hg lamp and an appropriate filter, and grounded in the course of irradiation to allow maximum exposure. This process induced a color change from deep orange to royal blue. The spectrum was then again recorded at room temperature (Figure 10.2). The ring closure of the diarylethene unit in *o*-L led to less distorted Fe(II) octahedra in the

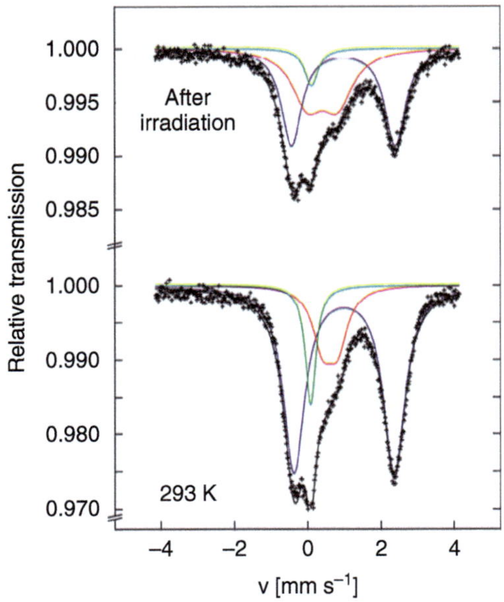

Figure 10.2 ^{57}Fe Mössbauer spectra of $[^{57}Fe_{0.1}{}^{56}Fe_{0.9}(L)_2(NCS)_2]\cdot 2CH_3OH)_2$ at 293 K (bottom) and at 293 K after $\lambda = 365$ nm (top). Source: Ref. [30]/with permission from Elsevier.

coordination polymer with $\Delta E_Q{}^{HS} = 2.82(1)$ mm.s^{-1}, but somewhat more distorted in the case of HS Fe(III) ions with $\Delta E_Q = 0.81(1)$ mm.s^{-1}. Interestingly, the Fe(III) ratio increased from 19% to 42%, upon UV irradiation, thus indicating a more rigid lattice, because the Debye temperature for HS Fe(III) ions is known to be higher compared to one of HS Fe(II) ions [34]. This is also translated by the presence of more diffraction peaks in the X-ray powder diffraction pattern, after irradiation. This is consistent with the photo-induced transformation from *o*-L to the closed form of *c*-L. This phenomenon was interpreted as a light-induced electron transfer involving three partners: Fe(II) and Fe(III) ions, as well as a ligand-radical L* [30].

In order to exploit the enticing properties of Fe(II) SCO compounds for application purposes, it is necessary that the spin switching be accompanied by both a color change and magnetic detection. Because LIESST and LD-LISC examples were limited so far [27, 35, 36], there is a current need for innovative approaches in order to observe efficient light-induced spin transitions.

In this respect, our laboratory has proposed a novel approach to address the spin state of Fe(II) complexes by light irradiation, using a photo-active component introduced at a remote position from the spin center. In this methodology, photo-responsive anionic organic molecules were incorporated into the crystal lattice of cationic Fe(II) complexes. Such incorporation of photo-responsive anions would not only introduce photo-responsive properties to the host complex in a controlled manner but also leads to the observation of a light-induced spin transition, through a long-range effect. We termed such phenomenon "Anion Driven Light-Induced Spin Change" (AD-LISC) as the spin transition would be induced by photo-responsive changes occurring in an anion. The idea behind this approach differs from the LD-LISC one, which is based on a local molecular effect, through a photo-induced change of ligand conformation. This effect can however be affected by steric hindrance due to the needed structural reorganization upon photoisomerization [20]. We expect SCO to ensue in synergy with photo-switching of non-coordinated anions because non-coordinated anions may modulate the ligand field strength and intermolecular lattice pressure, therefore causing a significant change in SCO characteristics such as $T_{1/2}$ (transition temperature where HS and LS are equally populated), different transition curves, etc. There are several reported cases which show that even subtle changes in non-coordinated anions can result in different SCO behavior [37–42]. The term "AD-LISC" was first tossed a few years ago as a concept by our research team [43]. Though the concept has been around for 10 years, it has only been observed experimentally [44]. This phenomenon offers more perspectives compared to the GD-LISC effect [45], for which the concentration of photo-responsive molecules is randomly included in the solid, limited to porous crystal lattices, and which has not yet been observed at room temperature.

10.2 Introduction to a New Photo-responsive Anion: *psca*

To realize AD-LISC at ambient conditions, we needed such a photo-responsive anion which could undergo photo-switching at room temperature in a solid state without requiring large spatial voids for photo-switching. In this context, we successfully designed and synthesized a photo-responsive anion, using a sulfonation reaction applied to *trans*-cinnamic acid to produce *para*-sulfocinnamic acid (***psca***). Two molecules of ***psca*** were found to undergo a [2 + 2] photodimerization from the vinyl bond in a solid state when irradiated with UV-light to give 4,4′-disulfonate-truxillic acid (***dsta***) as shown in Scheme 10.1 [46]. The [2 + 2] photodimerization of ***psca*** follows Schmidt's topochemical rules of [2 + 2] photodimerization which states that the photodimerization takes place provided: (i) the distance between the two vinyl bonds to be dimerized is $d \leq 4.2$ Å, and; (ii) the two vinyl bonds are parallel to each other in order to ensure a good overlap of their π-orbitals [47].

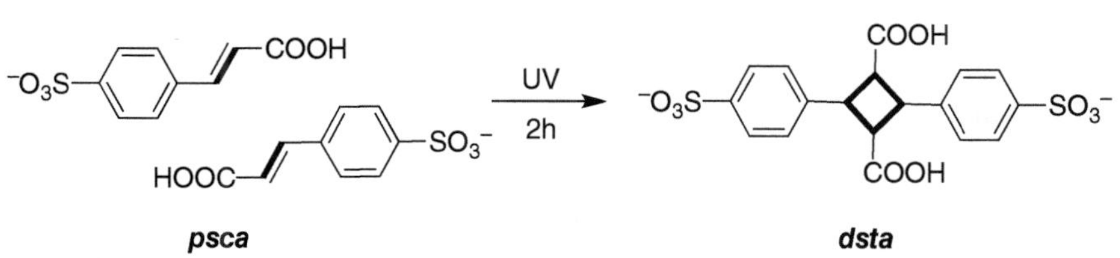

Scheme 10.1 [2 + 2] photodimerization of two molecules of ***psca***. The photodimerizable vinyl bonds and cyclobutane ring thus formed are emphasized in bold. Source: Ref. [46]/with permission from American Chemical Society.

The [2 + 2] photodimerization of ***psca*** is an irreversible reaction, which is an asset for a clear observation of the AD-LISC effect, in order to sustain the changes in properties followed by light irradiation, in contrast to reversible photoreactions where

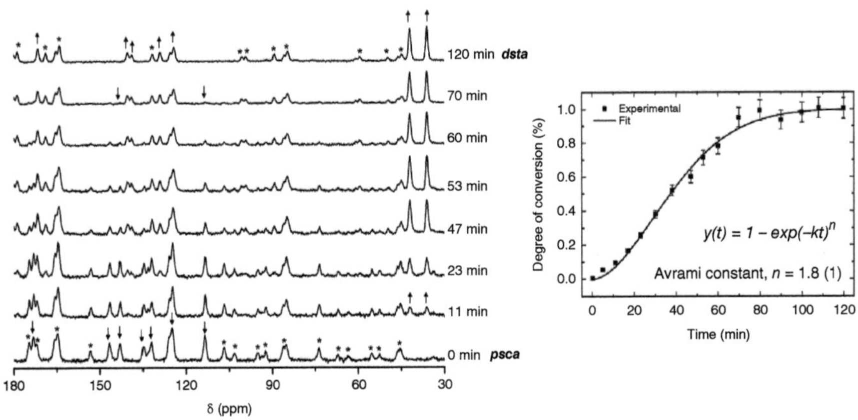

Figure 10.3 (a) Solid-state ^{13}C CP-MAS (cross-polarization magic angle spinning) NMR spectra of photoconversion from psca to dsta recorded at different irradiation times. The symbols (↓) and (↑) represent signals from psca and dsta respectively, whereas the asterisks (*) represents the spinning sidebands. The irradiation time is written beside each spectrum; (b) Degree of conversion plot with irradiation time to explain [2 + 2] photoconversion kinetics from psca to dsta. The curve is fitted according to the JMAK model of nucleation and growth. Source: Ref. [46]/with permission from American Chemical Society.

natural light usually interferes with the photoswitching of the molecules and makes it difficult to follow the changes in detail [30].

The detailed characterization of ***psca*** and ***dsta***, along with the kinetic study of the photo-conversion from ***psca*** to ***dsta*** was recently reported [46]. The time dependence of the photoreaction was monitored using ^{13}C solid-state NMR spectroscopy. NMR spectra of the photoreaction with respect to time are given in Figure 10.3(a). The degree of conversion and time plot which were used to explain the kinetics of the photoreaction are shown in Figure 10.3(b). The kinetics results were interpreted on the basis of Johnson–Mehl–Avrami–Kolmogorov (JMAK) model of nucleation and growth [48–52], which revealed that ***psca*** gets converted into ***dsta*** through a 1D heterogeneous growth.

10.3 Combining Fe(II) and *psca* Together in a Single Compound

After asserting the photo-properties of ***psca***, the next step toward AD-LISC observation was to combine ***psca*** and Fe(II) ions in a single coordination compound. It was successfully achieved by synthesizing a hexaquairon(II) complex, **$[Fe(H_2O)_6](psca)_2{\cdot}2H_2O$**, as a template, in which six water molecules are coordinated to Fe(II) ions and two ***psca*** molecules remain outside the coordination sphere as non-coordinated anions. **$[Fe(H_2O)_6](psca)_2{\cdot}2H_2O$** is an iron precursor which can ease the synthesis of a wide range of coordination complexes, by replacing H_2O ligand molecules with heterocyclic ligands. A new in situ synthesis procedure was thus devised to obtain hexaquairon(II) complex with aromatic sulfonate anions [46].

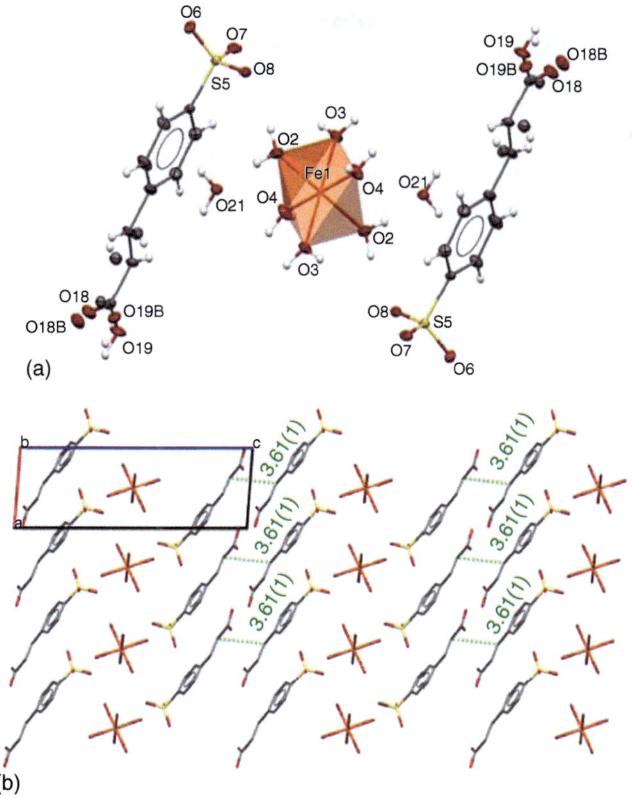

Figure 10.4 Crystal structure of $[Fe(H_2O)_6](psca)_2{\cdot}2H_2O$. (a) Formula unit, showing thermal displacement ellipsoids at the 30% probability level. The disorder of the vinyl and carboxylic group is shown by the corresponding atom color. (b) Crystal packing when viewed along the b-axis. Hydrogen atoms and solvent molecules are removed for clarity. Color code: Fe (orange), C (dark grey). O (red), S (yellow), and H (light grey). Source: Ref. [46]/ with permission from American Chemical Society.

The crystal structure of $\mathbf{[Fe(H_2O)_6](psca)_2{\cdot}2H_2O}$ shown in Figure 10.4 reveals that the distance between the concerned vinyl bonds to be dimerized is 3.6 Å, and these bonds are parallel to each other. Thus, ***psca*** molecules are packed in favorable arrangement for photodimerization reaction to take place according to Schmidt's topochemical rules. Hence, we proceeded to irradiate $\mathbf{[Fe(H_2O)_6](psca)_2}$ with UV light in solid state at room temperature and found that ***psca*** got partially dimerized in ***dsta***, as concluded from IR spectroscopy and powder X-ray diffraction [53].

To observe an eventual modification of the spin state distribution after UV irradiation, ^{57}Fe Mössbauer spectroscopy was used as it can efficiently give a compositional analysis of iron-based compounds, even at room temperature. ^{57}Fe Mössbauer spectra of $\mathbf{[Fe(H_2O)_6](psca)_2{\cdot}2H_2O}$, recorded at ambient conditions, before and after UV irradiation were fitted to the sum of Lorentzian lines by a least-squares refinement (Figure 10.5). A doublet (highlighted in red) corresponding to HS Fe(II) ions was distinguished, before and after UV irradiation, with identical parameters (Table 10.1). Another doublet (highlighted in yellow) and corresponding

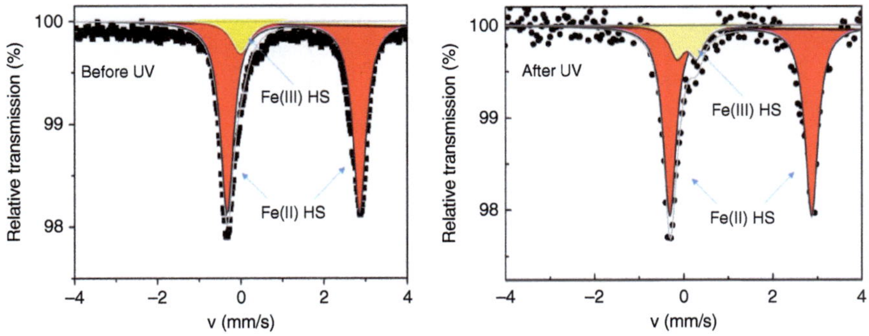

Figure 10.5 Room temperature ^{57}Fe Mössbauer spectra of $[Fe(H_2O)_6)](psca)_2{\cdot}2H_2O$ before irradiation (left) and after irradiation (right).

Table 10.1 ^{57}Fe Mössbauer parameters $[Fe(H_2O)_6](psca)_2{\cdot}2H_2O$ at room temperature before and after irradiation.

	Sites	δ (mm/s)	ΔE_Q (mm/s)	$\Gamma/2$ (mm/s)	Fraction (%)
Before irradiation	HS Fe(II)	1.26(2)	3.18(4)	0.20(1)	91
	HS Fe(III)	0.14(1)	0.43(4)	0.19(1)	9
After UV irradiation	HS Fe(II)	1.28(2)	3.18(3)	0.18(2)	85
	HS Fe(III)	0.07(1)	0.45(2)	0.20(2)	15

Isomer shifts (δ) are given with respect to metallic α-Fe. $\Gamma/2$ = half width of the lines [46].

to HS Fe(III) ions [54], with $\delta = 0.14$ mm/s and $\Delta E_Q = 0.43$ mm/s was also observed before irradiation. After UV irradiation, the Fe(III) HS content was found to increase to 15%, as a result of air oxidation of this material. Because of the weak ligand field strength set up by H_2O molecules in $\mathbf{[Fe(H_2O)_6](psca)_2{\cdot}2H_2O}$, UV irradiation did not have any effect on the spin population of this material, which remains in the HS state.

10.4 Fe(II) Mononuclear Complexes with DMPP and *psca* Ligands

After successfully synthesizing and studying $\mathbf{[Fe(H_2O)_6](psca)_2{\cdot}2H_2O}$, the next step toward AD-LISC was to replace all H_2O ligands in $\mathbf{[Fe(H_2O)_6](psca)_2{\cdot}2H_2O}$ by a heterocyclic nitrogen donor ligand of suitable ligand field strength to obtain an Fe(II) SCO compound. Following this, the final SCO compound would be irradiated with UV light to observe any changes in the magnetic and optical properties.

In this context, we chose a bidentate ligand, **DMPP** (3,5-dimethyl-1-(2′-pyridyl) pyrazole), which is known to give Fe(II) SCO mononuclear complexes having the coordination sphere $[Fe(DMPP)_3]^{2+}$ [55]. The Fe(II) mononuclear complexes with **DMPP** can be synthesized in solution, by mixing $\mathbf{[Fe(H_2O)_6](psca)_2}$ and

DMPP in a 1 : 3 molar ratio [56]. However, the homoleptic coordination complex **[Fe(DMPP)$_3$](psca)$_2$** was found to be unstable at ambient conditions and could not be isolated as a solid phase. To counter this problem, diethyl ether (Et_2O) was thought to be diffused into the complexation reaction to isolate **[Fe(DMPP)$_3$](psca)$_2$** in single crystalline or at the least in powder form for further investigation. Surprisingly, we ended up with single crystals of the heteroleptic complex **[Fe(DMPP)$_2$(psca)$_2$]**, in which two ***psca*** molecules were found to be coordinated to the Fe(II) ion along with two **DMPP** ligands. Upon investigating the heteroleptic formation further, we figured out that it was Et_2O which caused this heteroleptic coordination layout. The whole process of expected and obtained complexes is summed up in Scheme 10.2.

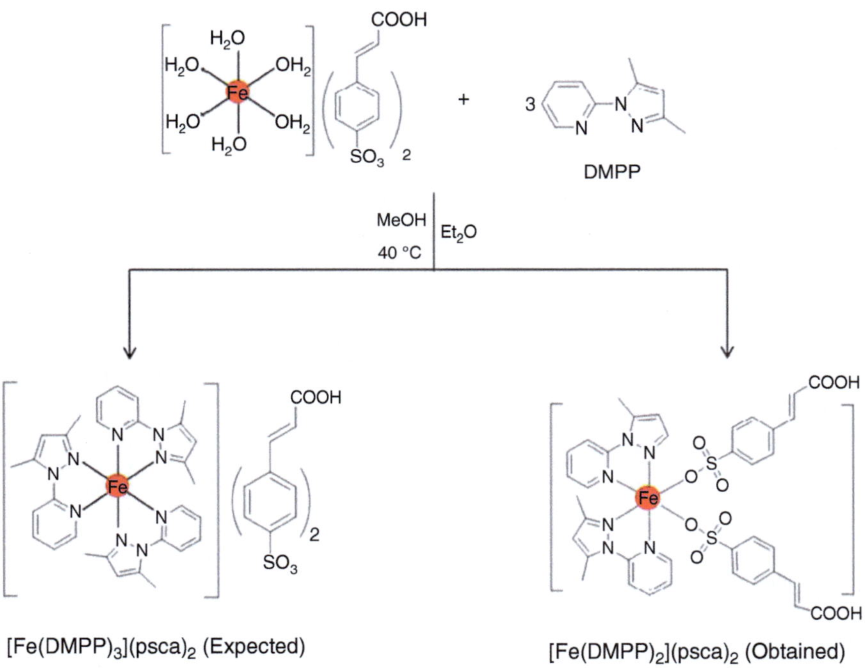

Scheme 10.2 Complexation reaction scheme of [Fe(H_2O)$_6$](psca)$_2$ with DMPP showing the expected and obtained coordination complexes. Source: Ref. [56]/with permission from Royal Society of Chemistry.

We further ascertained the role of Et_2O in dictating the heteroleptic coordination layout by replacing ***psca*** anion with other similar aromatic sulfonate anions, namely ***ps*** (phenyl sulfonate), ***tos*** (*p*-toluenesulfonate), and ***nitps*** (*m*-nitrophenylsulfonate) in the complexation reaction. Interestingly, all the complexation reactions developed exactly in the same manner and yielded heteroleptic coordination complexes, *viz.* **[Fe(DMPP)$_2$(ps)$_2$]**, **[Fe(DMPP)$_2$(tos)$_2$]**, and **[Fe(DMPP)$_2$(nitps)$_2$]** in which the sulfonate anions were found to coordinate to the Fe(II) ions along with two **DMPP** ligands. The crystal structure of **[Fe(DMPP)$_2$(psca)$_2$]** is shown in Figure 10.6(a), two ***psca*** molecules can be seen coordinated to Fe(II) centre through SO_3 groups.

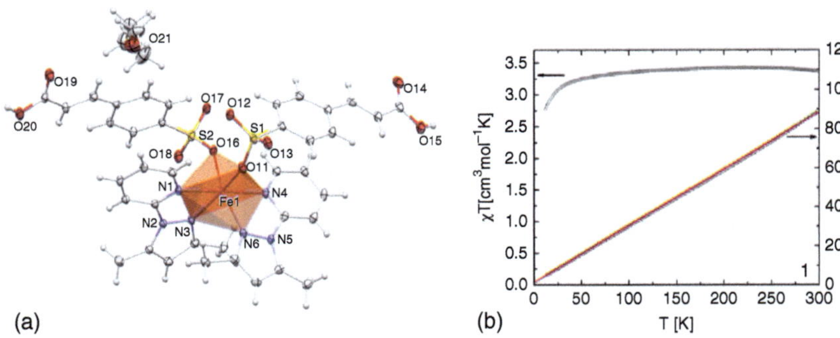

Figure 10.6 (a) Asymmetric unit of $[Fe(DMPP)_2(psca)_2]$, showing thermal displacement ellipsoids at the 30% probability level. Color code: Fe (orange), N (blue), C (grey), H (light grey), S (yellow), O (red); (b) χT vs. T, and χ^{-1} vs. T plots of $[Fe(DMPP)_2(psca)_2]$. The red line represents a Curie-Weiss law fit of the magnetic data. Source: Ref. [56]/with permission from Royal Society of Chemistry.

Magnetic susceptibility measurements revealed that all the complexes remained HS at all temperature ranges because the aromatic sulfonate anionic ligands contribute toward a weak ligand field strength (Figure 10.6b) [56].

^{57}Fe Mössbauer spectra of these complexes were found to be in accordance with susceptibility measurements (Figure 10.7). A quadrupole doublet (in red) with

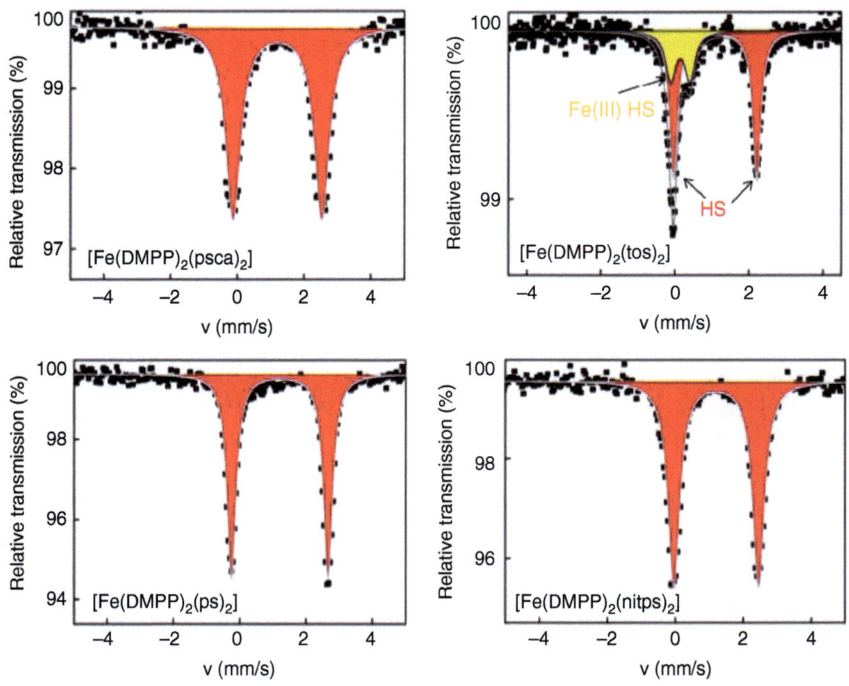

Figure 10.7 Room temperature ^{57}Fe Mössbauer spectra of **$[Fe(DMPP)_2(psca)_2]$**, **$[Fe(DMPP)_2(ps)_2]$**, **$[Fe(DMPP)_2(tos)_2]$**, and $[Fe(DMPP)_2(nitps)_2]$. Source: Ref. [56]/with permission from Royal Society of Chemistry.

Table 10.2 ^{57}Fe Mössbauer parameters of **[Fe(DMPP)$_2$(psca)$_2$]**, **[Fe(DMPP)$_2$(ps)$_2$]**, **[Fe(DMPP)$_2$(tos)$_2$]**, and **[Fe(DMPP)$_2$(nitps)$_2$]** at 298 K.

Compound	Sites	δ (mm/s)	ΔE_Q (mm/s)	$\Gamma/2$ (mm/s)	Fraction (%)
[Fe(DMPP)$_2$(psca)$_2$]	HS Fe(II)	1.20(1)	2.50(1)	0.21(1)	100
[Fe(DMPP)$_2$(ps)$_2$]	HS Fe(II)	1.20(1)	2.66(1)	0.27(1)	100
[Fe(DMPP)$_2$(tos)$_2$]	HS Fe(II)	1.09(1)	2.25(1)	0.13(1)	72
	HS Fe(III)	0.14(2)	0.51(5)	0.16(3)	28
[Fe(DMPP)$_2$(nitps)$_2$]	HS Fe(II)	1.21(1)	2.91(1)	0.16(5)	100

Isomer shifts (δ) are given with respect to metallic α-Fe. $\Gamma/2$ = half width of the lines.
Source: Adapted from Ref. [56].

δ = 1.1–1.2 mm/s and a large ΔE_Q = 2–3 mm/s, typical of HS Fe(II) ions is detected in all spectra. However, an additional yellow doublet was observed in the spectrum of **[Fe(DMPP)$_2$(tos)$_2$]** centered at δ = 0.14 mm/s with a quadrupole splitting of ΔE_Q = 0.51 mm/s, which corresponds to HS Fe(III) ions [30]. The absence of any HS Fe(III) in the SQUID measurement leads to the conclusion that **[Fe(DMPP)$_2$(tos)$_2$]** was synthesized as an Fe(II) HS compound but got partially oxidized to Fe(III) over a period of time needed to record the Mössbauer spectrum in air. The reason that only **[Fe(DMPP)$_2$(tos)$_2$]** got partially oxidized with time and not the other complexes could be due to the distortion degree of the coordination sphere of this complex. Indeed, quadrupole splitting can not only provide information about the spin state but also about the degree of distortion. When comparing ferrous compounds, the higher the distortion degree, the lower is the quadrupole splitting, due to the lattice contribution to the quadrupole splitting [57], which is indeed the case with **[Fe(DMPP)$_2$(tos)$_2$]** as can be seen in Table 10.2. Regarding a potential photodimerization in **[Fe(DMPP)$_2$(psca)$_2$]**, ***psca*** molecules were found to not meet Schmidt's topochemical criteria for [2 + 2] photodimerization [47] because the concerned vinyl bonds were found to be pointing away in opposite directions from each other (Figure 10.6a). Therefore, **[Fe(DMPP)$_2$(psca)$_2$]**, was acknowledged as a non-photoresponsive complex. However, on the positive side, this study allowed us to establish a new procedure to synthesize heteroleptic Fe(II) complexes with **DMPP** and aromatic sulfonate ligands [56].

10.5 1D Fe(II) Coordination Polymer with *psca* as Non-Coordinated Anions

After unsuccessful attempts on with **[Fe(DMPP)$_2$(psca)$_2$]**, we thought to incorporate ***psca*** in a more confined space where ***psca*** molecules could be close to each other for photodimerization. In this regard, we chose a 1D Fe(II) coordination polymer built from **NH$_2$trz** (4-amino-1,2,4-triazole). 1,2,4-triazole ligands are known to triply bridge Fe(II) ions to form coordination polymers which assemble in 1D chains [58–63]. In this regard, we synthesized the 1D chain, **$[Fe(NH_2trz)_3](psca)_2{\cdot}2H_2O$**, following the general complexation procedure of 1,2,4-triazole based ligands with Fe(II) salts [64]. The compound was synthesized as a white color powder but turned to pink color during the drying process by absorbing atmospheric water. This behavior is a common observation for Fe(II) coordination polymers with 1,2,4-triazole-based ligand and aromatic sulfonate anions [64, 65]. Due to the poor crystallinity of **$[Fe(NH_2trz)_3](psca)_2{\cdot}2H_2O$**, no information regarding ***psca*** arrangement for photodimerization could be obtained prior to irradiation. However, as in Fe(II) 1,2,4-triazole coordination polymers, non-coordinated anions occupy the positions in between 1D chains [66], we presumed that ***psca*** molecules in **$[Fe(NH_2trz)_3](psca)_2{\cdot}2H_2O$** were situated close to each other. Keeping this structural fact in mind, we irradiated with confidence **$[Fe(NH_2trz)_3](psca)_2{\cdot}2H_2O$** with UV light (Scheme 10.3). As a first positive sign of any structural change, the color of **$[Fe(NH_2trz)_3](psca)_2{\cdot}2H_2O$** switched from pink to brown after ~12 h of irradiation, as shown by the photographs in Figure 10.8(b).

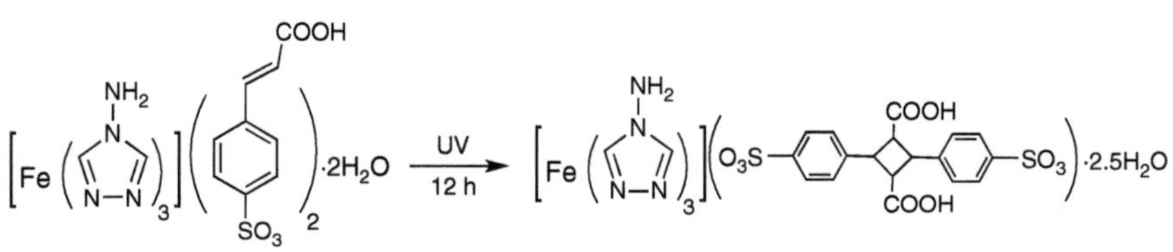

$[Fe(NH_2trz)_3](psca)_2{\cdot}2H_2O$

Scheme 10.3 [2 + 2] photodimerization reaction of **$[Fe(NH_2trz)_3](psca)_2{\cdot}2H_2O$** leading to **$[Fe(NH_2trz)_3](dsta)$**$\cdot 2.5H_2O$. Source: Ref. [44]/Royal Society of Chemistry.

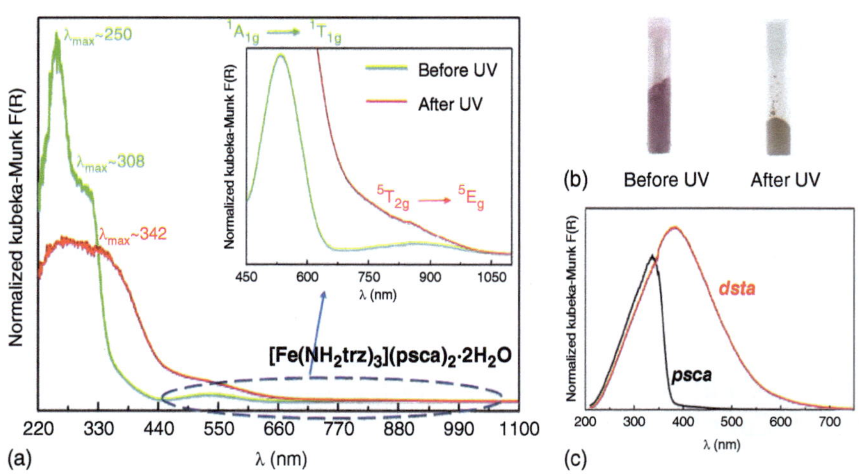

Figure 10.8 (a) Diffuse reflectance spectra of **$[Fe(NH_2trz)_3](psca)_2{\cdot}2H_2O$** before (in green) and after UV irradiation (in red); (b) Photographs of **$[Fe(NH_2trz)_3](psca)_2{\cdot}2H_2O$** taken before and after UV irradiation; (c) Bottom: Diffuse reflectance spectra of **psca** and **dsta.** Source: Ref. [44]/with permission from Royal Society of Chemistry.

The photodimerization reaction was followed using diffuse reflectance spectroscopy (DRS) at room temperature. Diffuse reflectance spectra of **[Fe(NH$_2$trz)$_3$](psca)$_2$·2H$_2$O** before (in green) and after irradiation (in red) are shown in Figure 10.8(a). Diffuse reflectance spectra of pristine ***psca*** and ***dsta*** are also given in Figure 10.8(c), for reference purpose. Before irradiation, a band is observed at $\lambda_{max} \sim 250$ nm, which corresponds to NH$_2$trz [67], accompanied by a shoulder at $\lambda_{max} \sim 308$ nm, corresponding to ***psca*** (Figure 10.8(c)). A closer view of the spectrum from 450 to 1050 nm is given in the inset plot of Figure 10.8(a). The absorption band at $\lambda_{max} \sim 530$ nm (pink color) is due to the $^1A_{1g} \rightarrow {}^1T_{1g}$ d–d transition of LS Fe(II) ions. Another band, much broader, lies in the near infra-red region at $\lambda_{max} \sim 870$ nm and is attributed to the $^5T_{2g} \rightarrow {}^5E_g$ d–d transition of HS Fe(II) ions. Thus, DRS suggests that in **[Fe(NH$_2$trz)$_3$](psca)$_2$·2H$_2$O**, a mixed spin state situation occur, with a major fraction of LS Fe(II) ions.

After UV irradiation, significant changes were observed (Figure 10.8(a)). The absorption band in the UV region became broader and the shoulder at $\lambda_{max} \sim 308$ nm was red-shifted to $\lambda_{max} \sim 342$ nm. Such a red shift is attributed to the formation of ***dsta*** (Figure 10.8(c)). Thus, DRS indicated that ***psca*** in **[Fe(NH$_2$trz)$_3$](psca)$_2$·2H$_2$O** underwent photodimerization to **[Fe(NH$_2$trz)$_3$](*dsta*)·2.5H$_2$O**, the number of water molecules being determined by thermogravimetry and CHN analysis. The intensity of the absorption band in the near infra-red region which is attributed to the $^5T_{2g} \rightarrow {}^5E_g$ d–d transition of HS Fe(II) ions increased after irradiation, which indicates that the HS state has become more populated. Whereas the absorption band in the visible region, which is assigned to the $^1A_{1g} \rightarrow {}^1T_{1g}$ d–d transition of LS Fe(II) ions, became broader. Two important developments were thus inferred by DRS as an effect of UV irradiation: (i) the [2+2] dimerization of ***psca*** was evident by the red shift of the absorption band ($\lambda_{max} \sim 308$ nm) in the UV region; (ii) an increase in the HS Fe(II) ion population was suggested by the increase in the intensity of the absorption band in the near infra-red region.

The change in spin state in **[Fe(NH$_2$trz)$_3$](psca)$_2$·2H$_2$O** upon UV irradiation was followed too by ^{57}Fe Mössbauer spectroscopy at room temperature (Figure 10.9). Before irradiation, a quadrupole doublet (in red) with $\delta = 1.10$ mms^{-1} and $\Delta E_Q = 3.03$ mms^{-1} was observed (Table 10.3). Such values fit well with those of the previously [Fe(NH$_2$trz)$_3$]$^{2+}$ 1D chains, e.g. for [Fe(NH$_2$trz)$_3$](2ns)$_2$·2H$_2$O (2ns = 2-napthalene sulfonate) with $\delta = 1.07$ mm.s^{-1} [65]. Another quadrupole doublet (in blue) was observed at $\delta = 0.4$ mms^{-1} with $\Delta E_Q = 0.16$ mms^{-1}, which corresponds to LS Fe(II) ions in [Fe(NH$_2$trz)$_3$]$^{2+}$. It matches the parameters recorded for [Fe(NH$_2$trz)$_3$](NO$_3$)$_2$ with $\delta = 0.4$ mms^{-1} and $\Delta E_Q = 0.19$ mm.s^{-1} [68]. Such a low quadrupole splitting is in agreement with a slight distortion of the octahedra which are constrained within a 1D chain [69]. The area fraction corresponding to HS and LS ions is 29/71% (Table 10.3). Similarly, after irradiation, the Mössbauer spectrum of **[Fe(NH$_2$trz)$_3$](*dsta*)·2.5H$_2$O** shows two doublets, highlighted in red ($\delta = 1.07$ mms^{-1}, $\Delta E_Q = 3.01$ mms^{-1}) and blue ($\delta = 0.43$ mms^{-1}, $\Delta E_Q = 0.19$ mms^{-1}) which correspond to HS and LS Fe(II) ions, respectively. The difference in isomer shift and quadrupole-splitting values before and after irradiation is negligible,

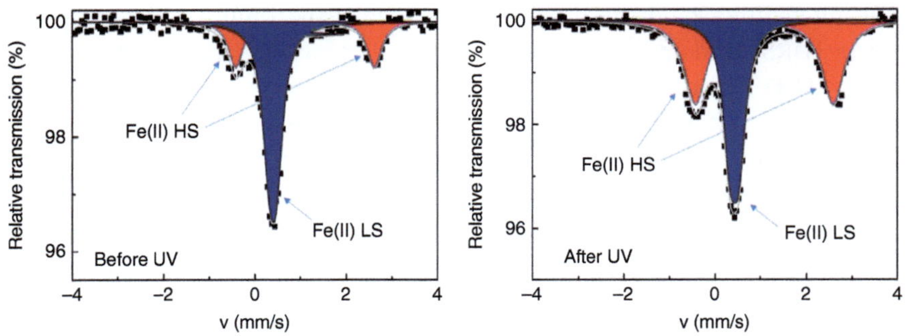

Figure 10.9 Room temperature ^{57}Fe Mössbauer spectra of **[Fe(NH$_2$trz)$_3$](psca)$_2$·2H$_2$O** before (left) and after UV irradiation leading to **[Fe(NH$_2$trz)$_3$](dsta)·2.5H$_2$O** (right). Source: Ref. [44]/with permission from Royal Society of Chemistry.

Table 10.3 ^{57}Fe Mössbauer parameters of **[Fe(NH$_2$trz)$_3$](psca)$_2$·2H$_2$O** before and after UV irradiation.

[Fe(NH$_2$trz)$_3$](psca)$_2$·2H$_2$O	**Sites**	δ **(mm/s)**	ΔE_Q **(mm/s)**	$\Gamma/2$ **(mm/s)**	**Fraction (%)**
Before irradiation	Fe(II) HS	1.10(4)	3.03(1)	0.19(4)	29
	Fe(II) LS	0.40(1)	0.16(1)	0.17(1)	71
After UV irradiation	Fe(II) HS	1.07(1)	3.01(2)	0.29(2)	53
	Fe(II) LS	0.43(1)	0.19(2)	0.18(1)	47

Isomer shifts (δ) are given with respect to metallic α-Fe. $\Gamma/2$ = half width of the lines.
Source: Adapted from Ref. [44].

which signifies that the FeN$_6$ octahedron geometry remains unperturbed upon light irradiation.

Most interestingly, after UV irradiation, the relative area fraction of the HS doublet increased from 29% to 53%. Such results fit well with DRS results which indicated an increase in the HS state, as well as with magnetic susceptibility measurements. To confirm such behavior, UV light irradiation was applied to [Fe(NH$_2$trz)$_3$](tos)$_2$·H$_2$O, which is known to present spin crossover behavior, associated with water release around the room temperature region [70]. No change was observed in the Mössbauer spectra recorded before and after light irradiation [44]. Thus, photodimerization of ***psca*** within **[Fe(NH$_2$trz)$_3$](psca)$_2$·2H$_2$O** caused a partial spin transition, thus indicating the first observation of an AD-LISC. Magnetic susceptibility measurements supported well ^{57}Fe Mössbauer spectroscopy results [44].

10.6 Conclusions and Perspectives

We have shown, in this chapter, how conditions for the first observation of the AD-LISC could be met following a stepwise approach. Though the light-induced spin transition was achieved by UV irradiation, it holds great significance as the transition occurred at room temperature and in the solid state. In this respect, the use of ^{57}Fe Mössbauer spectroscopy was crucial for the clear observation of the AD-LISC at room temperature. The developed methodology actually goes much far from the introduction of a [2+2] photodimerization in a molecular switch to control the spin state, as recently applied for the LD-LISC effect [31, 32], because it could be applied to any type of molecular switch, delocalized on the counteranion, for a given coordination complex or inorganic network toward the design of hybrid materials.

References

1 Gütlich, P., Garcia, Y., and Goodwin, H.A. (2000). Spin crossover phenomena in Fe(II) complexes. *Chem. Soc. Rev.* 29: 419–427.

2 Gütlich, P., Gaspar, A.B., and Garcia, Y. (2013). Spin state switching in iron coordination compounds. *Beilstein J. Org. Chem.* 9: 342–391.

3 Kahn, O., Kröber, J., and Jay, C. (1992). Spin transition molecular materials for displays and data recording. *Adv. Mater.* 4: 718–728.

4 Lefter, C., Davesne, V., Salmon, L. et al. (2016). Charge transport and electrical properties of spin crossover materials: towards nanoelectronic and spintronic devices. *Magnetochemistry* 2: 18–26.

5 Kumar, K.S. and Ruben, M. (2017). Emerging trends in spin crossover (SCO) based functional materials and devices. *Coord. Chem. Rev.* 346: 176–205.

6 Kipgen, L., Bernien, M., Tuczek, F., and Kuch, W. (2021). Spin-crossover molecules on surfaces: from isolated molecules to ultrathin films. *Adv. Mater.* 33: 2008141.

7 Molnar, G., Rat, S., Salmon, L. et al. (2018). Spin crossover nanomaterials: from fundamental concepts to devices. *Adv. Mater.* 30: n/a.

8 Linares, J., Codjovi, E., and Garcia, Y. (2012). Pressure and temperature spin crossover sensors with optical detection. *Sensors* 12: 4479–4492.

9 Jureschi, C.M., Linares, J., Rotaru, A. et al. (2015). Pressure sensor via optical detection based on a 1D spin transition coordination polymer. *Sensors* 15: 2388–2398.

10 Jureschi, C.M., Linares, J., Boulmaali, A. et al. (2016). Pressure and temperature sensors using two spin crossover materials. *Sensors* 16: 187.

11 Boukheddaden, K., Ritti, M.H., Bouchez, G. et al. (2018). Quantitative contact pressure sensor based on spin crossover mechanism for civil security applications. *J. Phys. Chem. C.* 122: 7597–7604.

12 Guo, Y., Xue, S., Dîrtu, M.M., and Garcia, Y. (2018). Versatile iron(II)-based colorimetric sensor for the vapor-phase detection of alcohols and toxic gases. *J. Mater. Chem. C* 6: 3895–3900.

13 Xue, S., Wang, L., Naik, A.D. et al. (2021). Iron(II) pillared-layer responsive frameworks via "kagomé dual" (kgd) supramolecular tessellations. *Inorg. Chem. Front.* 8: 3532–3546.

14 Sun, L., Rotaru, A., Robeyns, K., and Garcia, Y. (2021). A colorimetric sensor for the highly selective, ultra-sensitive and rapid detection of volatile organic compounds and hazardous gases. *Ind. Eng. Chem. Res.* 60: 8788–8798.

15 Li, W., Sun, L., and Garcia, Y. (2021). Subcomponent self-assembly of a Fe(II) based mononuclear complex for potential NH_3 sensing applications. *Hyperfine Interact.* 242: 7.

16 Li, W., Sun, L., Liu, C. et al. (2022). Supramolecular $Fe^{II}{}_4L_4$ cage for fast ammonia sensing. *J. Mater. Chem. C* 10: 9216–9221.

17 Sun, L., Rotaru, A., and Garcia, Y. (2022). A non-porous Fe(II) complex for the colorimetric detection of hazardous gases and the monitoring of meat freshness. *J. Hazardous Mater.* 437: 129364.

18 Gütlich, P. and Garcia, Y. (2016). Photomagnetism of molecular systems. In: *Reference Module in Materials Science and Materials Engineering* (ed. S. Hashmi), 1–5. Oxford: Elsevier.

19 Decurtins, S., Gütlich, P., Köhler, C.P. et al. (1984). Light-induced excited spin state trapping in a transition-metal complex: the hexa-1-propyltetrazole-iron(II) tetrafluoroborate spin-crossover system. *Chem. Phys. Lett.* 105: 1–4.

20 Gütlich, P., Garcia, Y., and Woike, T. (2001). Photoswitchable coordination compounds. *Coord. Chem. Rev.* 219–221: 839–879.

21 Garcia, Y., Ksenofontov, V., and Gütlich, P. (2002). Spin transition molecular materials: new sensors. *Hyperfine Interact.* 139/140: 543–551.

22 Garcia, Y., Renz, F., and Gütlich, P. (2016). LIESST effect in Fe(II) 1,2,4-triazole chains. *Curr. Inorg. Chem.* 6: 4.

23 Roux, C., Zarembowitch, J., Gallois, B. et al. (1994). Toward ligand-driven light-induced spin changing. Influence of the configuration of 4 styrylpyridine (stpy) on the magnetic properties of $Fe^{II}(stpy)_4(NCS)_2$ complexes. Crystal structures of the spin-crossover species Fe(*trans*-stpy)$_4$(NCS)$_2$ and of the high-spin species Fe(*cis*-stpy)$_4$(NCS)$_2$. *Inorg. Chem.* 33: 2273.

24 Boillot, M.-L., Roux, C., Audière, J.-P. et al. (1996). Ligand-driven light-induced spin change in transition-metal complexes: selection of an appropriate system and first evidence of the effect, in $[Fe^{II}(4\text{-styrylpyridine})_4(NCBPh_3)_2]$. *Inorg. Chem.* 35: 3975–3980.

25 Boillot, M.-L., Zarembowitch, J., and Sour, A. (2004). Ligand-driven light-induced spin change (LD-LISC): a promising photomagnetic effect. *Top. Curr. Chem.* 234: 261–276.

26 Sour, A., Boillot, M.L., Rivière, E., and Lesot, P. (1999). First evidence of a photoinduced spin change in an Fe^{III} complex using visible light at room temperature. *Eur. J. Inorg. Chem.* 2117–2119.

27 Brachňaková, B. and Šalitroš, I. (2018). Ligand-driven light-induced spin transition in spin crossover compounds. *Chem. Papers* 72: 773–798.

28 Rösner, B., Milek, M., Witt, A. et al. (2015). Reversible photoswitching of a spin-crossover molecular complex in the solid state at room temperature. *Angew. Chem. Int. Ed.* 54: 12976–12980.

29 Nihei, M., Suzuki, Y., Kimura, N. et al. (2013). Bidirectional photomagnetic conversions in a spin-crossover complex with a diarylethene moiety. *Chem. Eur. J.* 19: 6946–6949.

30 Garcia, Y., Ksenofontov, V., Lapouyade, R. et al. (2011). Synthesis and magnetic properties of an iron 1,2-bis thienyl perfluorocyclopentene photochromic coordination compound. *Opt. Mater.* 33: 942–948.

31 Wang, L.-F., Zhuang, W.-M., Huang, G.-Z. et al. (2019). *Chem. Sci.* 10: 7496–7502.

32 Fei Wang, L., Lv, B.-H., Wu, F.-T. et al. (2022). Reversible on-off switching of spin-crossover behavior via photochemical [2+2] cycloaddition reaction. *Sci. China Chem.* 65: 120–127.

33 Reiff, W.M., Frankel, R.B., Little, B.F., and Long, G.J. (1974). Mössbauer, electronic, and structural properties of bis(pyridine)dithiocyanatoiron(II). Ferromagnetically ordered polymer. *Inorg. Chem.* 13: 2153–2158.

34 Gibb, T.C. and Greenwood, N.N. (1971). *Mössbauer Spectroscopy*. London: Chapman & Hall.

35 Unruh, D., Homenya, P., Kumar, M. et al. (2016). Spin state switching of metal complexes by visible light or hard X-rays. *Dalton Trans.* 45: 14008–14018.

36 Chastanet, G., Lorenc, M., Bertoni, R., and Desplanches, C. (2018). Light-induced spin crossover—solution and solid-state processes. *C. R. Chimie* 21: 1075–1094.

37 Dîrtu, M.M., Rotaru, A., Gillard, D. et al. (2009). Prediction of the spin transition temperature in Fe^{II} one-dimensional coordination polymers: an anion based database. *Inorg. Chem.* 48: 7838–7852.

38 Zhao, X.-H., Zhang, S.L., Shao, D., and Wang, X.-Y. (2015). Spin crossover in $[Fe(2\text{-picolylamine})_3]^{2+}$ adjusted by organosulfonate anions. *Inorg. Chem.* 54: 7857–7867.

39 Klug, C.M., McDaniel, A.M., Fiedler, S.R. et al. (2012). Anion dependence in the spin-crossover properties of a Fe(II) podand complex. *Dalton Trans.* 41: 12577–12585.

40 Dîrtu, M.M., Naik, A.D., Rotaru, A. et al. (2016). Fe(II) spin transition materials including an amino-ester 1,2,4-triazole derivative, operating at, below and above room temperature. *Inorg. Chem.* 55: 4278–4295.

41 Shen, F.-X., Pi, Q., Dhi, L. et al. (2019). Spin crossover in hydrogen-bonded frameworks of Fe^{II} complexes with organodisulfonate anions. *Dalton Trans.* 48: 8815–8825.

42 Milin, E., Benaicha, B., El Hajj, F. et al. (2016). Magnetic bistability in macrocycle-based Fe^{II} spin-crossover complexes: counter ion and solvent effects. *Eur. J. Inorg. Chem.* 34: 5305–5314.

43 Robert, F., Naik, A.D., Tinant, B., and Garcia, Y. (2012). *N*-Salicylidene anil anions as thermo-sensitive components of organic–inorganic hybrid materials. *Inorg. Chim. Acta* 380: 104–113.

44 Kumar, V., Rotaru, A., and Garcia, Y. (2022). Room temperature light induced spin state switching change in a Fe(II) coordination polymer featuring a photo responsive anion. *J. Mater. Chem. C* 10: 14128–14134.

45 Xie, K.-P., Ruan, Z.-Y., Lyu, B.-H. et al. (2021). Guest-driven light-induced spin change in an azobenzene loaded metal–organic framework. *Angew. Chem. Int. Ed.* 60: 27144.

46 Kumar, V., Fusaro, L., Aprile, C. et al. (2020). [2+2] photodimerization of sulfonate derivative of *trans*-cinnamic acid: kinetics study using solid state ^{13}C NMR, and hybrid material inclusion. *Crystal Growth Des.* 20: 7850–7861.

47 Schmidt, G.M.J. (1971). Photodimerization in the solid state. *Pure Appl. Chem.* 27: 647–678.

48 Kolmogoroff, A. (1937). Zur Statistik der Kristallisationsvorgange in Metallen. *Izvestiya Rossiiskoi Akademii Nauk. Seriya Matematicheskaya* 1: 355–359.

49 Johnson, W.A. and Mehl, R.F. (1939). Reaction kinetics in processes of nucleation and growth. *Am. Inst. Mining. Metall. Eng. Inst. Met. Div. Tech. Publ.* 1089: 1–27.

50 Avrami, M. (1941). Granulation, phase change, and microstructure kinetics of phase change. III. *J. Chem. Phys.* 9: 177–184.

51 Avrami, M. (1940). Kinetics of phase change. II Transformation-time relations for random distribution of nuclei. *J. Chem. Phys.* 8: 212–224.

52 Avrami, M. (1939). Kinetics of phase change. I General theory. *J. Chem. Phys.* 7: 1103–1112.

53 Kumar, V. and Garcia, Y. (2021). Anion driven light induced spin crossover systems: towards multifunctional coordination compounds. *Möss. Effect Data Ref. J.* 44: 135–143.

54 Robert, F., Naik, A.D., and Garcia, Y. (2010). Influence of a thermochromic anion on the spin crossover of iron(II) trinuclear complexes probed by Mössbauer spectroscopy. *J. Phys. Conf. Ser.* 217: 012031.

55 Baker, A.T., Ferguson, N.J., Goodwin, H.A., and Rae, A.D. (1989). Structural, magnetic and spectral properties of iron(II) and nickel(II) complexes of N^1-(pyridin-2-yl)-3,5-dimethylpyrazole. *Aust. J. Chem.* 42: 623–638.

56 Kumar, V., El-Massaoudi, M., Radi, S. et al. (2020). Iron(II) coordination pyrazole complexes with aromatic sulfonate ligands: the role of ether. *New J. Chem.* 44: 13902–13912.

57 Ingalls, R. (1962). Variational calculation of the Sternheimer factors for the ferrous ion. *Phys. Rev.* 128: 1155–1158.

58 Michalowicz, A., Moscovici, J., Ducourant, B. et al. (1995). EXAFS and X-ray powder diffraction studies of the spin transition molecular materials $[Fe(Htrz)_2(trz)]BF_4$ and $[Fe(Htrz)_3](BF_4)_2 \cdot H_2O$ (Htrz = 1,2,4-4H-triazole; trz = 1,2,4-triazolato). *Chem. Mater.* 7: 1833–1842.

59 Michalowicz, A., Moscovici, J., Garcia, Y., and Kahn, O. (1999). Polymeric spin transition compounds: EXAFS and thermal behaviour. *J. Synchr. Rad.* 6: 231–232.

60 Garcia, Y., Moscovici, J., Michalowicz, A. et al. (2002). A spin transition molecular material with a wide bistability domain. *Chem. Eur. J.* 8: 4992–5000.

61 Seredyuk, M., Gaspar, A.B., Ksenofontov, V. et al. (2008). One-dimensional iron(II) compounds exhibiting spin crossover and liquid crystalline properties in the room temperature region. *Inorg. Chem.* 47: 10232–10245.

62 Grosjean, A., Daro, N., Kauffmann, B. et al. (2011). The 1D polymeric structure of the $[Fe(NH_2trz)_3](NO_3)_2 \cdot nH_2O$ (with n = 2) spin crossover compound proven by single crystal investigations. *Chem. Commun.* 47: 12382–12384.

63 Grosjean, A., Négrier, P., Bordet, P. et al. (2013). Crystal structures and spin crossover in the polymeric material $[Fe(Htrz)_2(trz)](BF_4)$ including coherent-domain size reduction effects. *Eur. J. Inorg. Chem.* 796–802.

64 Garcia, Y., van Koningsbruggen, P.J., Lapouyade, R. et al. (1998). Influences of temperature, pressure, and lattice solvents on the spin transition regime of the polymeric compound $[Fe(hyetrz)_3]A_2 \cdot 3H_2O$ (hyetrz = 4-(2′-hydroxyethyl)-1,2,4-triazole and A^- = 3-nitrophenylsulfonate). *Chem. Mater.* 10: 2426–2433.

65 van Koningsbruggen, P.J., Garcia, Y., Codjovi, E. et al. (1997). Non-classical Fe^{II} spin-crossover behaviour in polymeric iron(II) compounds of formula $[Fe(NH_2trz)_3]X_2 \cdot xH_2O$ (NH_2trz = 4-amino-1,2,4-triazole; X = derivatives of naphthalene sulfonate). *J. Mater. Chem.* 7: 2069–2075.

66 Garcia, Y., Niel, V., Muñoz, M.C., and Real, J.A. (2004). Spin crossover in 1D, 2D and 3D polymeric Fe(II) networks. *Top. Curr. Chem.* 233: 229–257.

67 Lynch, A.M., Robinson, S.A., Wilcox, P. et al. (2008). Cycloheximide and disulfoton are positive in the photoclastogencity assay but do not absorb UV irradiation: another example of pseudophotoclastogenicity? *Mutagenesis* 23: 111–118.

68 Dîrtu, M.M., Neuhausen, C., Naik, A.D. et al. (2010). Insights into the origin of cooperative effects in the spin transition of $[Fe(NH_2trz)_3](NO_3)_2$: the role of supramolecular interactions evidenced in the crystal structure of $[Cu(NH_2trz)_3](NO_3)_2 \cdot H_2O$. *Inorg. Chem.* 49: 5723–5736.

69 Dîrtu, M.M., Garcia, Y., Nica, M. et al. (2007). Iron(II) spin transition 1,2,4-triazole chain compounds with novel inorganic fluorinated counteranions. *Polyhedron* 26: 2259–2263.

70 Codjovi, E., Sommier, L., Kahn, O., and Jay, C. (1996). A spin transition molecular material with an exceptionally large thermal hysteresis loop at room temperature. *New. J. Chem.* 20: 503–505.

11

^{57}Fe Mössbauer Spectroscopy as a Prime Tool to Explore a New Family of Colorimetric Sensors

Li Sun, Weiyang Li, and Yann Garcia

Institute of Condensed Matter and Nanosciences, Molecular Chemistry, Materials and Catalysis (IMCN/MOST), Université catholique de Louvain, Place Louis Pasteur 1, 1348 Louvain-la-Neuve, Belgium

List of Abbreviations

HGs	hazardous gases
HS	high-spin
LS	low-spin
MOFs	metal organic frameworks
PCPs	porous coordination polymers
SCO	spin crossover
VOCs	volatile organic compounds

11.1 Introduction and General Context

In recent years, with the continuous improvement of the quality of life, the demand for a healthier environment has been increasing, thus making air-quality monitoring a constant concern [1]. The monitoring can also be performed by applying **^{57}Fe Mössbauer spectroscopy** to collected dusts from roads and polluted environments [2–4]. The technique can be applied even to amorphous dust which cannot be distinguished by X-ray powder diffraction [5]. Gas pollution, however, not only originates from factories and car exhaust emissions but also from many indoor decorative items (e.g. home furniture and coatings), which can also emit a large number of harmful gases [6]. Toxic gases are dangerous to human life and health, leading to skin redness, infection, and, in more serious cases, possible cancer development [7]. Therefore, in order to live in a healthier environment, it is important to accelerate the development of methods to detect and control toxic and harmful gases. Traditional detection methods include gas chromatography [8], gas chromatography-mass spectrometry [9], and fluorescence detection method [10, 11], to name a few techniques etc. [8, 9] However, these methods rely on expensive and unportable large equipment, so that they cannot

Mössbauer Spectroscopy: Applications in Chemistry and Materials Science,
First Edition. Edited by Yann Garcia, Junhu Wang, and Tao Zhang.

achieve *in-situ* and *real-time* detection. An alternative has been proposed through electrochemical methods, which are known to be relatively low-cost and portable, less time-consuming, and used for the detection of liquid phase components. Part of electrochemical sensors require however energy input [12] and therefore cannot be considered green sensors [12]. Therefore, the preparation of sensors for detecting toxic gases with the advantages of easy portability, fast response time, high selectivity and sensitivity, low cost and energy less is an urgent problem to be solved.

Research efforts have been directed towards **colorimetric sensors**, due to their simple and rapid detection, intuitive and accurate results and excellent detection performance. Generally speaking, a colorimetric sensor is a method of determining the content of a component to be measured by comparing or measuring the color depth of the colored substance based on the color reaction of the generated colored compound [13, 14]. One approach to achieve this kind of material is through the construction of porous coordination polymers (PCPs) [15] (or metal-organic frameworks, MOFs) [16, 17], which by nature contain voids necessary to absorb guest molecules, thereby changing their properties such as change their color [18–21]. In addition, there exist some molecular materials without any pores, which can be used as colorimetric sensors. Insights into the color change mechanism have been recently disclosed for these systems. This includes: (i) direct coordination of guest molecules to the metal center, leading to significant changes to the d–d absorptions [22–29]; (ii) Weak interaction of guest molecules with the sensor (such as H-bonding), causing the ligand field strength change of central ions [30, 31]; (iii) modulation of the molecular structure inside the sensor crystal *via* guest molecules [32]; (iv) guest-exchange of guest molecules in the crystal lattice [33–35]. Regardless of the color change mechanism, the current challenge is to design and synthesize complex molecules that are sensitive to small changes and capable of making color changes. Spin crossover (SCO) materials could be considered great candidates to act as molecular chemical sensors provided their spin state switching is associated to a color change as signal detection. Among this family of materials, some Fe(II) SCO complexes show stable color sets in ambient conditions corresponding to their high-spin (HS) and low-spin (LS) states [36, 37]. The spin state modification actually depends on subtle changes in their chemical environment, which can be tracked by **^{57}Fe Mössbauer spectroscopy** which can quantitatively determine the respective spin state population, i.e. HS and LS states [38]. Research efforts have focused in our laboratory on the detection and discrimination among a wide range of toxic and hazardous guest vapor molecules by colorimetric sensors based on Fe(II) complexes, which we describe in this chapter. ^{57}Fe Mössbauer spectroscopy was applied in transmission mode at room temperature on our newly designed sensors.

11.2 Colorimetric Gas Sensors Based on Fe(II) Complexes

Recently, Garcia *et al.* reported on a series of Fe(II) complexes chemically responsive pigment with the ability to differentiate different hazardous gases [24–31].

Three novel Fe(II) mononuclear complexes, namely $[Fe(trz\text{-}tet)_2(H_2O)_4]\cdot nH_2O$ (**F1**) (*trz-tetH* = 4H-1,2,4-triazolyl-2H-tetrazole) [24–26] $[Fe(H_2btm)_2(H_2O)_2]Cl_2$ (**F2**) (H_2btm = di(1H-tetrazol-5-yl)methane) [27, 28], and $[Fe(Hbta)_2(H_2O)_2]\cdot 2H_2O$ (**F3**) (H_2bta = bis(1H-tetrazol-5-yl)amine) [29] were synthesized and characterized. These systems include an Fe(II) center coordinated to water molecules, being either neutral (for **F1** and **F3**) or including non-coordinated anion in the crystal lattice (for **F2**) (Scheme 11.1). These complexes were found to act as colorimetric sensors for a wide range of Volatile Organic Compounds (VOCs) including toxic gases, such as methanol, hydrazine, ammonia and various amines, etc.).

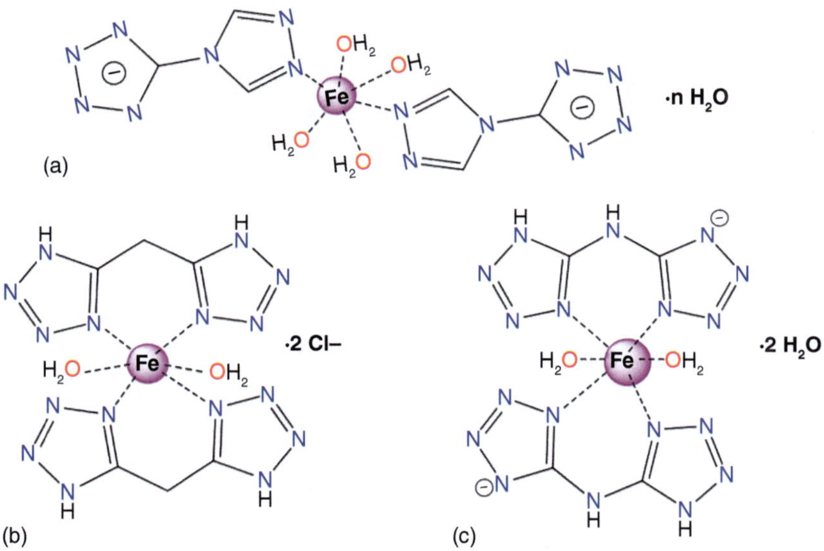

Scheme 11.1 View of the molecular formula of (a) **F1**, (b) **F2**, and (c) **F3**.

Interestingly, **F1** was found to detect methanol among alcohols and toxic industrial chemicals in the gas phase based on molecular sieving and spin-state change. Neighboring ligands were suspected to coordinate with Fe(II) ions starting from an HS octahedral FeN_2O_4 coordination sphere to form an LS FeN_6 unit thus provoking a remarkable color change, from colorless to red (Figure 11.1) [24–26]. Diffuse reflectance spectroscopy as well as ^{57}Fe Mössbauer spectroscopy were applied to investigate the driving spin state change of the guest molecules in **F1**. ^{57}Fe Mössbauer spectroscopy was applied on powdered samples in transmission mode using a conventional Wissel spectrometer operating at room temperature and equipped with a ^{57}Co(Rh) radioactive source.

The spectrum of **F1** (Figure 11.1a) was fitted with one quadrupole doublet revealing an isomer shift $\delta = 1.17(2)$ mm/s and a quadrupole splitting $\Delta E_Q = 3.18(3)$ mm/s, which are typical for HS Fe(II) ions. After $MeOH_{(g)}$ adsorption, a new quadrupole doublet (19%) appears at $\delta = 0.37(1)$ mm/s and $\Delta E_Q = 0.20(2)$ mm/s, for **F1**@MeOH (Figure 11.1b). Such a signal which is characteristic of an LS FeN_6 core with azole ligands is a signature of the modification of the coordination sphere of iron

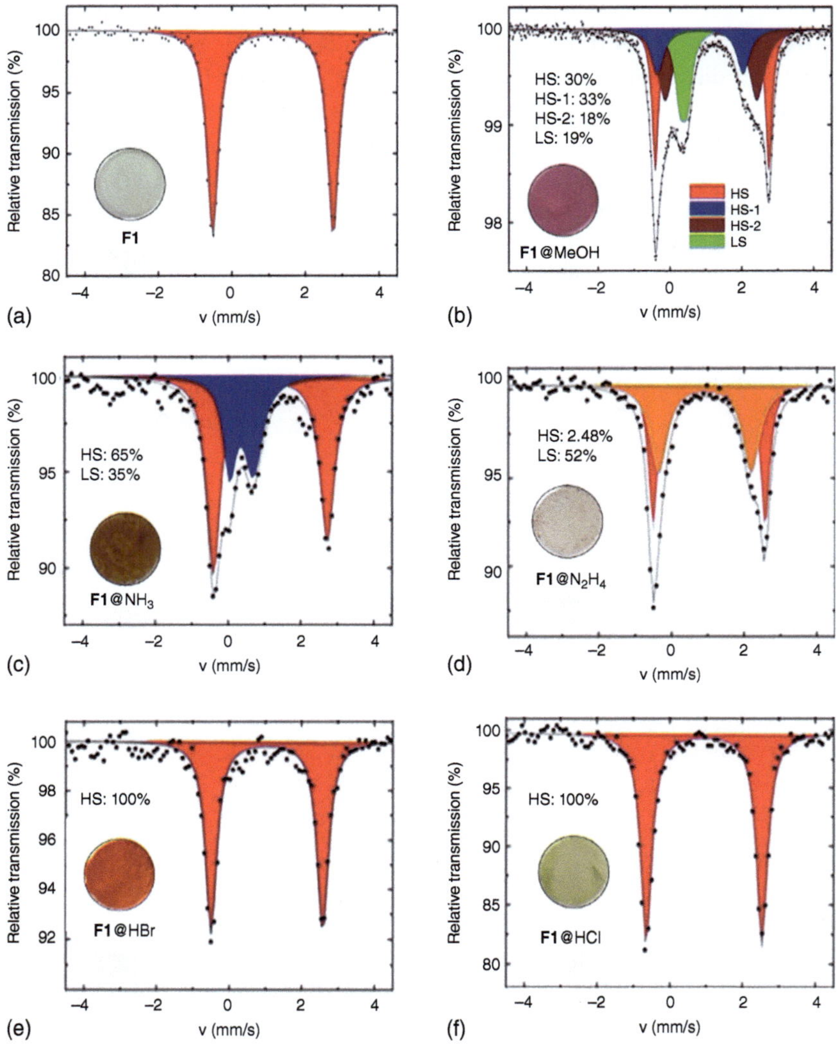

Figure 11.1 Spin-state tracking by ^{57}Fe Mössbauer spectroscopy of **F1** before and after adsorption of toxic gases at room temperature. Source: Ref. [25]/with permission from Royal Society of Chemistry.

and calls for a direct coordination of nitrogen atoms to the metal ion. The two remaining contributions correspond to HS Fe(II) sites: HS-1 [$\delta = 1.14(2)$ mm/s; $\Delta E_Q = 2.56(5)$ mm/s; 33%] and HS-2 [$\delta = 0.83(3)$ mm/s; $\Delta E_Q = 2.40(5)$ mm/s; 18%]. This situation corresponds to an incomplete transformation in the crystal lattice. The spectrum is much simpler when **F1** is in contact with $NH_{3(g)}$ with a single Fe(II) LS contribution indicating an incomplete spin state switching. For **F1**@HBr and **F1**@HCl, no change is observed in their Mössbauer spectra in contrast to **F1**@Hydrazine, with the detection of two HS Fe(II) sites (Figure 11.1). Relevant ^{57}Fe Mössbauer parameters for sensing $MeOH_{(g)}$ and selected toxic industrial chemicals are listed in Table 11.1 for completeness.

Table 11.1 Overview of ^{57}Fe Mössbauer parameters of **F1** for sensing $MeOH_{(g)}$ and toxic industrial chemicals in the gas phase.

	Spin state	Mössbauer parameters			% of species
		δ (mm s^{-1})	ΔE_Q (mm s^{-1})	$\Gamma/2$ (mm s^{-1})	
F1	HS	1.17(2)	3.18(3)	0.20(1)	100
F1@NH_3	HS	1.17(1)	3.13(3)	0.25(2)	65
	LS	0.36(3)	0.65(5)	0.30(6)	35
F1@HCl	HS	1.06(1)	3.26(1)	0.20(1)	100
F1@HBr	HS	1.16(1)	3.14(2)	0.20(2)	100
F1@N_2H_4 (hydrazine)	HS	1.14(1)	3.16(2)	0.16(2)	48
	HS-2	1.03(2)	2.61(8)	0.29(4)	52
F1@MeOH	HS	1.17(1)	3.17(1)	0.12(1)	30
	HS-1	1.14(2)	2.56(5)	0.27(4)	33
	HS-2	0.83(3)	2.40(5)	0.22(4)	18
	LS	0.37(1)	0.20(2)	0.17(2)	19

δ: isomer shift (r.t., *vs.* α-Fe); ΔE_Q: quadrupole splitting, $\Gamma/2$: half width at half maximum.
Source: Adapted from Ref. [25].

The investigations were extended to another chemosensor $[Fe(H_2btm)_2(H_2O)_2]Cl_2$ (**F2**) which was found to display different colors after adsorption for a set of various toxic gases (Figure 11.2) [27]. Interestingly, **F2** can detect not less than 14 different VOCs and HGs at real time, with high selectivity and ultra-sensitivity. With this system, colors can be distinguished by the naked eye without the need for any cryogenic equipment or temperature pre-treatment at room temperature, which is an invaluable asset compared to other materials.

In particular, amines are detected very quickly (<2 min) and with a very high sensitivity. In addition, toxic gases could be readily identified using a standard chemometric approach (hierarchical clustering analysis) to record and analyze collected data on the color of the picture [27]. In particular, the purple and red color detection (e.g. after adsorption 1,2-diaminopropane$_{(g)}$ and ammonia$_{(g)}$ molecules, respectively) were systematically associated to the emergence of $Fe^{II}N_6$ LS ions as shown by ^{57}Fe Mössbauer spectroscopy (Figure 11.3), due to guest molecules replacement of H_2O molecules coordinated to the metal center, leading to significant changes to the d–d absorptions. More importantly, the sensor itself has a high sensitivity detection toward toxic gases even at very low gas concentration, especially for ammonia$_{(g)}$.

Furthermore, a friendly smartphone-based analytical method was developed to make **F2** accessible for *in-field* applications. ^{57}Fe Mössbauer spectroscopy of **F2** recorded in transmission mode displayed a symmetric quadrupole doublet with parameter characteristics of HS Fe(II) ions ($\delta = 1.15(1)$ mm/s, $\Delta E_Q = 3.66(1)$ mm/s) (Figure 11.3). After adsorption of $NH_{3(g)}$ (Figure 11.3b), and 1,2-diaminopropane

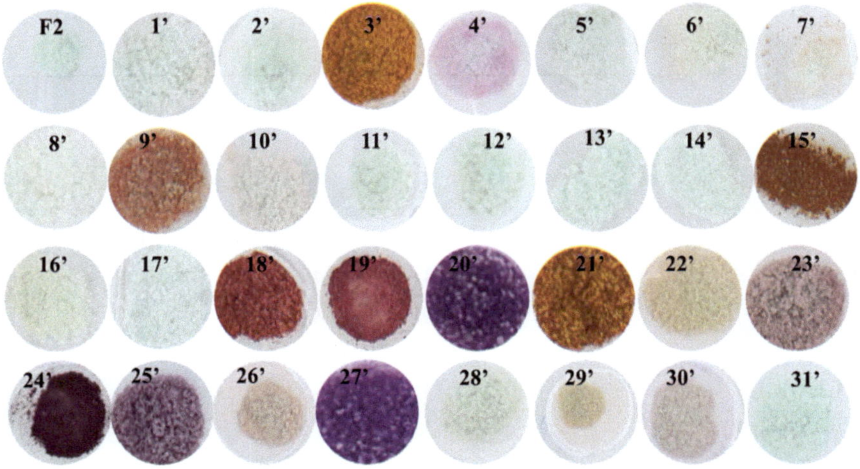

Figure 11.2 Images depicting vapochromic transformation on the sensing of VOCs by complex **F2** for 120 min at room temperature on ground (**F2**; 1′ = water; 2′ = acetonitrile; 3′ = pyridine; 4′ = formaldehyde; 5′ = acetone; 6′ = chloromethane; 7′ = dichloromethane; 8′ = tetrahydrofuran; 9′ = methanol; 10′ = ethanol; 11′ = 1-pentanol; 12′ = 1-butanol; 13′ = hydrochloric acid; 14′ = sulfuric acid; 15′ = nitric acid; 16′ = bromic acid; 17′ = acetic acid; 18′ = ammonia; 19′ = hydrazine; 20′ = ethylenediamine; 21′ = isobutylamine; 22′ = 1,2-diaminocyclohexane; 23′ = furfurylamine; 24′ = hexylamine; 25′ = diethylamine; 26′ = 3-picolylamine; 27′ = 2-diaminopropane; 28′ = aniline; 29′ = triethylamine; 30′ = diethylenetriamine; 31′ = 2-chloroaniline). Source: Ref. [27], From American Chemical Society, 2021.

(Figure 11.3d) molecules, in addition to major quadrupole doublet attributed to HS Fe(II) ions, a new quadrupole doublet characteristic of LS Fe(II) ions was observed at [$\delta = 0.38(1)$ mm/s, $\Delta E_Q = 0.23(1)$ mm/s] with 70% and [$\delta = 0.40(1)$ mm/s, $\Delta E_Q = 0.22(2)$ mm/s] with 53%, respectively.

The same situation is observed for **F2**@ethylenediamine (Figure 11.3c), **F2**@N_2H_4 (Figure 11.3f), **F2**@isobutylamine (Figure 11.3e) and **F2**@hexylamine (Figure 11.4d), with similar isomer shift and similar quadrupole splitting except for **F2**@isobutylamine with $\Delta E_Q = 0.08(3)$ mm/s, which thus reveals a more symmetric octahedron. These parameters are typical for LS Fe(II) ions, indicating that **F2** absorbs $NH_{3(g)}$, 1,2-diaminopropane, ethylenediamine, $N_2H_{4(g)}$ and isobutylamine$_{(g)}$ molecules to build an LS FeN_6 core with nitrogen-based ligands. These materials are very stable after guest adsorption with no sign of oxidation to Fe(III) having been observed. No quadrupole splitting was found for LS Fe(II) signals for **F2**@triethylamine (Figure 11.3g), **F2**@diethylamine (Figure 11.3h) **F2**@MeOH (Figure 11.4b), **F2**@diethylenetriamine (Figure 11.4c), **F2**@3-picolylamine (Figure 11.4f), **F2**@1,2-diaminocyclohexane (Figure 11.4e) and **F2**@furfurylamine (Figure 11.4h), which indicates that the iron(II) sites are completely symmetrical after guest adsorption. Interestingly, a lower isomer shift $\delta^{LS} = 0.14(1)$ mm s^{-1} was found for **F2**@pyridine (Figure 11.4g) with a larger quadrupole splitting $\Delta E_Q{}^{LS} = 0.62(2)$ mm s^{-1} (Table 11.2). Indeed, the replacement of oxygen atoms by nitrogen atoms, which are less electronegative, increases the electronic density

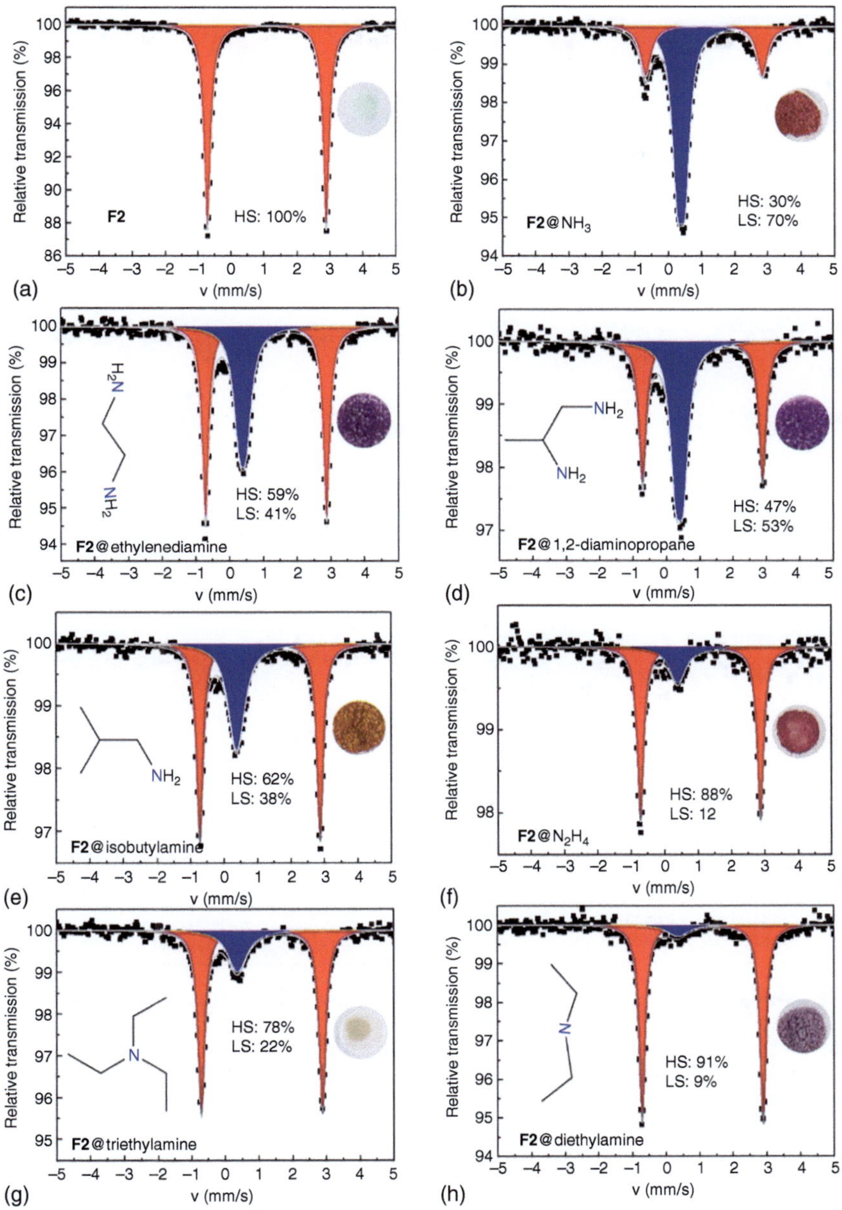

Figure 11.3 Spin-state tracking by ^{57}Fe Mössbauer spectroscopy of **F2** before and after adsorption of toxic gases at room temperature. Source: Ref. [27]/with permission from American Chemical Society.

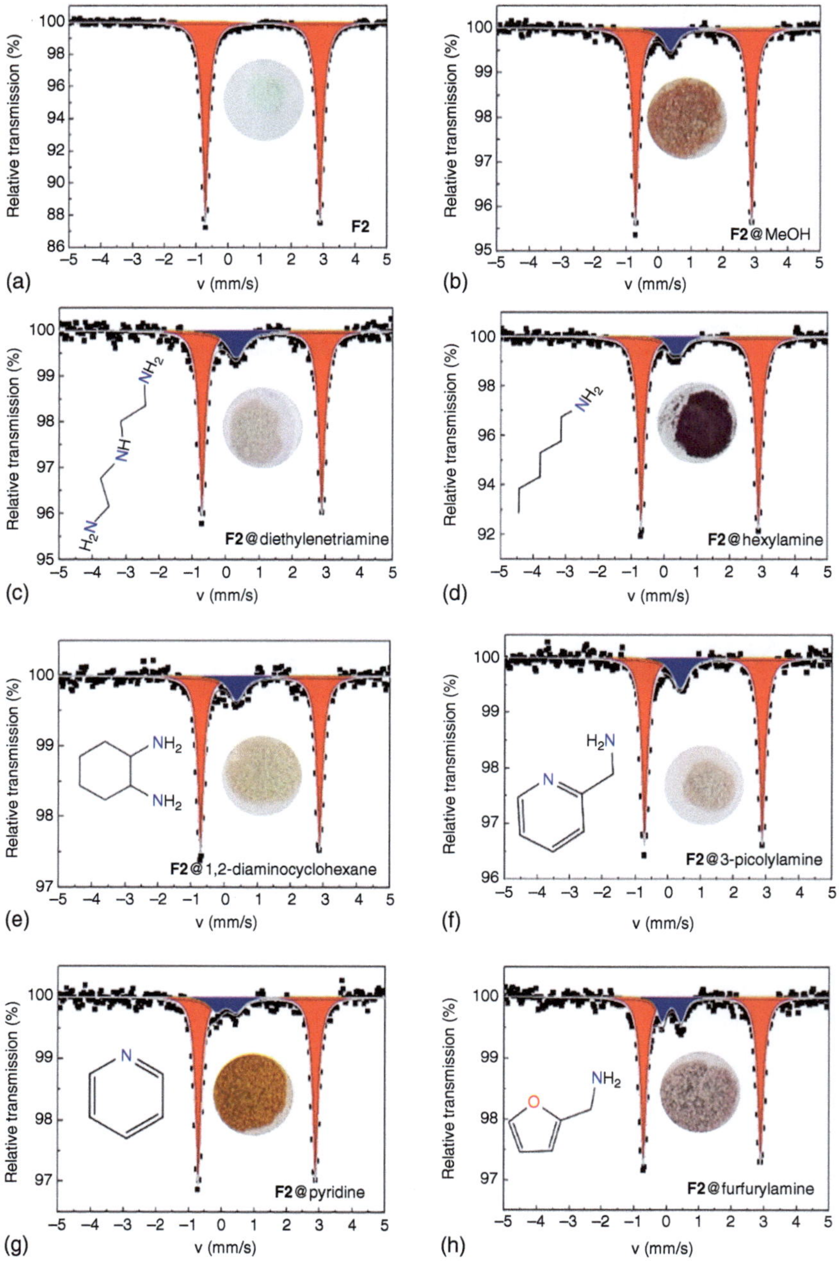

Figure 11.4 Spin-state tracking by ^{57}Fe Mössbauer spectroscopy of **F2** before and after adsorption of toxic gases at room temperature. Source: Ref. [28]/with permission from Springer Nature.

Table 11.2 Overview of ^{57}Fe Mössbauer parameters for **F2** and guest adsorbed species after measurements in transmission mode at room temperature.

	Spin state	Mössbauer parameters			
		δ (mm s^{-1})	ΔE_Q (mm s^{-1})	$\Gamma/2$	% of species
F2	HS	1.15(1)	3.66(1)	0.14(1)	100
F2@NH_3	HS	1.08(1)	3.51(3)	0.20(2)	30
	LS	0.38(1)	0.23(1)	0.17(1)	70
F2@N_2H_4	HS	1.00(1)	2.57(1)	0.17(1)	88
	LS	0.39(3)	0.20(5)	0.14(1)	12
F2@isobutylamine	HS	1.08(3)	3.59(1)	0.12(1)	62
	LS	0.38(1)	0.08(3)	0.28(5)	38
F2@1,2-diaminopropane	HS	1.09(1)	3.60(1)	0.14(1)	47
	LS	0.40(1)	0.22(2)	0.19(2)	53
F2@ethylenediamine	HS	1.08(1)	3.59(1)	0.12(1)	59
	LS	0.39(1)	0.20(3)	0.20(1)	41
F2@triethylamine	HS	1.09(3)	3.60(1)	0.13(1)	78
	LS	0.36(1)	0	0.34(1)	22
F2@diethylamine	HS	1.08(1)	3.59(1)	0.12(1)	91
	LS	0.35(1)	0	0.37(6)	9
F2@MeOH	HS	1.08(1)	3.60(1)	0.12(1)	89
	LS	0.36(4)	0	0.26(1)	11
F2@diethylenetriamine	HS	1.08(1)	3.60(1)	0.13(1)	85
	LS	0.30(1)	0	0.31(3)	15
F2@hexylamine	HS	1.09(1)	3.59(1)	0.12(1)	91
	LS	0.36(4)	0.26(1)	0.19*	9
F2@3-picolylamine	HS	1.08(1)	3.60(1)	0.12(1)	86
	LS	0.41(5)	0	0.26(2)	14
F2@1,2-diaminocyclohexane	HS	1.08(1)	3.60(1)	0.12(1)	86
	LS	0.36(1)	0	0.28(2)	14
F2@pyridine	HS	1.08(4)	3.60(1)	0.11(1)	86
	LS	0.14(1)	0.62(2)	0.27(1)	14
F2@furfurylamine	HS	1.09(1)	3.60(1)	0.12(1)	86
	LS	0.40(1)	0	0.25(2)	14

δ: isomer shift (with respect to α-Fe at 298 K); ΔE_Q: quadrupole splitting; $\Gamma/2$: half width at half maximum; * fixed parameter.
Source: Adapted from Refs. [27, 28].

of s electrons in the vicinity of the iron nucleus and thereby decreases the isomer shift (the spin states of Fe(II) from HS to LS). Last but not least, our studies have also shown that **F2** has excellent regeneration performance of adsorption and de-adsorption of VOCs. It operates at ambient conditions, is low cost, environmentally friendly, easy to use, and shows excellent and fast detection performances. Such features offer attractive prospects for **F2**, which could be used for *in-field* detection and food safety control in environmental conditions.

In order to study the aerochromic properties of such mononuclear Fe(II) functional molecules, we synthesized another sensor with a related structure compared to **F2** with $[Fe(Hbta)_2(H_2O)_2]{\cdot}2H_2O$ (**F3**) [29], which shows fast and different colors after contacting different toxic gas molecules at room temperature. As can be seen in Figure 11.5, the color change of **F3** is particularly pronounced after absorption of amine noxious gas molecules, although other pyridine and methanol can also induce a color change.

Similarly, we have used ^{57}Fe Mössbauer spectroscopy as a tool to conduct an in-depth study of the color change mechanism of **F3** sensor [29]. Studies have shown that there are significant differences in the ^{57}Fe Mössbauer spectra before and after adsorption of toxic gas molecules, as shown in Figure 11.6 and Table 11.3.

Figure 11.5 Colorimetric sensor responses for 32 common analytes at saturation vapor pressure at room temperature with **F3**.1a = H_2O; 2a = acetonitrile; 3a = formaldehyde; 4a = dichloromethane; 5a = chloroform; 6a = acetone; 7a = tetrahydrofuran; 8a = pyridine; 9a = MeOH; 10a = ethanol; 11a = 1-butanol; 12a = 1-pentanol; 13a = hydrochloric acid; 14a = hydrogen bromide; 15a = acetic acid; 16a = sulfuric acid; 17a = nitric acid; 18a = NH_3; 19a = isobutylamine; 20a = diethylamine; 21a = 1,2-diaminopropane; 22a = ethylenediamine; 23a = N_2H_4; 24a = diethylenetriamine; 25a = furfurylamine; 26a = hexylamine; 27a = triethylamine; 28a = aniline; 29a = 3-picolylamine; 30a = 2-(trifluoromethyl)aniline; 31a = 2-chloroaniline; 32a = 1,2-diaminocyclohexane. Source: Ref. [29], From ELSEVIER, 2022.

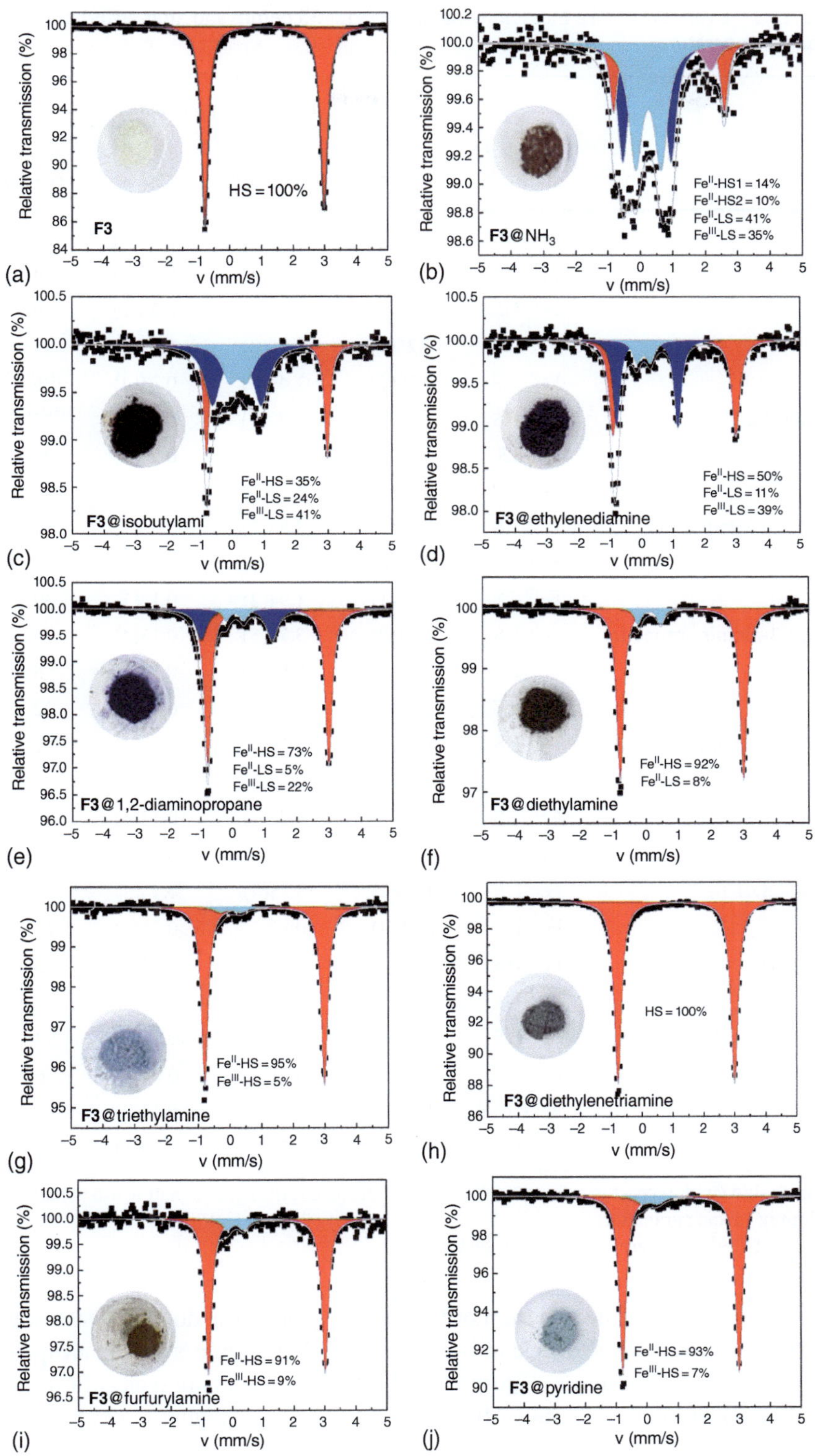

Figure 11.6 ^{57}Fe Mössbauer spectra of **F3** before and after the detection of HGs with amino groups at room temperature. Source: Ref. [29]/with permission from Elsevier.

Table 11.3 Overview of ^{57}Fe Mössbauer parameters for **F3** before and after adsorbing HGs at room temperature.

		Mössbauer parameters			
	Spin state	**δ (mm s^{-1})**	**ΔE_Q (mm s^{-1})**	**$\Gamma/2$**	**% of species**
F3	Fe^{II}-HS	1.10(2)	3.78(3)	0.15(3)	100
F3@NH_3	Fe^{II}-HS1	0.89(2)	3.42(5)	0.14(4)	14
	Fe^{II}-HS2	1.18(9)	2.00(2)	0.28(2)	10
	Fe^{II}-LS	0.25(2)	0.78(8)	0.25(5)	41
	Fe^{III}-LS	0.20(2)	1.49(6)	0.21(4)	35
F3@isobutylamine	Fe^{II}-HS	1.10(9)	3.78(2)	0.13*	35
	Fe^{II}-LS	0.22(5)	0.57(8)	0.29(6)	24
	Fe^{III}-LS	0.15(3)	1.53(5)	0.29(4)	41
F3@ethylenediamine	Fe^{II}-HS	1.04(1)	3.88(2)	0.15(1)	50
	Fe^{II}-LS	0.07(5)	0.43(8)	0.16(7)	11
	Fe^{III}-LS	0.17(1)	1.94(2)	0.13(1)	39
F3@1,2-diaminopropane	Fe^{II}-HS	1.11(2)	3.78(5)	0.13(4)	73
	Fe^{II}-LS	0.08(4)	0.55(7)	0.14(5)	5
	Fe^{III}-LS	0.13(2)	2.21(3)	0.20(2)	22
F3@diethylamine	Fe^{II}-HS	1.11(2)	3.79(5)	0.13(4)	92
	Fe^{II}-LS	0.11(3)	0.70(5)	0.13(3)	8
F3@triethylamine	Fe^{II}-HS	1.11(2)	3.78(5)	0.13(4)	95
	Fe^{II}-LS	0.04(1)	0.64(2)	0.24(1)	5
F3@MeOH	Fe^{II}-HS	1.10(2)	3.78(4)	0.14(3)	89
	Fe^{II}-LS	0.36(4)	0	0.31(1)	11
F3@pyridine	Fe^{II}-HS	1.10(2)	3.78(3)	0.14(3)	93
	Fe^{II}-LS	0.11*	0.5*	0.26*	7
F3@diethylenetriamine	Fe^{II}-HS	1.11(1)	3.78(2)	0.15(1)	94
	Fe^{II}-LS	0.11*	0.51*	0.21*	6
F3@furfurylamine	Fe^{II}-HS	1.11(5)	3.79(1)	0.13(7)	91
	Fe^{II}-LS	0.10(1)	0.54(2)	0.19(1)	9

δ: isomer shift (with respect to α-Fe at 298 K); ΔE_Q: quadrupole splitting; $\Gamma/2$: half width at half maximum; * fixed parameter.
Source: Ref. [29]/with permission from Elsevier.

At room temperature, the spectrum of **F3** shows a quadrupole doublet which was fitted by Lorentzian lines to yield to isomer shift $\delta = 1.10(2)$ mm/s and quadrupole splitting $\Delta E_Q = 3.78(3)$ mm/s (Figure 11.6a). Such parameters are typical for octahedral Fe(II) ions in the HS state. After adsorption of $NH_{3(g)}$ (Figure 11.6b) and 1,2-diaminopropane (Figure 11.6e) molecules, a new quadrupole doublet characteristic of LS Fe(II) ([$\delta = 0.25(2)$ mm/s, $\Delta E_Q = 0.78(8)$ mm/s]; [$\delta = 0.08(4)$ mm/s, $\Delta E_Q = 0.55(7)$ mm/s]) was observed. In addition, the main quadrupole doublet of HS Fe(II) ions was preserved and observed at [$\delta = 0.89(2)$ mm/s, $\Delta E_Q = 3.42(5)$ mm/s] and [$\delta = 0.11(2)$ mm/s, $\Delta E_Q = 3.78(5)$ mm/s], respectively. A quadrupole splitting was found for LS Fe(II) signals for **F3**@ethylenediamine (Figure 11.6d), **F3**@1,2-diaminopropane (Figure 11.6e) **F3**@diethylamine (Figure 11.6f), **F3**@triethylamine (Figure 11.6g), **F3**@diethylenetriamine (Figure 11.6h), **F3**@pyridine (Figure 11.6j) and **F3**@furfurylamine (Figure 11.6i), respectively. Therefore, the color change mechanism of **F3** could be similar to that of **F2**. And the mechanism of color change of **F3** is inferred to be that the toxic gas molecules replace the coordinated water molecules, and then the Fe(II) ions of **F3** form a FeN_6 coordination sphere starting from FeN_4O_2. Overall, such colorimetric material (**F3**) could provide broad selectivity towards various HGs with amino groups and high-sensitivity NH_3 molecules, opening up an opportunity to develop a new generation of room temperature-portable electronic nose systems for real-time assessment and in-situ measurement of the freshness of meat, e.g. by inserting our sensor into food packaging, via processing a thin film [29].

Recently, we have extended our Mössbauer investigations to a new mononuclear Fe(II) complex (**F4**) with formula $[FeL_2](BF_4)_2$ (L = (E)-2-((pyridin-2-ylmethylene)amino)-1Hbenzo[*de*]-isoquinoline-1,3(2H)-dione), which was prepared in powdered state (Scheme 11.2).

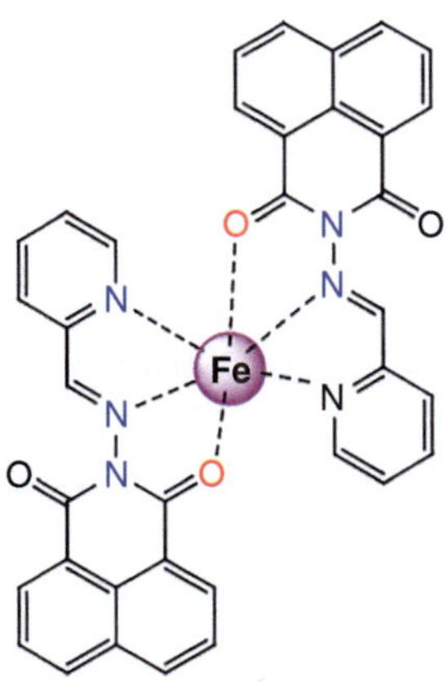

Scheme 11.2 View of the molecular formula of **F4**.

The interest of this compound from a chemical point of view is that it does not contain any coordinated water molecules contrary to **F1**, **F2**, and **F3**. This compound shows a mixed spin state population at room temperature [25]. Indeed, a quadrupole doublet characteristic of HS Fe(II) ions is observed (74%) along with LS Fe(II) ions (26%) (Figure 11.7 and Table 11.4). The remarkable and quick color change from pink to black after exposure to $NH_{3(g)}$, at room temperature, (Figure 11.7) led us to record the Mössbauer spectrum again. After adsorbing NH_3 gas molecules, the ^{57}Fe Mössbauer spectrum is obviously modified (Figure 11.7b). **F4**@NH_3 exhibits a new

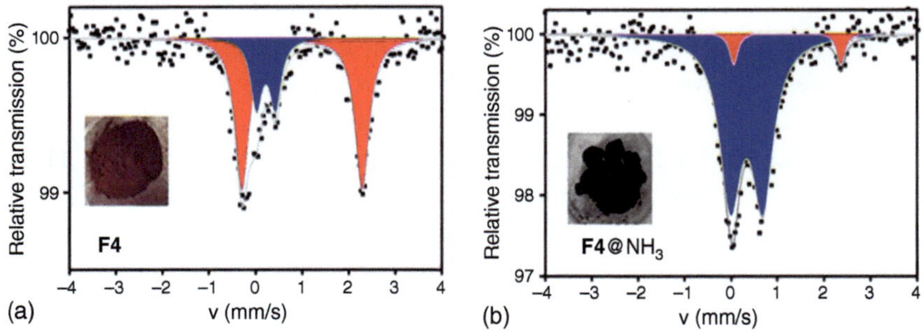

Figure 11.7 ^{57}Fe Mössbauer spectrum of **F4** (a) before and (b) after adsorption of $NH_{3(g)}$ recorded at 298 K. The red and blue quadrupole doublets correspond to HS and LS Fe(II) ions, respectively. The inset shows the pink and black colors of the powder before and after $NH_{3(g)}$ adsorption, respectively. Source: Ref. [30]/with permission from Springer Nature.

Table 11.4 Overview of selected ^{57}Fe Mössbauer parameters for **F4** and **F4@NH$_3$**.

		Mössbauer parameters			
	Spin state	**δ (mm s^{-1})**	**ΔE_Q (mm s^{-1})**	**$\Gamma/2$**	**% of species**
F4	HS	1.00(2)	2.58(4)	0.18(3)	74
	LS	0.22(5)	0.42(1)	0.16(1)	26
F4@NH$_3$	HS	1.21(1)	2.31(2)	0.11(1)	7
	LS	0.33(2)	0.70(4)	0.28(3)	93

δ: isomer shift (with respect to α-Fe at 298 K); ΔE_Q: quadrupole splitting; $\Gamma/2$: half width at half maximum.
Source: Ref. [30] / Springer Nature.

quadrupole doublet characteristic of LS Fe(II) ions centered at $\delta = 0.33(2)$ mm/s and $\Delta E_Q = 0.70(4)$ mm/s with a large area fraction of 93%. Such spin state change could be caused by weak interactions (e.g. H-bonding) between the oxygen-rich ligands and the guest NH_3 molecules, resulting in a modification of the ligand field strength thereby leading to a spin state change [30]. This spin state configuration applies to the powdered state only.

Continuing our investigations of non-aqua coordinated complexes, we have expanded our research to a new tetranuclear Fe(II) cage compound (**F5**) with formula $C_{144}H_{120}N_{48}Fe_4B_8F_{32}{\cdot}9H_2O$ (Scheme 11.3) [31]. **F5** can detect $NH_{3(g)}$ molecules at room temperature very quickly (10 s) and the detection process is accompanied by a dramatic colour change from light brown to purple. ^{57}Fe Mössbauer spectroscopy of **F5** and **F5**@NH$_3$ was used to study the sensing mechanism (Figure 11.8 and Table 11.5). The spectrum of **F5** shows three quadrupole doublets (Figure 11.8a). The first one [isomer shift $\delta = 0.20(3)$ mm s^{-1}; quadrupole splitting $\Delta E_Q = 0.70(1)$ mm s^{-1}] is characteristic of LS Fe(II) ions and the area fraction (A) of 29%. The two other doublets HS-1 [$\delta = 0.99(1)$ mm s^{-1}; $\Delta E_Q = 1.39(1)$ mm s^{-1}; A = 44%] and HS-2 [$\delta = 1.10(4)$ mm s^{-1}; $\Delta E_Q = 2.60(1)$ mm s^{-1}; A = 27%] were

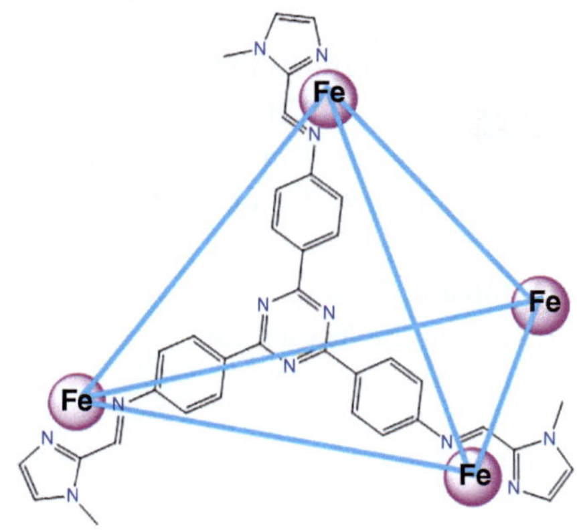

Scheme 11.3 View of the molecular formula of **F5**.

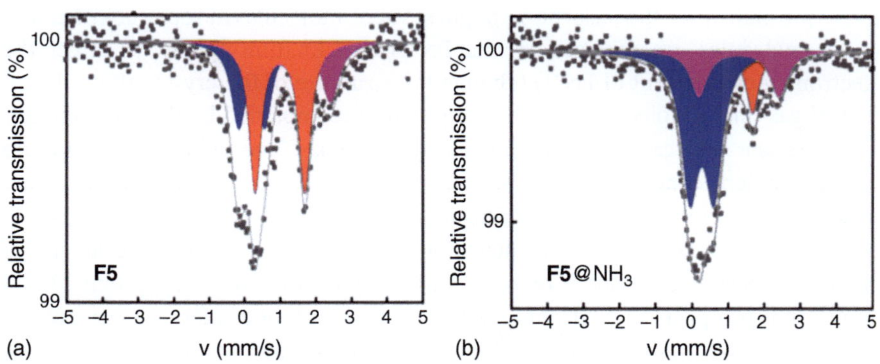

Figure 11.8 ^{57}Fe Mössbauer spectra of **F5** (a) and **F5**@NH_3 (b) recorded at 298 K. The red, magenta and blue quadrupole doublets correspond to HS-1, HS-2 and LS Fe(II) ions, respectively.

Table 11.5 ^{57}Fe Mössbauer parameters for **F5** and **F5**@NH_3.

			Mössbauer parameters		
Sample	**Spin state Fe(II)**	**A/A_{tot} (%)**	**δ (mm s^{-1})**	**ΔE_Q (mm s^{-1})**	**$\Gamma/2$**
F5	LS (blue)	29	0.20(3)	0.70(1)	0.25(1)
	HS-1 (red)	44	0.99(1)	1.39(1)	0.20(1)
	HS-2 (magenta)	27	1.10(4)	2.60(1)	0.30(1)
F5@NH_3	LS (blue)	60	0.27(1)	0.66(1)	0.28(1)
	HS-1 (red)	20	0.96(1)	1.45(2)	0.21(1)
	HS-2 (magenta)	20	1.29(4)	2.19(1)	0.28(1)

assigned to HS Fe(II) ions. The LS/HS Fe(II) ratio is 29 : 71%. After absorption of NH_3 molecules, the Mössbauer spectrum of **F5**@NH_3 also shows three quadrupole doublets: [$\delta = 0.27(1)$ mm s^{-1}; $\Delta E_Q = 0.66(1)$ mm s^{-1}; A = 60%]; [$\delta = 0.96(1)$ mm s^{-1}; $\Delta E_Q = 1.45(2)$ mm s^{-1}; A = 20%]; [$\delta = 1.29(4)$ mm s^{-1}; $\Delta E_Q = 2.19(1)$ mm s^{-1}; A = 20%], corresponding to LS, HS-1 and HS-2 Fe(II) ions, respectively (Figure 11.8b). The LS/HS ratio changed significantly to 60 : 40%. Thus, the sensing mechanism of **F5** was ascribed to the transition of the spin state of Fe(II) ions after the intrusion of $NH_{3(g)}$ molecules. Such molecules are presumably hydrogen-bonded and induce a spin state change [31].

11.3 Conclusions and Perspectives

We have shown that colorimetric sensors based on Fe(II) complexes can change their spin state after the adsorption of guest vapor molecules at room temperature. Guest molecules are replaced in the Fe(II) coordination sphere or affect the electronic environment of Fe(II) (change the structure of the crystal after adsorption of gas-phase molecules). Such changes are accompanied by the color change of the sensor and can be distinguished by the naked eye at room temperature and in ambient conditions. Usually, with a fast response, intuitive and accurate results of a colorimetric sensor, not only achieve the specific detection of toxic gases but also can be used for *in-field* detection, provided an association to an image database processing method is realized, which can be used by a smartphone. Broad application prospects in environmental protection and food safety assessment are foreseen, given the potential of developments from the coordination chemistry side of such pioneering systems, in particular for meat freshness assessment [29, 31]. **^{57}Fe Mössbauer spectroscopy** is expected to play a key role in this growing area.

References

1 Nicolaou, L. and Checkley, W. (2021). Inequities in air pollution exposure and gaps in air quality monitoring. *J. Allergy Clin. Immun.* 148: 64–66.

2 Szumiata, T., Gawronski, M., Gorka, B. et al. (2013). Chemical, magnetic and Mössbauer effect analysis of road dust from expressway. *Nukleonika* 58: 109–112.

3 Kierlik, P., Kuczkowska, A.H., Rachwal, M. et al. (2020). Application of Mössbauer spectroscopy for identification of iron-containing components in upper Silesian topsoil being under industrial anthropopressure. *Materials* 13: 5206.

4 Yang, D., Wang, M., Liu, J. et al. (2020). Occurrence and characteristics of iron-bearing minerals in surface road dusts: a case study in the coastal areas of southern Fujian, China. *J. Soils Sediments* 20: 3406–3416.

5 Kierlik, P., Kuczkowska, A.H., Meczynski, R. et al. (2019). Phase composition of urban soils by X-ray diffraction and Mössbauer spectroscopy analysis. *Arch. Metall. Mater.* 64: 1029–1032.

6 Soni, V., Singh, P., Shree, V., and Goel, V. (2018). *Air Pollution and Control-Effects of VOCs on Human Health*, 119–142. Singapore: Springer Nature.

7 Colman Lerner, J.E., Gutierrez, M.L.A., Mellado, D. et al. (2018). Characterization and cancer risk assessment of VOCs in home and school environments in gran La Plata, Argentina. *Environ. Sci. Pollut. Res. Int.* 25: 10039–10048.

8 Ligor, M., Ligor, T., Bajtarevic, A. et al. (2009). Determination of volatile organic compounds in exhaled breath of patients with lung cancer using solid phase microextraction and gas chromatography mass spectrometry. *Clin. Chem. Lab Med.* 47: 550–560.

9 Colomb, A., Yassaa, N., Williams, J. et al. (2008). Screening volatile organic compounds (VOCs) emissions from five marine phytoplankton species by head space gas chromatography/mass spectrometry (HS-GC/MS). *J. Environ. Monit.* 10: 325–330.

10 Kim, S.H., Kirakosyan, A., Choi, J., and Kim, J.H. (2017). Detection of volatile organic compounds (VOCs), aliphatic amines, using highly fluorescent organic-inorganic hybrid perovskite nanoparticles. *Dyes Pigm.* 147: 1–5.

11 Bagheri, M., Masoomi, M.Y., and Morsali, A. (2017). Highly sensitive and selective ratiometric fluorescent metal–organic framework sensor to nitroaniline in presence of nitroaromatic compounds and VOCs. *Sensors Actuators B Chem.* 243: 353–360.

12 Mori, M., Nishimura, H., Itagaki, Y. et al. (2017). Detection of sub-ppm level of VOCs based on a Pt/YSZ/Pt potentiometric oxygen sensor with reference air. *Sensors Actuators B Chem.* 143: 56–61.

13 Lim, S.H., Feng, L., Kemling, J.W. et al. (2009). An optoelectronic nose for the detection of toxic gases. *Nat. Chem.* 1: 562–567.

14 Janzen, M.C., Ponder, J.B., Bailey, D.P. et al. (2006). Colorimetric sensor arrays for volatile organic compounds. *Anal. Chem.* 78: 3591–3600.

15 Garcia, Y. (2020). Selected polyazole based coordination polymers displaying functional properties. *Adv. Inorg. Chem.* 76: 121–153.

16 Naik, A.D., Dîrtu, M.M., Railliet, A. et al. (2011). Coordination polymers and metal organic frameworks derived from 1,2,4-triazole amino acid linkers. *Polymers* 3: 1750–1775.

17 Li, H.Y., Zhao, S.N., Zang, S.Q., and Li, J. (2020). Functional metal–organic frameworks as effective sensors of gases and volatile compounds. *Chem. Soc. Rev.* 49: 6364–6401.

18 Urien, E.R., López, L.P., Bartolome, E.F. et al. (2020). Covalent post-synthetic modification of switchable iron-based coordination polymers by volatile organic compounds: a versatile strategy for selective sensor development. *Dalton Trans.* 49: 7315–7318.

19 Homayoonnia, S. and Zeinali, S. (2016). Design and fabrication of capacitive nanosensor based on MOF nanoparticles as sensing layer for VOCs detection. *Sensors Actuators B Chem.* 237: 776–786.

20 Wenger, O.S. (2013). Vapochromism in organometallic and coordination complexes: chemical sensors for volatile organic compounds. *Chem. Rev.* 113: 3686–3733.

21 Li, E., Jie, K., Liu, M. et al. (2020). Vapochromic crystals: understanding vapochromism from the perspective of crystal engineering. *Chem. Soc. Rev.* 49: 1517–1544.

22 Bartolome, E.F., Urien, E.R., Vidal, M.M. et al. (2021). Sequential single-crystal-to-single-crystal vapochromic inclusion in a nonporous coordination polymer: unravelling dynamic rearrangement for selective pyridine sensing. *Inorg. Chem. Front.* 8: 2426–2432.

23 Supriy, S. and Das, S.K. (2007). Reversible single crystal to single crystal transformation through Fe-O(H)Me/Fe-OH_2 bond formation/bond breaking in a gas-solid reaction at an ambient condition. *J. Am. Chem. Soc.* 129: 3464–3465.

24 Naik, A.D., Robeyns, K., Meunier, C.F. et al. (2014). Selective and reusable iron(II)-based molecular sensor for the vapor-phase detection of alcohols. *Inorg. Chem.* 53: 1263–1265.

25 Guo, Y., Xue, S., Dîrtu, M.M., and Garcia, Y. (2018). A versatile iron(II)-based colorimetric sensor for the vapor-phase detection of alcohols and toxic gases. *J. Mater. Chem. C* 6: 3895–3900.

26 Xue, S., Wang, L., Naik, A.D. et al. (2021). Iron(II) pillared-layer responsive frameworks via "kagomé dual" (kgd) supramolecular tessellations. *Inorg. Chem. Front.* 8: 3532–3546.

27 Sun, L., Rotaru, A., Robeyns, K., and Garcia, Y. (2021). A colorimetric sensor for the highly selective, ultra-sensitive, and rapid detection of volatile organic compounds and hazardous gases. *Ind. Eng. Chem. Res.* 60: 8788–8798.

28 Sun, L., Rotaru, A., and Garcia, Y. (2021). ^{57}Fe Mössbauer study of an iron(II) sensor for the detection of toxic gases at room temperature. *Hyperfine Interact.* 242: 23.

29 Sun, L., Rotaru, A., and Garcia, Y. (2022). A non-porous Fe(II) complex for the colorimetric detection of hazardous gases and the monitoring of meat freshness. *J. Hazard. Mater.* 437: 129364.

30 Li, W., Sun, L., and Garcia, Y. (2021). Subcomponent self-assembly of a Fe(II) based mononuclear complex for potential NH_3 sensing applications. *Hyperfine Interact.* 242: 7.

31 Li, W., Sun, L., Liu, C. et al. (2022). Supramolecular $Fe^{II}{}_4L_4$ cage for fast ammonia sensing. *J. Mater. Chem. C* 10: 9216–9221. https://doi.org/10.1039/D2TC01655A.

32 Jimenez, S.R., Feltham, H.L., and Brooker, S. (2016). Non-porous iron(II)-based sensor: crystallographic insights into a cycle of colorful guest-induced topotactic transformations. *Angew. Chem. Int. Ed.* 55: 15067–15071.

33 Costa, J.S., Jiménez, S.R., Craig, G.A. et al. (2020). Selective signalling of alcohols by molecular lattice and mechanism of single-crystal-to-single-crystal transformation. *Inorg. Chem. Front.* 7: 3165–3175.

34 Costa, J.S., Jiménez, S.R., Craig, G.A. et al. (2014). Three-way crystal-to-crystal reversible transformation and controlled spin switching by a nonporous molecular material. *J. Am. Chem. Soc.* 136: 3869–3874.

35 Develioglu, A., Urien, E.R., Poloni, R. et al. (2021). Tunable proton conductivity and color in a nonporous coordination polymer via lattice accommodation to small molecules. *Adv. Sci.* 8: 2102619.

36 Gütlich, P., Gaspar, A.B., and Garcia, Y. (2013). Spin state switching in iron coordination compounds. *Beilstein J. Org. Chem.* 9: 342–391.

37 Xue, S., Guo, Y., and Garcia, Y. (2021). Spin crossover crystalline materials engineered via single-crystal-to-single-crystal transformations. *CrystEngComm* 23: 7899–7915.

38 Gütlich, P. and Garcia, Y. (2013, chapter 2). Chemical applications of Mössbauer spectroscopy. In: *Mössbauer Spectroscopy – Tutorial Book* (ed. Y. Yoshida and G. Langouche), 23–89. Springer.

Index

Mössbauer Spectroscopy: Applications in Chemistry and Materials Science,
First Edition. Edited by Yann Garcia, Junhu Wang, and Tao Zhang.